The Biology of

THE BIOLOGY OF HABITATS SERIES

This attractive series of concise, affordable texts provides an integrated overview of the design, physiology, and ecology of the biota in a given habitat, set in the context of the physical environment. Each book describes practical aspects of working within the habitat, detailing the sorts of studies which are possible. Management and conservation issues are also included. The series is intended for naturalists, students studying biological or environmental science, those beginning independent research, and professional biologists embarking on research in a new habitat.

The Biology of Disturbed Habitats

Lawrence R. Walker

OXFORD

UNIVERSITY PRESS

OXFORD

UNIVERSITY PRESS

Great Clarendon Street, Oxford, OX2 6DP,
United Kingdom

Oxford University Press is a department of the University of Oxford.
It furthers the University's objective of excellence in research, scholarship,
and education by publishing worldwide. Oxford is a registered trade mark of
Oxford University Press in the UK and in certain other countries

Published in the United States of America by Oxford University Press
198 Madison Avenue, New York, NY 10016, United States of America

British Library Cataloguing in Publication Data
Data available

Library of Congress Cataloguing in Publication Data
Data Available

ISBN 978-0-19-957530-5

Preface

Disturbances are an integral part of the environment. Large-scale geological forces such as volcanoes and earthquakes sculpted the earth and, combined with glaciers, floods, and other natural disturbances, continue to shape it. Organisms survive only when such forces are not dominant, and can be destroyed by them. Natural disturbances can also occur at much smaller scales, as exemplified by insect herbivores defoliating a tree or storm waves eroding a coral reef. Large and small disturbances combined become part of a suite of interacting disturbances that constitute the disturbance regime of a particular area. Life, including humanity, has evolved to survive and even thrive in the presence of disturbances. Even when there is widespread destruction from larger disturbances, plants, animals, and humans recolonize the disturbed patches where they often benefit from new habitats and a renewal of nutrients and light.

The growing human population has now disrupted this dynamic equilibrium between natural disturbances and ecosystem recovery. Not only do humans create anthropogenic disturbances that mimic natural ones, but we also create novel disturbances. Both types of disturbance can augment the effects of natural disturbances, sometimes leading to a cascade of events that magnify formerly moderate natural disturbances to levels that threaten ecosystem processes. For example, the Smokey the Bear campaign to prevent forest fires in the U.S. led to occasional, but hotter and more destructive, fires when the forests eventually did burn. Humans have also built an extensive infrastructure including dwellings, factories, and transportation corridors that did not exist before, thereby increasing our vulnerability to disturbances. The novel disturbances that humans create challenge the adaptive mechanisms of organisms. The biggest challenge, however, is the extent to which anthropogenic disturbances have impacted the world's ecosystems and reduced their ability to recover. The intensification and expansion of disturbance regimes has become so pervasive that humans now control the fate of the biosphere. We have become engineers at a global scale of ecology (e.g. restoration), hydrology (e.g. dams), geology (e.g. mines), and even meteorology (e.g. climate change). Yet we are still learning the limits to our manipulative powers and what disturbances we can or cannot change.

I wrote this book to help explain to ecologists, naturalists, land managers, and others interested in disturbances what effects disturbances have on habitats, including both the initial disruption and subsequent habitat responses. I examine the parallels and differences between natural and anthropogenic disturbances to better understand the contributions of humans to present disturbance regimes. I also explore the constructive responses we can have to disturbances that will preserve the biodiversity and ecosystem functions that are vital to our well-being. Novel and shifting conditions in the future will require both understanding and foresight about disturbance effects and responses. The dramatic effects that humans are having on climate and natural resources constitute a grand experiment that we are all enmeshed in. With increased awareness and understanding of the effects of humans on disturbance regimes, coupled with a better understanding of natural responses to disturbance, we may be able to avert the worst consequences of disturbances. For example, possible lessons to draw from the triple disaster in Japan in March 2011 are: (1) we can do little about our vulnerability to earthquakes because of their unpredictability and spatial scale, except to minimize our risks by not building in earthquake zones and maximizing earthquake-resistant construction; (2) we have slightly more control over our vulnerability to tsunamis because they are more localized and we can decide not to live or build on vulnerable coastlines; and (3) we have near total control over our vulnerability to nuclear disasters by choosing whether or not to develop nuclear energy and where to place nuclear reactors. Understanding disturbances and our variable relationship to them is therefore essential to minimize disruptions of our communities and even loss of lives.

The scope of this book is broader than many others in the Biology of Habitat Series because disturbance is present in almost every habitat. Such breadth necessitates a less comprehensive coverage than one might like on any given topic, but I hopefully provide the necessary introduction and theoretical structure for readers to explore topics of interest in related literature. I address both terrestrial and aquatic disturbances of natural and anthropogenic origin. In addition, I cover the consequences of disturbance on ecosystem processes, biodiversity, and spatial and temporal patterns of ecosystems. Finally, I tackle how humans manage disturbances and the future consequences of global climate change and continued human population growth.

I thank many people for inspiration and insight, including such wonderful "disturbance" colleagues as Richard Bardgett, Peter Bellingham, Klaus Mehltreter, Roger del Moral, Duane Peltzer, Joanne Sharpe, Aaron Shiels, David Wardle, and Mike Willig. Roger del Moral and I have co-authored two previous books on disturbance and succession and he was a constant intellectual companion as I constructed this book. Contributors to an earlier book on disturbance ecology that I edited were also *de facto* writers of this book, as their chapters provided essential background in a wide range of topics. I also thank my undergraduate and graduate students for continual inspiration and feedback. Students in my Conservation Biology classes have been particularly helpful foils in the preparation of this book because they helped me grapple with the ecological tragedies of habitat loss and disruption without becoming too pessimistic. Two book series, The State of the World and Plan B, both by Lester Brown and his colleagues at the Worldwatch Institute deserve our collective gratitude for regularly summarizing the status of critical resources and offering practical roadmaps on how to respond to global problems. Working for the first time with Oxford University Press, I also acknowledge their professional and courteous assistance, particularly from Helen Eaton. Finally, I acknowledge with much love and admiration all four generations of my family, and in particular, my wife, Elizabeth Powell. I hope that my grandchildren can enjoy the strong connection with the natural world that I have been privileged to experience. Such connections provide the underpinnings of a value system that will hopefully lead us to improve our stewardship of the earth.

I was supported in part by grant DEB-0620910 from the U.S. National Science Foundation to the Puerto Rico Long-Term Ecological Research Program, by the U.S. Fish and Wildlife Foundation for work in Alaska, and by the Wilder Chair Program in the Botany Department at the University of Hawaii at Manoa that graciously supported me during the 2009–10 academic year. David Smee and particularly Paula Garrett ably assisted with the graphics and Peter Bellingham, Karl Gunnarsson, Elizabeth Powell, Ariel Lugo, and V. J. Neldner kindly provided photographs. I acknowledge a huge debt to the following people who helped improve this book by reviewing one or more chapters: Richard Bardgett, Peter Bellingham, Nick Brokaw, Robin Chazdon, Bruce Clarkson, Roger del Moral, Klaus Mehltreter, Karel Prach, Joanne Sharpe, Liz Walker, Margery Walker, and David Wardle. Special acknowledgement is due to my mother, Margery Walker, who has been a particularly diligent reader, reviewing every one of my books.

Contents

1 Introduction

1.1 Why study disturbance?

Disturbances are such an integral part of every habitat that any detailed study of habitats incorporates them. A **disturbance** is defined as a relatively abrupt loss of **biomass** or alteration of ecosystem structure or ecosystem function (Pickett and White 1985). Through their effects on growth, reproduction, and mortality (Wardle 2004), disturbances impact the diversity and evolution of organisms (Huston 1994). Disturbances also influence the spatial and temporal patterns of ecosystems across a landscape (Cale and Willoughby 2009) and ecosystem processes such as nutrient cycling. Humans are also affected by disturbances, most directly when they disrupt our work and travel, destroy homes, and kill people. We also try to predict and manage disturbances (Birkmann 2006) and manipulate disturbances through **restoration** (Hobbs and Suding 2009b).

There are many reasons for studying disturbances. Foremost is the curiosity that humans have about their environment, particularly when the environment is being disrupted in a dramatic way. This curiosity, imbued with a dose of anxiety about our own safety, involves wanting to know at what intervals the disturbances will arrive and how destructive they will be (Davis 2003). There is nothing like a dramatic natural disturbance such as a volcanic eruption, an earthquake, a fire, or a flood to command one's attention. Equally widespread, but sometimes less threatening, are **anthropogenic** (human-caused) disturbances such as transformation of land through agriculture, transportation corridors, urban sprawl, mining, and oil spills. Human history is full of disasters (disturbances that humans view as having negative consequences). Deaths and displacements of humans just in the last 80 years have been caused by volcanoes (U.S. 1980, Philippines 1991), floods (China 1931, Pakistan 2010), earthquakes (Pakistan 2005, Haiti 2010, New Zealand 2011), tsunamis (Indonesia 2004, Chile 2010, Japan 2011), oil spills (France 1978, USA 1989 and 2010), nuclear accidents (U.S. 1979, Soviet Union 1986, Japan 2011), industrial explosions (India 1984, Nigeria 1998), and bombings (England 1941, Japan 1945). However, disturbances can also benefit humans; for example, forest fires, flooding, or volcanic eruptions can result in improved soil fertility. With nearly instant global communication, we learn about recent dramatic disturbances wherever and whenever they occur. In that sense, we have all become amateur "disasterologists," following the disruptions of nature and their consequences for humanity. Unfortunately, as human populations increase, the impacts of disturbances on our property, livelihoods, and lives also increase (del Moral and Walker 2007) (Fig. 1.1).

A second reason for studying disturbances is to attempt to understand them (Walker 1999a). One way is to examine repeated patterns of damage or pathways of recovery across types of disturbances (Willig and Walker 1999). Questions arise about how similar one flood or one cyclone is to the last one in terms of damage to and responses of the biota and their ecosystems. If clear patterns emerge, can they be applied to interpreting and perhaps ameliorating similar disturbances at different locations? Ultimately, are there any general patterns that could apply to responses across types of disturbances (Glenn-Lewin *et al.* 1992; White and Jentsch 2001)?

A third reason for studying disturbances is to conserve biodiversity, in part because of an apparent increase in the damage caused by disturbances (Mittermeier *et al.* 2005). In this view, humans have so dramatically altered the earth by the disturbances that we have initiated or exacerbated, that we may be endangering not only other species but our own health and survival (Diamond 2005; del Moral and Walker 2007; but see Yoffee, 2009).

A fourth reason for studying disturbances is to be able to better manipulate and restore damaged ecosystems and recover the services that healthy

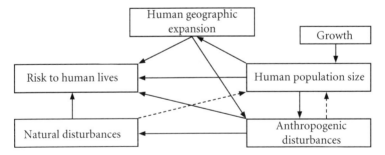

Figure 1.1   Human population growth leads to geographic expansion, and both of these changes increase anthropogenic disturbances that precipitate and intensify natural disturbances. Human lives are then increasingly at risk from the direct effects of larger population size and indirect effects of expansion and increased disturbance. Disturbances also kill people, but not in numbers sufficient to offset population growth. Solid lines indicate a positive influence, dashed lines a negative influence. (From Walker and del Moral (2009a) with permission from Wiley-Blackwell.)

ecosystems provide humans, such as clean air and water or healthy soils. To do this, we need to learn enough about how both natural and anthropogenic disturbances disrupt ecosystems and natural recovery processes so that we can assist in the recovery process. This challenge is compounded by the need to adapt to novel conditions that will arise due to combinations of natural and anthropogenic disturbances. Human disturbances are so pervasive that we now live in what has been called the Anthropocene or Age of Man (Crutzen 2002). **Biomes** have been so altered that they can now be called "**anthromes**" (Ellis *et al.* 2010). Sometimes humans attribute natural disturbances to divine causes or just bad luck. Praying, leading less sinful lives, or the purchase of good luck charms (Fig. 1.2) are seen as antidotes. Now, with the growing recognition that our own activities are causing many of these disturbances, actions that could lead to possible amelioration are our best antidote.

The concept of a disturbance is complex because it includes a cause (trigger event), the physical characteristics of that event (frequency, intensity, extent) and the damage that results (severity), and the consequences that are demonstrated by ecosystem responses (**succession**). For example, a volcano is considered a disturbance because eruptions with characteristic frequencies, intensities, and extent damage pre-existing ecosystems, which then respond and change over time. Additional complexity arises because loss of biomass (e.g. living or dead organisms) or structure (e.g. canopy) can be minimal to total and there are many types of ecosystem functions that can be altered (e.g. nutrient cycling or productivity). Furthermore, disturbances interact at

Figure 1.2   Japanese good-luck charm available in a temple in Kyoto along with charms "for happiness," "for lucky money," "for child," "for conception," and "for easy delivery."

a given site in what is called a disturbance regime (see Section 1.4). I use a broad definition of disturbance in this book that includes causes, characteristics, and consequences, but when discussing a particular disturbance, it is important to clarify which of these is the focus and to specify the relevant spatial and temporal contexts under consideration (Walker and Willig 1999).

Disturbances are relatively abrupt events in time, but actually occur across a range of temporal (and spatial) scales. Ecological responses to large disturbances are measured over longer periods than responses to small ones. For example, dinosaurs and half of all other organisms living 65 million years ago were probably annihilated due to the consequences of a meteorite impact in Mexico. That event, although short in duration, permanently altered the long-term course of evolution on earth. Alternatively, very small-scale disturbances can impact microhabitats of seedlings or insects and occur repeatedly during a single day. To interpret the effects of a given disturbance, it is important to choose the appropriate temporal and spatial scales that are both context- and organism-dependent.

Responses to disturbances are also measured on a variety of spatial and temporal scales (Fig. 1.3). That disturbance is considered a relatively abrupt

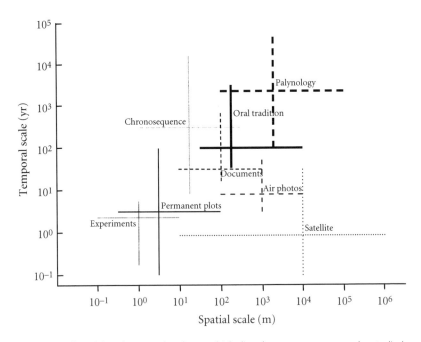

Figure 1.3 The range of spatial and temporal scales at which disturbance responses may be studied. Where the horizontal and vertical lines cross is where the method is most commonly used. (From Walker and del Moral (2003) with permission from Cambridge University Press.)

process distinguishes it from both long-term, gradual changes to an environment induced by **stress** and repeated, but minor, **fluctuations** such as seasonal changes. However, such distinctions are not always clear and depend largely on the chosen perspective. For example, a fire that is unusual, destructive, and resets successional trajectories is clearly a disturbance. Annual fires that burn only some species such as understory herbs, leaving forest canopies intact, are not clearly disturbances unless one focuses on local damage to the herbs or associated fauna in that particular year. With a broader spatial or temporal perspective, the annual fires are just fluctuations. Similarly, plants evolved with herbivores so herbivory at background levels could be considered a stress. However, a sudden insect outbreak inflicting unusual damage on a single species could be considered a disturbance.

1.2 Types of disturbance

1.2.1 Natural

Natural disturbances are independent of human influences. This definition includes any disturbances that occurred before human contact and any that are currently not affected by human activities such as volcanoes, tsunamis, and most earthquakes. Natural disturbances will be examined for a variety of terrestrial habitats in Chapter 2 and for aquatic habitats in Chapter 3.

1.2.2 Anthropogenic

Anthropogenic disturbances are those that are primarily due to humans. Some are uniquely human creations such as pavement or mine tailings. Others resemble natural disturbances (Table 1.1) or form complex mixes with natural disturbances (e.g. when humans remove rocks from a newly flooded river basin or logs from a recently burned forest). Such novel mixes of types of disturbance, often followed by colonization by novel assemblages of native and non-native species, provide new perspectives for studies of **competition**, succession, and even evolution (Hobbs *et al.* 2006). Anthropogenic disturbances will be explored in Chapter 4.

1.2.3 Allogenic and autogenic

Allogenic disturbances originate from outside the system of interest. A cyclone that blows across the Atlantic Ocean to the Caribbean Sea is clearly allogenic, as is a caribou that tramples a particular patch of lichen during its long seasonal migration. Autogenic disturbances originate within the system. Examples include local erosion triggered when a senescent tree falls and the digging, burrowing, and other activities of resident animals. Note

Table 1.1 Analogous surfaces resulting from either natural or potentially analogous anthropogenic disturbances, listed in order of decreasing severity, from the complete removal of all soil and biomass to very little disruption. (From Walker (in press), with permission from Wiley-Blackwell.)

Surfaces from natural disturbances	Surfaces from anthropogenic disturbances
Lava flow	Pavement
Cliff face	Wall of building
Volcanic ash	Mine tailing
Glacial moraine	Large construction site
Landslide (triggered by rain)	Landslide (triggered by road building)
Floodplain	Mine slurry
Burn (triggered by lightning)	Burn (triggered by matches)
Grazed meadow	Pasture
Forest (felled by hurricane)	Forest (felled by logging)
Treefall (triggered by windstorm)	Treefall (felled by humans)
Gopher mound	Local construction site
Animal trail	Vehicle track

that both allogenic and autogenic disturbances are often combined and can have either **abiotic** or **biotic** causes (Glenn-Lewin & van der Maarel 1992). A tree disease may begin in one forest patch (autogenic) then spread to others (allogenic). Similarly, a moth may fly into an area from some distance (allogenic) and deposit eggs; the resulting larvae then grow by feeding on leaves of a local tree (autogenic). Allogenic and autogenic are also terms applied to succession (Tansley 1935) (see Chapter 8).

1.2.4 Disturbance by addition

Disturbances are usually defined by loss of biomass or alteration of structure, but disturbances can also be caused by the addition of species or nutrients. For example, the invasion (addition) of a non-native organism can lead to the exclusion of native species, especially when diseases or predators (e.g. brown tree snake, *Boiga*) are introduced to islands (Davis 2003). Similarly, additions of nutrients to an ecosystem can also be considered a disturbance, particularly when the addition leads to dominance by species adapted to high-nutrient conditions that exclude other organisms and to altered successional trajectories. For example, the introduction of the nitrogen (N)-fixing tree *Morella* to low-nutrient volcanic soils in Hawaii locally excluded native trees and promoted alternative communities that were adapted to the increased nutrient levels (Walker and Vitousek 1991). In both cases, overall biomass might decline, increase, or remain constant, but the original community structure and ecosystem function have been altered.

1.3 Causes of disturbance

The initial cause that triggers a disturbance (such as increased ocean temperature) is distinguished from the characteristics of the disturbance itself (such as wind speed) and the response of the ecosystem to the disturbance (such as recovery of biota from damage caused during a cyclone). A cause can be abiotic, as with a cyclone, or biotic (e.g. gophers build mounds and tunnels or insect outbreaks defoliate a stand of trees). Sometimes the immediate cause of a disturbance cannot be determined, although the larger context can usually be identified. For example, a slope may be gradually eroding but the cause of a particular slippage may be obscure. Disturbances can occur when accumulated stresses reach a **tipping point**. When a particular cause has been identified, further disruption of the system can be avoided (e.g. by not building roads on steep slopes because they can trigger landslides).

1.4 Characteristics of disturbance

1.4.1 Frequency

Disturbance frequency is the number of similar disturbance events at a given location per unit time. Return intervals are often used as a measure of disturbance frequency (Glenn-Lewin and van der Maarel 1992). For example, a treefall gap in a forest can have a gap return interval, or the time between gap formation events for a particular forest. Forest gap return intervals range from 50 to 100 years (Veblen 1992), but return intervals of fires in grasslands can be every few years. Return intervals of earthquakes or volcanic eruptions can reach thousands of years. The size distribution of gaps can be more important in determining species composition than mean return interval (Brokaw 1987). Minor disturbances that return on regular intervals within the life span of an organism, such as seasonal changes for perennial plants and long-lived animals, are often considered fluctuations. For annual organisms, the onset of winter can be fatal and certainly can be considered a disturbance. Species may evolve adaptations to disturbances that occur several times during the life span of an individual.

1.4.2 Intensity

Intensity is a measure of the force of the disturbance. Examples of measurements of intensity include volcanic explosivity, earthquake magnitude, fire temperature, wind speed, and rainfall accumulation per hour (Table 1.2). Intensity can be altered by past disturbances in the area as well as current conditions. For example, fire temperature depends on current fuel loads, which are in turn dependent on the intensity and biological response to

Table 1.2 Measurements of disturbance intensity for a variety of natural disturbances, listed in approximate order of increasing extent (see Chapter 2).

Disturbance	Measure of intensity
Volcano	Volcanic explosivity index (height and volume of tephra plume)
Earthquake	Horizontal displacement (Richter and moment magnitude scales)
Dune	Depth of sand added
Erosion	Depth of soil displaced
Cyclone	Wind speed (five categories); atmospheric pressure
Flood	Depth above mean level; rate of rise; flow rate; mm rain per hour
Drought	Days without rain; departure from mean precipitation
Fire	Temperature, duration
Tsunami	Wave height; distance traveled inland

past fires. Wind speed, temperature, and humidity at the time of the fire will also affect fire temperature. As instrumentation improves, measurements of intensity become more sophisticated. Networks of sensors allow extensive and real-time measurements not previously available.

1.4.3 Severity

Severity is the amount of biomass lost or the degree of alteration of ecosystem structure or function caused by a disturbance. Severity can be related to intensity (higher wind speeds usually cause more damage) and both together are considered as the magnitude of a disturbance. Typical measures of severity include percentage mortality of species or kilograms of biomass removed. When disturbances are frequent, net loss of biomass may be reduced because the ecosystem does not recover in the interlude between disturbances. When disturbances are so severe that there is little or no **biological legacy**, the recovery process is called primary succession. When there is a substantial biological legacy (i.e. substantial amounts of soil and many organisms survive), the process is called secondary succession.

Substrate characteristics such as fertility, stability, and texture following a disturbance (Fig. 1.4) are another way to measure severity and to assess the initial conditions for succession. Disturbances that remove most fertile soil, for example, can be considered more severe than those that have little effect on soil fertility. Relative loss of fertility (compared with initial, pre-disturbance fertility) is one measure of severity, but absolute fertility that remains (regardless of pre-disturbance conditions or percentage loss) is more relevant for succession. Substrate stability, a measure of resistance to erosion, is, like fertility, often reduced by a disturbance and can reduce rates of colonization (Fig. 1.5). However, reduction in either

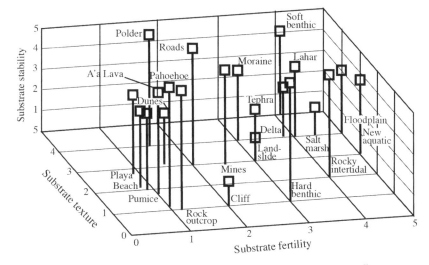

Figure 1.4 Substrate stability, texture, and fertility in different types of primary succession. A score of 5 indicates substrates that are very stable, fine-textured, or fertile. From Walker and del Moral (2003) with permission from Cambridge University Press.

Figure 1.5 Hydroids at 20 m below sea level on relatively stable, hard volcanic slopes of Surtsey Volcano, Iceland. Colonization was greater on stable than unstable slopes. (Photograph by Karl Gunnarsson.)

fertility or stability by a disturbance may favor organisms adapted to unstable or infertile substrates. Substrate texture is also frequently altered by a disturbance. For example, river floods can change floodplain substrates on a given stretch of a river from fine-textured, clay-rich soils to sands and gravel or can inundate sands and gravels with fine sediments. Severity in this case would be relative to the organism of interest, as most organisms have a preferred substrate texture (Grubb 1986). Alterations of texture by disturbance therefore also alter post-disturbance colonization dynamics (see Chapter 2), particularly through their influence on substrate fertility and spatial heterogeneity of resources (Dale *et al.* 2001; Walker in press).

1.4.4 Extent

Extent is the area impacted by a disturbance and it varies from global to local. Volcanic gases can spread into the upper atmosphere and affect the entire world (Fig. 1.6). Likewise, diseases and certain animals and weedy plants have invaded many parts of the world. Dunes, floodplains, and road corridors can also impact large areas of land. Disturbances of

Figure 1.6 A minor steam eruption (in August 2010) on Anak Krakatau, a volcano that first emerged above the ocean in the Sunda Strait, Indonesia in 1927. The tourist boat in the foreground is traversing the area of the collapsed and flooded caldera from the major eruption of the original island in 1883, which caused a decline in global temperatures.

intermediate extent include cyclones, glacial moraines, and large mines. Animal burrows, herbivory, single treefalls, and whale carcasses on the sea floor are examples of disturbances that can be mostly local. Within a disturbed habitat there can also be undisturbed patches, resulting in spatial heterogeneity and variable disturbance responses (see Chapter 7).

1.4.5 Interactions

Ecosystems are shaped by complex arrays of interacting disturbances and the sum of all disturbances at a given site forms the disturbance regime (Fig. 1.7; Walker and del Moral 2003). Examination of just one disturbance at a time therefore gives an unrealistic perspective on the conditions that local organisms must adapt to. A better but unusual approach is to determine the array of disturbances impacting a site and partition the proportion of the damage that was caused by each disturbance. An easier task is to measure the net responses of the organisms to the disturbance regime (Zimmerman *et al.* 1996). Interactions among disturbances generally occur when certain disturbances trigger other ones. For example, the melting of a glacier produces an unstable moraine prone to landslides and flooding. An earthquake can trigger volcanic activity, tsunamis, landslides, and flooding. A cyclone can lead to drought, floods, treefalls, landslides, and herbivory (Fig. 1.8). Disturbance effects can also cascade down food chains or trophic levels, especially when a keystone species such as a wolf (D. Smith *et al.* 2003), a sea star (Paine 1966), or a dominant plant species (Bentz *et al.* 2010) is removed.

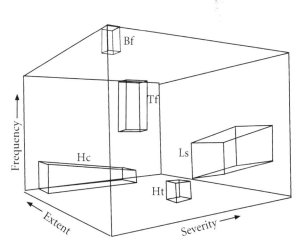

Figure 1.7 Disturbances in the Luquillo Experimental Forest, Puerto Rico vary in frequency, extent, and severity. Bf = branchfall; Tf = treefall; Hc = hurricane; Ls = landslide; Ht = tree harvest. (From Waide and Lugo (1992) with permission from Springer.)

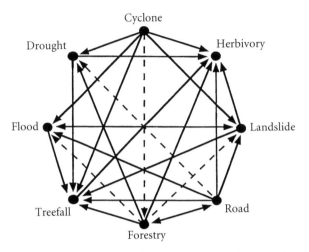

Figure 1.8   Disturbance interactions in a lowland rainforest in Puerto Rico. Arrows indicate when the occurrence of one disturbance enhances the probability of another disturbance. Weaker links are shown as dotted lines. (From Willig and Walker (1999) with permission from Elsevier.)

1.5 Theory of disturbance

1.5.1 History

Disturbances have shaped the earth and molded evolution, including human evolution. Early human societies, in turn, caused disturbances by setting forest fires, clearing forests, hunting, fishing, and plowing fields to ensure adequate supplies of food. As human societies became more centralized they also became increasingly vulnerable to natural disturbances (e.g. the Minoan civilization collapsed after Thera Volcano erupted in 1623 BC and two Mongol invasion fleets were destroyed when Japan defended itself with the aid of timely cyclones in 1274 and 1281; del Moral and Walker 2007). Early writings about disturbance ecology include those by Theophrastus (300 BC) who examined plants along floodplains, Linnaeus (1735) who noted succession in bog communities, and Thoreau (1860) who described succession following logging. Disturbance ecology had its formal birth with Warming's descriptions of plant succession on dunes, volcanoes, landslides, and other disturbed habitats (Warming 1895, 1909) and was supplemented by insights from Cowles (1899, 1901) about the role of abiotic disturbances. Davis (1909) noted more evidence of disturbance in younger versus older terrain. Clements (1928) explicitly addressed many types of disturbances in his seminal studies of succession by including volcanoes, dunes, flood-plains, and prairie fires. As the field of ecology developed during the 20th century, disturbance continued to be a central theme in the study of lakes

(Shelford 1911), abandoned agricultural fields (Odum 1950), animal populations (MacArthur 1955), forest cutting (Marks and Bormann 1972), and nutrient cycling (Rice and Pancholy 1972).

Modern disturbance ecology has seen a shift from the earlier view that ecosystems return to an equilibrial state following a disturbance (Odum 1969) to a view that disturbances constantly keep ecosystems in a state of disequilibrium, at least at scales relevant to the organisms inhabiting them (Pickett and White 1985). This trend is essentially one part of a larger shift from **holism** to **reductionism** that is summarized in McIntosh (1985) and Walker and del Moral (2003). Disturbances are now seen as essential drivers of changes in community structure and biodiversity (Connell 1978; Chapin *et al.* 2000). The recent emphasis has centered more on abiotic and biotic responses than on the characteristics of the disturbances themselves (but see Johnson and Miyanishi 2007). Modern disturbance ecology tackles issues of spatial heterogeneity (Turner 1989; 2010), hierarchical responses (Pickett *et al.* 1999), disturbance interactions (Dale *et al.* 2001), and hazard prediction and management (Restrepo *et al.* 2009). Instead of emphasizing natural disturbances in isolation from human activities, growing anthropogenic influences now direct disturbance ecologists to examine global warming, **novel ecosystems**, and feedbacks between human activities and disturbance characteristics (Turner *et al.* 2003; Hobbs *et al.* 2006; Turner 2010).

1.5.2 Resistance

Resistance is the ability of an ecosystem to persist in the presence of a disturbance without substantial alteration in biomass, biodiversity, or other characteristics. Resistance in a particular disturbance regime may not be equal among all species. For example, certain tree species are more susceptible to breakage in strong winds than are other species (Walker *et al.* 1992; Quine and Gardiner 2007). Among individuals of a given species, resistance may decrease and ultimate damage (severity) may increase as the intensity of the disturbance increases (e.g. first leaves, then small branches, and finally large branches and trunks break off in strong winds). This progression is not always followed, particularly if a disturbance is sudden. Under such circumstances, tree trunks can break while the leaves remain intact.

1.5.3 Resilience

Resilience is the ability of an ecosystem to recover following disturbance. The recovery can be slow or fast, partial, or complete, and can begin immediately or after a lag period (Fig. 1.9). Ecosystems tend to be either resistant (e.g. oak forest to fire) or resilient (e.g. grassland to fire), suggesting a close interaction between disturbance effects (loss of biomass) and disturbance response (amount and rate of increase of biomass; Walker in press). There

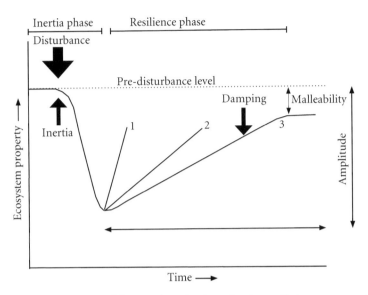

Figure 1.9 Resilience of an ecosystem following disruption by a disturbance that overcomes resistance (inertia). Hysteresis results in loss of elasticity with sequential perturbations. Damping modifies the final recovery of an ecosystem property that is then measured as the malleability of the ecosystem. (Modified from Westman (1986), with permission from Springer.)

are many characteristics of the pre-disturbance ecosystem that can recover, including physical structure, species composition, biomass, ecosystem function, and even resistance to future disturbances.

1.5.4 Stability

High resistance and high resilience are often considered to be aspects of stable ecosystems, once thought to be part of self-regenerating climax communities. The climax concept evolved into the concept of a dynamic equilibrium and then into the concept of multiple but spatially limited stables states (Beisner *et al.* 2003). Alternative stable states may exist during the recovery of an ecosystem from disturbance. They can arise through shifts in state variables (e.g. predator additions or deletions; overharvesting a resource) or shifts in ecosystem parameters or drivers (e.g. slope stability; minimum air temperature). **Hysteresis** results when a perturbed system cannot return to the original state (Fig. 1.10). The dominant paradigm now is that all systems are at some degree of disequilibrium and that disturbance frequently disrupts ecosystems before any equilibrium can be reached. **Stability** is affected by spatial scale. Local recovery of ecosystem structure or function may not occur even when stability is reached at larger spatial scales (Ingegnoli 2004).

Parameter shift

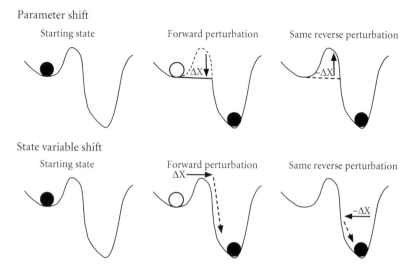

Starting state   Forward perturbation   Same reverse perturbation

State variable shift

Starting state   Forward perturbation   Same reverse perturbation

Figure 1.10   Alternative stable states can be achieved through shifts in ecosystem parameters or state variables, but those shifts are not necessarily reversible due to hysteresis. See text for further explanation. (From Beisner *et al.* (2003), with permission from the Ecological Society of America.)

1.5.5 Intermediate disturbance hypothesis

The intermediate disturbance hypothesis (IDH) has been a remarkably robust proposal that species diversity (see Chapter 6) is maximal when disturbance frequency and extent are intermediate. The IDH also suggests that species diversity peaks in mid succession (Grime 1973; Connell 1978). The implication is that at high disturbance frequency (and extent), many species die and that at low disturbance frequency (and extent) a few species with superior competitive abilities exclude most other species. At intermediate levels of disturbance, species adapted to both ends of the spectrum can co-exist, leading to maximal biodiversity (Huston 1979). Evidence for this general relationship has come from a variety of habitats including tropical forests (Molino and Sabatier 2001; Bongers *et al.* 2009), **boreal** forests (Biswas and Mallik 2010), coral reefs (Rogers 1993), intertidal zones (Sousa 1979a), riverbeds (Townsend *et al.* 1997), and planktonic habitats (Flöder and Sommer 1999). However, many questions have been raised by this deceptively simple hypothesis. Some of the many problems with the IDH include its lack of specificity about what scales, systems, or trophic levels are being considered, whether it can apply to disturbance severity (that may have different effects from disturbance frequency on biodiversity), whether species richness or evenness is being considered (Rogers 1993), and the impact of invasive species that might alter biodiversity (Hobbs and Huenneke 1992). Support for the IDH is countered by the evidence of many studies that do not show the expected relationship (Mackey and Currie 2001), particularly when addressing fire frequency (Collins *et al.* 1995; Schwilk *et al.* 1997) but also from additional habitats such as tropical forests (Sheil and Burslem 2003). Trophic levels can respond differently to disturbance as well. In microbial communities affected by anthropogenic disturbances, for example, the IDH applied to species diversity but not genotypic diversity (Parnell *et al.* 2009). The IDH has helped focus the attention of ecologists on the relationship between disturbance and species diversity but has also led to some confusion about its applicability, a scenario reminiscent of the debate surrounding the general applicability of Connell and Slatyer's (1977) three models of succession (see Chapter 8). The IDH actually summarizes a set of explanations for co-existence that requires patchy disturbances (Roxburgh *et al.* 2004), but within-patch disturbances are also possible. It appears that the links between disturbance and biodiversity are not easily represented by simple models (Whittaker 2010). Further discussion of disturbance/diversity relationships is given in Chapter 6.

1.6 Responses to disturbance

The focus of this book is the response of ecosystems and their constituent organisms that are disrupted by disturbances. After an introduction to particular types of disturbances, I examine changes in ecosystem processes, biodiversity, spatial patterns, and temporal dynamics across disturbance types. There are no prescribed patterns of how ecosystems respond to disturbances, making precise predictions difficult but leaving plenty of room for more study. Organisms do follow a general pattern through their life histories, however, of dispersing to a disturbed habitat, establishing in it, interacting with each other, and then dying (Walker and del Moral 2003). Clements (1928) recognized the first three of these responses (which he called migration, **ecesis**, and competition) as processes that drive succession.

1.6.1 Dispersal

Plants can disperse to a disturbed site as seeds, spores, or vegetative parts and via wind, water, gravity, or animal **dispersal**. After extensive disturbances that leave no biological legacy, colonization will depend on species able to disperse long distances (Wood and del Moral 1987). Most species have limited dispersal distances (Malanson and Cairns 1997), although only a few individuals need to spread widely to advance a population (Bullock and Clarke 2000). *In situ* survivors can spread out from surviving vegetation. Sometimes vegetation advances from the edges (phalanx dispersal) and sometimes from newly established pockets of vegetation within the disturbance (guerilla dispersal or **nucleation**; Clarkson and Clarkson 1983; Clark *et al.* 1999). Animals can disperse by wind (passively or actively by flying), water (passively or actively by swimming), soil (burrowing, tunneling), walking, or hitchhiking on other animals. Animals can also survive disturbances *in situ* and spread via phalanx or guerilla dispersal. The spatial aspects of dispersal are discussed in Chapter 7.

1.6.2 Establishment

Establishment is the successful colonization and growth of organisms following a disturbance. Successive waves of colonists find changing conditions as the original colonists alter the available space and resources. Establishment is slowest following severe or extensive disturbances (Miles 1979). Where there is a high degree of survivorship on a small affected area, establishment can be quite rapid (Hobbs and Cramer 2007). Factors that limit establishment include germination safe sites, surface stability, and unfavorable environmental conditions (e.g. drought, acidity, temperature extremes; Díaz *et al.* 1998). The temporal implications of establishment are discussed in Chapter 8.

1.6.3 Species interactions

Species using the same space and resources following a disturbance interact in positive (facilitative) and negative (competitive) ways or may not interact at all (Callaway 2007). These interactions, among other factors, determine competitive dominance and relative resource acquisition as well as species replacements when one species outcompetes another (Tilman 1988; Keddy 1989; Walker and del Moral 2003). Species interactions therefore influence both spatial patterns (see Chapter 7) and temporal dynamics (see Chapter 8).

1.6.4 Senescence

As organisms age their ecological roles change. When individuals die their absences sometimes create big gaps in an ecosystem, such as when a large tree falls or an alpha male wolf is killed by hunters. Death can strike many members of the same species at the same time. For example, pests such as bark beetles strike and kill many trees in large patches in forests (Ayres and Lombardero 2000; Fettig *et al.* 2007); alternatively, many trees die simultaneously due to cohort senescence (Mueller-Dombois 1986, 2006). Dead organisms provide fertile substrates for organisms that are decomposers (see Chapter 5) and support the establishment of colonists or growth of survivors.

1.7 Ecological and economic benefits of disturbance

Disturbance results in a loss of biomass, reduced community structure, and the removal of soil and nutrients. Yet one positive aspect of disturbance is that it provides new habitat for species adapted to conditions of higher light and lower nutrients (Bazzaz 1996). Nutrients such as phosphorus (P) are often exposed by disruptions of soil or bedrock (see Chapter 5) and made available to plants, while animals may benefit from newly exposed salts. All species evolved under some kind of disturbance regime so the absence of disturbance can actually be a problem. For example, occasional very intense fires destroy more biomass and soil than less severe but more frequent fires. Even anthropogenic disturbances can benefit certain species, as seen in the success of agricultural weeds under a regime that includes annual disruptions from plowing and harvesting. The relative value of a particular disturbance is therefore defined by the situation, much like intensity, severity, and other characteristics of disturbances.

1.8 Organization of the book

This book provides an overview of the topic of disturbance ecology, focusing on how organisms respond to disturbance. Although disturbance is broadly defined to encompass any relatively sudden loss of biomass or alteration of structure or function in an ecosystem, a specific disturbance can best be evaluated in terms of the spatial and temporal scales relevant to the organisms that are affected. I therefore examine disturbances first by habitat and increasing spatial extent (Chapters 2–4), then by general patterns and processes that are altered by disturbance (Chapters 5–9). Chapter 2 covers natural disturbances in terrestrial habitats, including the responses of soil communities. Chapter 3 considers the responses of marine and freshwater habitats to natural disturbances such as volcanoes, floods, tsunamis, and cyclones. Chapter 4 discusses responses to anthropogenic disturbances in both terrestrial and aquatic habitats and contrasts ones that are local (e.g. military activities) to global (e.g. agriculture) in spatial extent. Chapter 5 looks at how ecosystem processes such as nutrient cycling and productivity change when a habitat is disturbed. Chapter 6 explores how biodiversity of plants, animals, and microbes is altered by disturbances and examines the effects of invasive species on biodiversity. Chapter 7 addresses how both natural and anthropogenic disturbances alter the spatial patterns of organisms. Chapter 8 considers temporal responses to disturbance on gradients of decreasing severity that result in conditions triggering primary to secondary succession. Several theoretical perspectives on succession are also presented. Chapter 9 discusses how humans approach the management of disturbances by conserving biodiversity and maintaining disturbance regimes, as well as both passive and active efforts to restore biodiversity and ecosystem processes. Chapter 10 examines the links between global climate change, human population growth, and disturbance; summarizes lessons that can be learned from disturbance ecology; and ends by envisioning a future where ecosystems and humans would be less vulnerable to disturbance.

2 Terrestrial habitats

2.1 Introduction

Natural disturbances such as volcanoes and earthquakes have helped shape the earth's terrestrial surface since it was formed over 4 billion years ago. Sand dunes, floods, glaciers, and landslides have had local and regional influences on landscapes. Other disturbances including fire, insect outbreaks, droughts, and cyclones have altered the distribution and composition of the plants and animals living on the land. All of these natural disturbances continue, despite the growing influence of humans on the earth. Some are unaffected by humans (volcanoes, earthquakes), some are more frequent in human-altered than pre-human landscapes (landslides, and possibly cyclones), and the frequency of others is either decreased or increased by humans, depending on the local dynamics (drought, dunes, fire, insect outbreaks, and floods). Humans have also introduced novel disturbances (see Chapter 4). We have altered the biota through centuries of agriculture and horticulture, geology with mines, hydrology with dams, and now global climate with the release of greenhouse gases. Therefore, in the following discussion the focus will be on those natural disturbances that are least influenced by humans—but the pervasive influence of humans cannot be ignored.

Previous overviews have organized terrestrial disturbances by their affiliation with earth, air, water, or fire (Walker and del Moral 2003), by the severity of the disturbance (del Moral and Walker 2007), and by pairing the physical processes that characterize a disturbance with its biological implications (Johnson and Miyanishi 2007). In this chapter, terrestrial disturbances (and the terrestrial implications of floods) are organized by increasing spatial extent (percentage of the terrestrial land mass that they affect), with an emphasis on disturbances that are widespread, frequent, and with direct habitat implications (Table 2.1); natural disturbances in aquatic habitats are the subject of Chapter 3. This organization excludes terrestrial disturbances that are very localized (e.g. animal burrows), infrequent (e.g. meteorites), or with only indirect biological implications (e.g. plate tectonics). An overview of how soil structure and soil organisms are affected by disturbances precedes the discussion of 10 widespread and biologically relevant disturbances: volcanoes, earthquakes, dunes, erosion, insect outbreaks, glaciers, cyclones, floods, droughts, and fires. For each disturbance type the following topics are covered: a brief introduction to its relevance to humans, the characteristics of the disturbance, and the ecosystem responses. These topics are further developed in later chapters on ecosystem processes, biodiversity, spatial, and temporal patterns. This chapter and Chapters 3 and 4 focus on each disturbance while the remaining chapters address themes that are not disturbance-specific.

Table 2.1 Natural disturbances and their major habitat alterations for terrestrial (this Chapter) or aquatic (Chapter 3) disturbances. Within each category, disturbances are presented in approximate order of increasing spatial extent in that habitat (approximate percentage of habitat affected in parentheses). See Table 4.1 for anthropogenic disturbances (both terrestrial and aquatic).

Habitat	Name	Habitat alterations
Terrestrial	Volcanoes (1)	Destroy biota; create new landscapes; fertilize
	Earthquakes (1)	Damage biota; alter topography
	Dunes (7)	Bury soils; provide unstable, new environment
	Erosion (10)	Creates heterogeneity, fewer nutrients, more light
	Insect outbreaks (10)	Make plants vulnerable to other disturbances; promote species diversity
	Glaciers (10)	Expose new, unstable terrain; alter topography
	Cyclones (15)	Damage biota; cause erosion; create gaps
	Floods (15)	Disrupt topography; rearrange nutrients, soils
	Droughts (30)	Destroy susceptible, but promote tolerant, biota
	Fires (>50)	Recycle nutrients; create gaps; alter microsites
Aquatic	Carcasses (<1)	Introduce nutrients; support benthic fauna
	Volcanoes (1)	Destroy biota; create new surfaces for pioneers
	Hydrothermal vents (1)	Damage local biota; permit chemosynthesis; support unusual communities
	Earthquakes (1)	Bury or lift up sea floor, reefs, seagrass beds
	Erosion and deposition (10)	Bury biota; reduce photosynthesis; trigger eutrophication; favor pioneers
	Floods and droughts (15)	Erode riverbanks; scour riverbeds; sort and rearrange sediments; fertilize and destroy biota
	Tsunamis and other waves (15)	Churn up sediments; bury biota; damage reefs; create heterogeneity; promote pioneers
	Cyclones (15)	Disrupt wildlife; damage reefs; disperse corals

2.2 Soils

Most terrestrial disturbances affect soil structure, and soils are an essential factor in the response of ecosystems to disturbance. Soils provide a thin layer that covers most of the terrestrial surfaces of the earth and upon which plants, animals, and humans depend for their existence. Perhaps a world without soil is in our future, when all plants are grown hydroponically and we live packed into multi-story, urban dwellings. However, for now we depend on soil for agriculture, silviculture, nutrient cycling, decomposition, and the existence of nearly all terrestrial ecosystems. Soil is a complex mixture of minerals and organic matter in varying stages of decomposition (Bardgett 2005) that also includes gases, water, and a wide assortment of **archaea**, bacteria, and animals. Although many biological processes in soil (e.g. nutrient cycling through bacteria and other decomposers) occur so rapidly that it is hard for scientists to measure them, soil formation is a process that takes centuries to millennia. As humans cause over 77 billion tons of fertile soil to erode from the land every year (Myers 1993), we destroy a central and irreplaceable part of the ecosystems we depend on. The disturbance of soils is an essential part of any coverage of the disturbance of terrestrial habitats. Soils as sources of nutrients are discussed in Chapter 5.

2.2.1 Structure

Soil structure is the combination of physical characteristics that describe a soil, including fertility, aeration, stability, and texture. Soils are formed from parent material (rock) that is weathered by chemical, physical, and biotic means into its constituent minerals. These minerals, combined with decaying organic matter that collects on or in the soil, provide the substrate in which soil organisms live, interact, and die. The decaying bodies of the soil biota further enrich the developing soil by releasing nutrients that are quickly recycled by living organisms. Eventually, complex humic acids develop and soil structure reaches a state in which it retains water, has readily available nutrients, and can physically and nutritionally support the growth of plants. Soil horizons (layers) form over time with most organic matter near the surface. The number, thickness, and composition of these horizons vary depending on the local conditions of temperature, precipitation, drainage, and chemical composition of both the rocks and the decaying organic matter. Rudimentary soil horizons can develop within a few decades but continue to change for the life of the soil (which can reach millions of years) as minerals are gradually leached into lower layers. Nutrient availability (typically) increases then declines over geological time. In young soils, low levels of N can be limiting for plant growth, while in old soils low levels of P are more often limiting (Walker & Syers 1976, Peltzer *et al.* 2010).

Disturbances disrupt soil structure in a number of ways. Animals that dig up or burrow under soil normally loosen soil particles and improve the water percolation and soil aeration needed by most soil organisms. Soil compaction leaves soil horizons intact but reduces aeration. One example of a habitat disturbed by soil compaction is heavily grazed pastures, particularly along trails or around watering holes where herds of large grazing animals congregate. Flooding can also reduce soil aeration and lead to anoxic conditions as well as cause soil erosion upstream and deposition of unconsolidated organic matter downstream along floodplains and deltas. The eventual downstream deposition of eroded soils is a disturbance to *in situ* soils but, in many cases, these new deposits are quite fertile (Cassel and Fryrear 1990), especially if upstream soils were nutrient-rich, indicating that not all disturbances involve the removal of nutrients or soil. Colonization of these new surfaces depends on their stability, because they can be subject to renewed erosion. Trees that dominate floodplains are usually tolerant of some flooding. Wind is also a major contributor to soil erosion, and many land formations are created by the deposition of wind-borne soil, ash, and dust particles (e.g. **loess**, dunes). Biological soil crusts that form on soil surfaces from algae, lichens, mosses, and other organisms can reduce erosion in windy landscapes such as deserts where there is minimal vegetation cover (Belnap and Gillette 1998). Other disturbances can affect soil fertility, such as salt spray, ash deposition, or acid rain (from volcanoes or human activities).

2.2.2 Humans

Humans have had substantial effects on soil structure. Agriculture is essentially a manipulation of soil conditions to grow desired crops that involves improving fertility, aeration, and infiltration. In traditional, small, and organic farms this manipulation has included additions of manure, crop rotations, and multi-species crops in the same field that help to maintain active soil communities. Large-scale agriculture has sacrificed long-term soil health for short-term increases in production. Over time, a reliance on chemical fertilizers and frequent use of pesticides and herbicides to control undesirable species destroys natural soil communities. Frequent and deep plowing can improve soil aeration, decomposition, and infiltration, but also leads over the long term to compaction below the area of plowing. Plowing also results in accelerated soil erosion, especially where soils are left exposed between plantings. Agriculture and forestry have resulted in substantial increases in soil loss (Pimentel and Harvey 1999), but also in declines in numbers and species richness of soil organisms (Bardgett 2005).

In addition to agriculture, humans alternatively loosen and compact soils through surface mining, transportation, and construction of roads and buildings. Road (and railroad) construction has the opposite goal to agriculture for soil structure. Road engineers try to reduce aeration, organic matter, and infiltration to avoid the fluctuations in water content and root growth that would disrupt road surfaces (Sojka 1999). However, placing what are effectively impermeable barriers across the landscape can alter drainage patterns and result in both flooding and erosion.

2.3 Volcanoes

Volcanoes are perhaps the most intense natural disturbance on earth (del Moral and Grishin 1999). The magma comes from the depths of the earth, produces fiery displays, creates land, and builds mountains. Ash from a volcano can cause colorful sunsets around the world and cool global temperatures (e.g. Krakatau, Indonesia 1883; Pinatubo, Philippines, 1991). Millions of people live near or on volcanoes because decomposing ash and cinder provide fertile soils. Most volcanoes erupt episodically (with typical return intervals of centuries to millennia), and generations of people may no longer worry about an eruption (Scarth 1999), but volcanoes can be dormant for 10 million years or more, so threats to lives and property persist. The residents of Seattle, Washington (USA), for example, will be affected when Mount Rainier erupts again. Part of the fascination, perhaps, is our lack of control over volcanic eruptions. Only in rare instances have humans restricted the flow of lava. Icelanders on the small island of Heimaey kept a volcano from closing their irreplaceable harbor by spraying the flows day and night with hoses from fire boats (Williams and Moore 2008). Despite our lack of control over most volcanoes, we can study natural recovery following volcanic eruptions and use that knowledge to restore ecosystems damaged by volcanoes or disturbances that create similar surfaces (Walker and del Moral 2003).

Volcanoes, or the surface eruption of magma from the earth's interior, occur mostly along the edges of continental plates that are shifting in response to temperature changes and magma below. Seventy-five per cent of all active volcanoes (ones that have erupted in the last 10 000 years) are found around the edge of the Pacific Ocean (the Pacific Ring of Fire), particularly in Kamchatka (Russia), Japan, and Indonesia. Volcanoes occur either where the plates spread apart (Iceland, Cameroon) or where they collide with another plate (Japan, northwestern U.S.). Collisions result in one plate subducting below another plate that is therefore thrust upward. The dynamics of these plate movements (plate tectonics) can be seen in miniature on pools of molten lava in volcanic craters (e.g. Kilauea Volcano in Hawaii). The movements of the lava are analogous to a pot of oatmeal where the cooling surface crust sinks and is replaced by expanding hot oatmeal from below. Volcanoes also occur over hotspots that are places where magma swells upward; hotspots occur in the middle of plates (Hawaii) and on spreading zones (Iceland)

(Walker 1994). The location and timing of volcanic eruptions is still an imperfect science, but there may be some predictability to volcanic eruptions around the Pacific Ocean triggered by seasonal movements of water from land to sea (Mason *et al.* 2004).

The topography of areas disturbed by volcanic eruptions is partly determined by the type of lava. Volcanoes that produce basaltic lava that flows readily and has a low silica content have wide, gently sloped, domed-shaped outlines and are called shield volcanoes because of their resemblance to a warrior's shield. These volcanoes are characteristic of spreading zones (e.g. Hawaii and Iceland). Two textures of basalt are common, crinkly aa lava (Fig. 2.1, Box 2.1), which is more viscous and cools more quickly than the ropy pahoehoe, which is more fluid and cools more slowly. Volcanoes that produce andesitic or rhyolitic lavas, which are less fluid and contain more silica than basalt, tend to have more explosive eruptions and produce steep cones (Indonesia, Italy). Lava that is thrown into the air becomes tephra, which then lands as anything from car-sized lava bombs to cinder, ash, or even as very fine strands called Pele's hair. Lava reaching the ocean often builds black sand beaches (Fig. 2.2). Explosive volcanoes can also emit incandescent gas clouds that mix with solids to form **pyroclastic flows**, which descend the slopes at incredible speeds (del Moral and Grishin 1999).

The intensity of a volcanic disturbance is positively correlated with the amount of silica in the magma, so explosive ones are rated as more intense than shield volcanoes. The extent of a volcano is a function of the intensity of past eruptions and how widely volcanic ash is distributed, from local to global. Volcanoes occupy about 1% of the earth's terrestrial surface (Walker and Willig 1999). Sometimes heating of ice or snow on the slopes of volcanoes creates dry avalanches, debris flows, or **lahars** (mudflows). Volcanic surfaces are therefore quite variable in chemistry, texture, and stability, resulting in complex biotic responses. Volcanoes also trigger many secondary disturbances including earthquakes, landslides, and floods, each with its own disturbed habitat.

Volcanoes are perhaps the most severe of all disturbances because they generally destroy any previous ecosystem under lava or ash. Under such conditions, all post-disturbance colonists must disperse from surrounding, undamaged vegetation. Establishment is therefore slowest where lava flows are widest and most extensive and fastest when there are narrow flows flanked on both sides by surviving vegetation. Important early animal colonists include spiders and insects that arrive by aerial dispersal and nesting birds that bring plant propagules and nutrients to many volcanoes (Edwards 1988; Clarkson and Clarkson 1995; Crawford *et al.* 1995). However, at the edges of this disturbed habitat or where topographical features restrict ash deposits, erosion can remove ash and surviving plants and animals can respond quickly to the substantial input of minerals (Zobel and

Figure 2.1   Aa lava on the Island of Hawaii.

Figure 2.2   Black sand beach on the Island of Hawaii formed from lava. Note steam from the lava entering the ocean in the background.

Box 2.1 Long-term studies of volcanic ecosystems

Volcanoes create a variety of disturbances that vary in how they erupt, what they produce, and how they respond to subsequent erosion and colonization by organisms. Two of the best-studied examples of volcanoes are Krakatau (in the Sunda Strait just east of Java in Indonesia), and Surtsey (off the south coast of Iceland).

Krakatau's explosive 1883 eruption got the world's attention because it killed at least 21 000 people, was felt thousands of kilometers away, and cooled the world's climate for several years in addition to producing dramatic sunsets. Only a third of the original island was left after the eruption (this remnant is now called Rakata), although several nearby islands expanded in size. Studies of the colonization began immediately, with spiders and grasses found within the first year. It is still unclear if Rakata was completely sterilized. In 1927, a new island emerged and was named Anak Krakatau (or child of Krakatau). It has grown to a height of over 300 m and erupts fairly regularly. It has a fringe of sparse trees near part of its shoreline. When I visited Rakata and Anak Krakatau in 2010 I saw the fig trees (see Fig. 6.3) so important to animal colonization of denuded islands and the ongoing venting of steam from Anak Krakatau (see Fig. 1.6). Biologists and geologists have spent over a century monitoring the changes that have occurred and made many contributions to biogeography and succession.

Another volcano that is well-studied and, like Anak Krakatau, emerged from the ocean in the last century, is Surtsey. From 1963 to 1967, Surtsey grew to a height of 174 m and reached a maximum area of 2.5 km², although it has since eroded to about half that size. Biologists and geologists immediately began studying the island and still continue visiting Surtsey every year. Periodic collections of articles come out in Surtsey Research about colonization by terrestrial plants (e.g. Magnússon et al. 2009) and animals (Ólafsson and Ingimarsdóttir 2009) and by marine organisms (Gunnarsson and Hauksson 2009). When I visited in 2003, 14 species of birds had been found nesting on Surtsey, along with 69 species of vascular plants and several hundred species of invertebrates in both the terrestrial and marine habitats.

Both Krakatau and Surtsey have become valuable sources of information about volcanoes and the biotic and abiotic forces that affect them after they erupt. Ongoing research on volcanoes continues around the world and is particularly exciting when whole new surfaces are created with little or no biological legacy and research teams can study both terrestrial and marine ecosystems and the links between them (DeGange et al. 2010).

Antos 1997). For example, 1 year after the 2008 eruption on Kasatochi Volcano in Alaska that covered an Aleutian island in tens of meters of ash, plant and insect survivors emerged from buried soil on bluff edges and cliff bases due to a combination of shallow ash deposits and post-eruption erosion (DeGange et al. 2010). Biotic recovery following volcanic eruptions is highly variable, and volcanoes provide excellent opportunities to examine the relative importance of dispersal, gap size, nutrient cycles, survivors, and many other factors that influence colonization (del Moral et al. 2010).

2.4 Earthquakes

Earthquakes are sudden movements of the earth along faults where rocks slip past each other following a gradual increase in stress. The sudden release of energy sends seismic waves through the earth that can be felt for many kilometers in all directions. Earthquakes occur around the world but concentrate in the same areas of active plate boundaries where volcanoes are found (e.g. the Pacific Ring of Fire). Humans have a very rational fear of earthquakes because of their unpredictable nature and potential for extreme damage (Scarth 1999). Tens of thousands of people have been killed in major earthquakes such as in Portugal in 1755, Japan in 1923, or Pakistan in 2005. Today, millions of people live along major faults such as the Alpine Fault in New Zealand or the San Andreas Fault in California (U.S.), often in buildings that are not designed to resist the violent shaking of an earthquake.

Earthquakes occur in a variety of ways. Faults with mostly vertical displacement are called dip-slip faults, while mostly horizontal displacement occurs along strike-slip faults. Earthquakes are not just single events, but instead a composite of many movements as stresses build up and are then released by movement along the fault. We tend to label the biggest shock as the actual earthquake. Worldwide, there are about 50 earthquakes recorded per day or 20 000 per year (USGS Earthquake Hazard Program 2010). Many thousands more go undetected. Only about 150 earthquakes each year are intense enough to cause considerable damage. Modern measurements of earthquake intensity began in the 19th century and are based on readings from seismometers that record the earth's movements. The logarithmic Richter scale was useful for earthquakes near seismometers but is now mostly replaced by the moment magnitude scale that is based on the actual displacement and area of slippage along the fault. A subjective scale of what we experience and what types of structures get damaged during an earthquake (severity) roughly parallels the Richter scale of intensity and is called the Mercalli scale (Table 2.2). The extent of habitat disturbed by earthquakes is related to intensity, but strong shocks can sometimes be recorded around the world (e.g. Chile 1960). Earthquakes, like volcanoes, impact about 1% of the earth's terrestrial surface (Walker and Willig 1999). In addition to plate movements, earthquakes can be triggered by volcanoes, massive landslides, or floods. Humans can also trigger earthquakes by destabilizing the earth during blasting at mines or construction projects. Earthquakes, in turn, trigger a variety of other types of disturbances, including tsunamis and other floods, landslides (Garwood et al. 1979; Restrepo et al. 2009), or even fires (e.g. when gas lines are disrupted).

The direct biological effects of earthquakes have not been as well examined as the damage that earthquakes cause to physical structures. Nevertheless, earthquakes are major determinants of vegetation dynamics over large areas

Table 2.2 The Mercalli scale—a scale used for measuring the intensity of an earthquake. Values represent maximal intensity, but that will diminish with distance from the epicenter of the earthquake. Data are from the compilation of many individual observations. (From http://en.wikipedia.org/wiki/Mercalli_intensity_scale; accessed 15 February 2011.)

Mercalli intensity	Richter magnitude	Description of observations
I. Instrumental	1.0–3.0	Not usually felt
II. Weak	3.0–3.9	Suspended objects may swing; felt by very few
III. Slight	4.0–4.9	Felt by some people on upper floors; resembles vibrations from a passing truck
IV. Moderate	5.0–5.9	Felt by many on upper floors; felt by some outdoors; some awakened at night; dishes rattle; resembles a heavy truck striking building; standing cars rock noticeably
V. Rather strong	6.0–6.9	Felt outside by most; dishes and windows break; resembles large train passing close to house
VI. Strong	7.0+	Felt by all; many frightened and run outdoors; books fall off shelves; heavy furniture moves; damage slight
VII. Very strong	7.0+	Felt by motorists in cars; difficult to stand; furniture broken; some structural damage; some chimneys break
VIII. Destructive	7.0+	Structural damage moderate to substantial; chimneys, walls, monuments fall
IX. Violent	7.0+	General panic; structural damage considerable; buildings moved off foundations, even when well-built
X. Intense	7.0+	Most buildings destroyed, rails slightly bent
XI. Extreme	7.0+	Few buildings remain standing; rails bent; bridges collapse
XII. Cataclysmic	7.0+	Total destruction; objects thrown in air; large rock movements; landscape altered by several meters; rivers rerouted

as documented in New Guinea (Simonett 1967), Chile (Veblen and Ashton 1978), and Argentina (Kitzberger *et al.* 1995). In addition to widespread tree mortality from landslides, non-lethal damage can occur from smaller scale soil and rock movements. In a New Zealand forest, Allen *et al.* (1999) found that high levels of tree mortality (25%) from landslides and sedimentation triggered by intense earthquakes were accompanied by similar levels of injury from widespread, low-intensity damage (Fig. 2.3). Earthquakes clearly have long-term effects on tree population dynamics (Wells *et al.* 2001; Jacoby *et al.* 1988). Despite such evidence of direct earthquake damage, most biological consequences of earthquakes are measured indirectly through the impacts of secondary disturbances caused by earthquakes (e.g. landslides or tsunamis).

Figure 2.3   Effects of a 6.7 magnitude earthquake (in 1994) on a forest in the Avoca River drainage, South Island, New Zealand. (Photograph by Peter Bellingham.)

2.5 Dunes

Dunes are hills of wind-transported sand. They become disturbances when they advance across the landscape, burying ecosystems in their path. The sand in moving dunes and other wind-borne particles (e.g. volcanic ash, loess) are instrumental in shaping the topography of many landscapes. Red sand dunes that covered southwestern North America for about 10 million years (175 million years ago; Navajo Formation) now delight visitors as red sandstone formations (Fig. 2.4). Dunes cover about 7% of the earth's land mass and are particularly prominent in Africa (the Sahara and Namib deserts), the Middle East (Lut and Rub'al-Khali deserts), and China (Gobi and Taklamakan deserts). Large dunes are also found in coastal Peru (Sechura Desert) and Chile (Atacama Desert). Dunes are used for military training exercises, conservation reserves, and recreation, but dune expansion is part of a growing worldwide problem of **desertification** that affects a third of the earth's terrestrial surfaces. Heavily populated northeastern China is facing the eastern advances of the Gobi Desert in Inner Mongolia (Fullen and Mitchell 1994). The Sahara Desert periodically advances south into the semi-arid Sahel region in northern Africa, but there are also periods of retreat along this dynamic boundary (Tucker *et al.* 1991).

Dunes come in many shapes and sizes (Fig. 2.5). They can be shaped like crescents, lines, stars, domes, and parabolas. When wind directions shift, reversing dunes are created, with irregular shapes. Dunes can reach heights of over 1000 m (1176 m in Nazca, Peru) but most dunes are < 100 m high. The least relief is found in dune sheets that resemble sandy plains. Terrestrial dunes occur where wind-borne sand is deposited around an obstruction. Dunes advance by sand spilling down the leeward side of a dune (the slip face) and can migrate across the landscape at rates ranging from several meters to many kilometers each year (Hesp 2002). Underwater dunes are moved by water currents and form particularly in rivers and estuaries but also on the sea floor. Coastal dunes are the most extensive type of terrestrial dune and are built from beach sand and suspended sediments. Inland dunes form the largest

Figure 2.4   Highly eroded slopes in Bryce Canyon National Park, Utah (U.S.) highlighted by a recent snowfall.

dune complexes. Typical sources of sand for both coastal and inland dunes include tectonic movements (e.g. volcanoes and earthquakes), water erosion (e.g. glacial moraines and riparian sediments), terrestrial erosion (e.g. dried lake beds), and wind (e.g. cyclones). Coastal dunes are subject to disruption from wave erosion and salt spray, whereas all terrestrial dunes are subject to wind erosion, sand deposition, and droughts. Blowouts are depressions formed by wind erosion of dunes and feature higher wind speeds and more salt spray penetration inland than on intact coastal dunes. Dune intensity is measured by rate of advance, extent by the size of the active dune fields, and severity by damage to property and organisms, depending on their tolerance to burial by sand. However, few links have been made between dune geomorphology and biology (Hayden *et al.* 1995; Hesp and Martínez 2007). Human activities have accelerated dune formation by diverting lake water, removing vegetation, overgrazing pastures, and increasing global aridity.

Dunes can have a diverse flora and fauna especially adapted to the unstable conditions. Plants that thrive in shifting sands, such as the grasses *Ammophila* or *Leymus*, often accumulate sand but stabilize it with extensive roots and rhizomes. The roots tap into water tables that can be fairly high under dunes because water does not readily rise to the surface through coarse-textured sand (Danin 1991). In addition to tolerating sand inundation, many plants on coastal dunes can tolerate salt spray and saltwater immersion (Hesp and Martínez 2007). Once plants become established, animals

Figure 2.5   An aerial view from 1993 of the Shelburne Bay Dunefields, Cape York Peninsula, Queensland, Australia. (Photograph by V. J. Neldner.)

such as lizards, spiders, beetles, and snakes can live on dunes. Dunes can stabilize within decades if the supply of wind-blown sand does not overwhelm the colonists. Alternatively, dunes can remain unstable for thousands of years (Specht 1997). Ponds sometimes form between dune ridges in the dune slack, providing valuable surface water for plants and animals and furthering the process of soil and ecosystem development. Studies of changes in dune communities in Denmark (Warming 1909), Lake Michigan (U.S.; Cowles 1899; Olson 1958; Lichter 1998), and eastern Australia (Walker *et al.* 1981) have been fundamental to the development of the concept of ecological succession (Chapter 8). However, dune succession models do not always incorporate the many ongoing disturbances on dunes and their geomorphological consequences (Miyanishi and Johnson 2007).

2.6 Erosion

Erosion is any downslope movement of earth due to gravity, and it can be triggered by glacial melting, volcanoes, earthquakes, floods, and heavy or prolonged rainfall. Erosion affects about 10% of the earth's terrestrial surface at any given time. Human activities also contribute to erosion by removing vegetation (e.g. logging, fire) or construction activities (e.g. home building, road cuts). In a forest in eastern Puerto Rico, over half of all erosion was road-related (Larsen and Parks 1997). Landslides are a common and easily distinguishable erosional feature so they serve as the focus for this discussion (Fig. 2.6). The term landslide is used in its broadest sense to incorporate all mass movements from creeps and flows to rock falls and avalanches. Landslides destroy many properties and take hundreds of lives each year (Abbott 1996), so it is relevant to examine their biological features to further slope restoration efforts.

Landslides generally have well-defined edges and often feature relatively distinct vertical zonation from the upper slip face where the landslide began, to the chute where soil was transported, to the deposition zone where much of the original topsoil stopped sliding (Guariguata 1990, Walker *et al.* 1996b). Landslides can slide along a plane or rotate, but the term also includes earth that falls, topples (falls as a unit), flows, slumps, or spreads laterally (Sidle and Ochiai 2006). Landslides can include slumps that are several square meters in size to massive slope failures of several square kilometers. At landscape scales, studies have begun on multiple populations of landslides triggered by the same disturbance (e.g. an earthquake) that can also be seen as a community of disturbances across wider scales (Restrepo *et al.* 2009).

Landslides remove most of the soil, vegetation, and animals that formerly resided on the slope. The removal process is highly variable, so residual patches

Figure 2.6   Landslides and the ecologically important monkey-puzzle tree (*Araucaria*) in northwestern Argentina; the seeds feed many animals.

of the original ecosystems may remain intact or be rafted downslope to a new position. Recovery can be most rapid in the deposition zone with its net accumulation of organic matter, seeds, and plant parts. Convergence of plant biomass to values resembling those in adjacent mature forests may take as little as 55 yr (Zarin and Johnson 1995a) or as much as 500 yr (Dalling 1994). Nutrients in surface soils can recover rapidly when there is N-fixation by plants and ample allogenic litter inputs (Zarin and Johnson 1995b). Surviving rafts of vegetation and new colonists are also important contributors to ecosystem recovery on landslides. Secondary erosion can continue to influence succession for decades (Walker and Shiels 2008). Where landslides create gaps in a forest canopy, species tolerant of high light levels can invade and delay succession to forests (Walker *et al.* 2010a).

2.7 Insect outbreaks

Herbivory can be seen as a normal background stress for plants and its effects (e.g. stimulated growth of non-host plants) can be considered an integral part of an ecosystem (Schowalter 1985). Herbivory can also be considered a disturbance when significant loss of biomass results (e.g. a leaf that loses half its biomass in an afternoon), or when ecosystem structure is altered (e.g. the death of a stand of trees from a bark beetle infestation over the course of several months). At some threshold of frequency, intensity, and scale, levels of herbivory become a disturbance with disruptive consequences similar to other disturbances such as fire or drought (Schowalter and Lowman 1999; Heavilin *et al.* 2007). Insect herbivory affects approximately 10% of the earth's terrestrial surface. Insect outbreaks generally target a select subset of plant species because of the often specific biological relationships between the insects and their hosts. The selective damage differs from indiscriminate damage by volcanoes or dunes but resembles the selective tolerance of different species to fire or floods. Human-altered conditions can increase the frequency and intensity of insect outbreaks by increasing forest **fragmentation**, expanding agriculture, decreasing biodiversity, reducing fire frequency, introducing non-native insects, and reducing populations of native predators such as birds (Roland 1993; Schowalter and Lowman 1999).

Insect herbivory can result in damage to current-year foliage (budworms, *Choristoneura*, and tent caterpillars, *Malacosoma*), older foliage (tussock moths, family Lymantriidae, and hemlock loopers, *Lambdina*), or cambial tissues (bark beetles, family Curculionidae). Other herbivores form galls to induce abnormal growth of plant tissues, feed on sap, or eat seeds or seedlings. The severity of defoliation can be measured as loss of leaf area and can range from 1% to more than 50% (Schowalter and Lowman 1999). Initially non-lethal defoliation or cambial damage can reach a level of intensity that results in plant death, either directly from the herbivory or, more likely, by a complex interaction of insects with their hosts, predators, pathogens, and various abiotic and biotic site factors (Franklin *et al.* 1987; Morin *et al.* 2007). A resistant host plant is defended from herbivores with a complex arsenal of chemical and physical defenses (Coley *et al.* 1985). Insect herbivores have evolved many ways to overcome these defenses, but what usually makes a plant vulnerable to extensive damage is a stress (i.e. lack of resource) or disturbance (i.e. prior loss of biomass). Plants in monocultures or low-diversity stands are often more sensitive to insect outbreaks than mixtures of species (Schowalter and Turchin 1993) because insects are confused by multiple chemical cues (Visser 1986). Monocultures can result from previous selective mortality of other species, early successional dominance, or human management. Abiotic factors that impact herbivore success include disturbances such as fires (Miller and Wagner 1984) or cyclones (Willig and Camilo 1991) that can affect herbivore, predator, or pathogen populations directly or indirectly through alterations to the local environment. For example, some insect predators prefer cool, moist, understory conditions (Oboyski 1995) while some herbivores orient toward the green and yellow wavelengths of young leaves in well-lit conditions (Matthews and Matthews 1978).

Herbivory also alters community and ecosystem dynamics at a range of spatial and temporal scales. Spatial effects of insect outbreaks occur at the level of the individual plant, the stand, and the landscape. Temporal effects can be considered at daily, annual, and longer time scales. Interpretation of the effects of insect outbreaks on insect population dynamics, community-level processes, including plant–insect interactions and vegetation dynamics, primary productivity, and nutrient dynamics are all dependent on which spatial and temporal scales are selected. Such a web of effects and resultant feedbacks is understandably complex and difficult to study. Predictive models have been helpful in understanding these relationships. For example, linkages have been made between the recent and unusually severe outbreaks of bark beetles in western North America and several decades of warmer weather (Logan *et al.* 2003; Heavilin *et al.* 2007). Spatial patterns of an insect outbreak can be measured as losses of individuals, increased patchiness of a stand, or extensive and contiguous damage across a landscape. These spatial characteristics of an outbreak then impact future stand dynamics and influence where the next outbreak will occur. Temporal patterns of outbreaks include the periodicity of cyclical outbreaks of tent caterpillars (*Malacosoma*) in Ontario, Canada (every 13 yr; Fleming *et al.* 2000), jack pine budworms (*Choristoneura*) in Wisconsin, USA (between 5 and 14 yr cycles; Volney and McCullough 1994), spruce budworms (*Choristoneura*) in British Columbia, Canada (26 yr; Burleigh *et al.* 2002), and cicadas (families Tettigarctidae and Cicadidae) in many locations (various cycles; Williams and Simon 1995). Expected outbreaks do not always occur, however, and outbreaks can occur in unexpected regions and years, emphasizing our imperfect understanding of how temporal cycles are generated and controlled (Cooke *et al.* 2007).

2.8 Glaciers

Glacial advances shape landscapes by rounding mountains, carving valleys and fjords, transporting and pulverizing sediments, and isolating populations of plants and animals. Glacial retreats leave moraines, which are unconsolidated sediments that erode, form new drainages, open up new paths for animal migration, and expose new surfaces for colonization by plants, animals, and soil organisms (Fig. 2.7; Walker and del Moral 2003). At least 90% of all glaciers monitored worldwide are retreating and this rate has accelerated in the last decade, perhaps in part due to global warming (Dyurgerov 2002). Some tropical alpine glaciers (e.g. on Mount Kilimanjaro, Tanzania; Mote and Kaser 2007), are in danger of disappearing entirely. Further evidence of glacial retreats comes from icebergs that are the size of small countries calving from glaciers bordering the coastal waters of West Antarctica. Aerial views of the Greenland ice sheet are dotted with blue melt-water lakes, while Greenland's coastal bays are filled with icebergs. There are several concerns about the acceleration of glacial melting. First, sea levels would rise about 23 cm if all the world's glaciers were to melt, and about 100 cm if the Greenland and Antarctic ice caps were included (although estimates range from 18–220 cm; Rahmstorf 2010). Second, earthquakes might be triggered by the movement of so much water on the earth's surface. Third, if the Greenland ice-cap were to melt, the addition of a large amount of fresh water to the North Atlantic might alter cycling of the Gulf Stream with undetermined global climatic effects, including the possibility of cooler temperatures in Europe (Hewitt *et al.* 2006). Melting glaciers provide excellent opportunities for biologists studying succession because they expose and shape moraines that present new, disturbed habitats for organisms to colonize (Matthews 1992). Studies are warranted on what these moraines might mean to flora and fauna as they respond to global climate change, including their possible role as corridors for migration, loss of biodiversity, and reduction in water flows (Orlove *et al.* 2008).

Glaciers cover about 10% of the earth's surface, most of which (90%) is in Greenland and Antarctica (Matthews 1999). Of the 680 000 km² of glaciers in the world, 577 000 km² are in the Northern Hemisphere, principally on Arctic islands and in Alaska. The extent of glaciers has varied widely during

Figure 2.7   A melting glacier in Iceland. Note several layers of moraine exposed by the melting. (Photograph by Elizabeth Powell.)

the history of the earth. We are in only the third glacial age in the last 600 million yr. During our current glacial age, glaciers have advanced regularly during periods lasting 100 000 yr and retreated during shorter, 10 000 yr, interglacials. We are at the end of a rare interglacial in a rare glacial age. However, the likelihood of the earth cooling in the near future has perhaps been offset by the greenhouse gases that humans have been adding to the atmosphere during the last several thousand years (Vavrus *et al.* 2008).

Glaciers act as giant sanders because when the lower surface of ice is buried by a glacier at least 60 m thick it becomes plastic and picks up rock and other debris that scour the land as the stream of ice moves down slope. The effects of glacial scouring are seen in the landscape as lakes, U-shaped valleys, kettle holes (indentations formed when ice chunks melt), monadnocks (isolated, erosion-resistant hills), and rouches moutonnées (elongated, gradually ascending hills with a steep southern slope in the Northern Hemisphere). Deposits of rocks left by melting glaciers include a variety of moraines (lateral, central, terminal). These moraines help determine stream-flow patterns across the newly exposed land.

The disturbed habitat present when a glacier retreats is highly variable, depending on the regional climate, the rate of retreat, the type of substrate, and the closeness of potential colonists. Early colonists face a habitat mostly devoid of life that features some combination of melt-water lagoons with icebergs floating in them, rocky outcrops worn smooth by the ice, cliffs with water seepage, linear moraines, and less organized piles of sediments of a wide range of sizes from house-sized boulders to small, rounded pebbles and sand. Glaciers are actually not devoid of life, as they carry organisms in, on, and under the ice (Wynn-Williams 1993; Matthews 1999). Microorganisms dominate the glacial ecosystems, but non-vascular and vascular plants are often present as well, some of which can act as nuclei for colonization of recently exposed moraines. Aquatic organisms colonize the lagoons and nascent streams (Milner *et al.* 2007; Milner and Robertson 2010). Early terrestrial colonists include soil bacteria (Bardgett and Walker 2004) and mosses (Worley 1973), while vascular plants with N-fixing symbionts often dominate after several decades (Walker 1999b).

2.9 Cyclones

Cyclones (also known as hurricanes or typhoons) are widespread, spiraling storms that obtain their energy from warm, moist, rising air. Cyclones affect approximately 15% of the earth's terrestrial surface. They are characterized by strong winds, rain, and low atmospheric pressure. Cyclones occur in the temperate zone but are most common and destructive in the tropics, particularly in the Caribbean Sea (Lesser Antilles, Bahamas) and the western

Figure 2.8   Effects of a hurricane on tropical forests. (a) An uprooted tree in the lower cloud forest of Kauai, Hawaii, probably uprooted by Hurricane Iniki in 1992. (b) Defoliation and branch loss caused by Hurricane Hugo (1989) in a cloud forest in Puerto Rico. (Photograph by Ariel Lugo.)

Pacific Ocean (Taiwan, Japan). Cyclones are ranked on the Saffir–Simpson hurricane wind scale by average sustained wind speed, starting at 118 km hr⁻¹. Those that impact populated regions are widely covered in the news media and improvements in the prediction of cyclone trajectories have facilitated cyclone preparedness. Yet vulnerable populations such as people living in low-lying Bangladesh suffer enormous losses (at least 300 000 died there from flooding caused by a 1970 cyclone and 140 000 more in a 1990 cyclone). It is likely that human-induced warming of the Caribbean Sea has increased the intensity (wind speed) and possibly the severity (damage) of cyclones in that area (Goldenberg *et al.* 2001).

Cyclone damage to vegetation depends on wind speed, the duration of the storm, local and regional topography, and anatomical and stand traits of the vegetation. Plants can lose leaves or branches and tree trunks can snap or be uprooted (Fig. 2.8). When trees snap or are uprooted, mineral soils are exposed, organic soils are mounded, and increased availability of light in the gaps created by treefalls can speed recovery by fast-growing species and promote species diversity (Vandermeer *et al.* 1996). These gaps can impact 1–2% of tropical (Whigham *et al.* 1999) and temperate (Cogbill 1996) forests annually. Return intervals of cyclones determine what type of vegetation is able to re-establish before the next cyclone. Animals unable to fly or run away from a cyclone can suffer, while those that did escape might not thrive in the immediate post-cyclone environment, especially if they were dependent on arboreal habitats that were subsequently destroyed. If damage is severe, dead trees can provide fuel for fires (Whigham *et al.* 1991). Recovery of pre-cyclone plant community structure in mesic climates can occur within just a few years or decades, but recovery of the previous species composition can take many decades to several centuries (Walker *et al.* 1991, 1996a). Post-cyclone recovery of different aspects of a tropical ecosystem in Puerto Rico (e.g. stream and soil nutrients, plant and litter mass, abundance of shrimp, frogs, snails, bats) followed six distinct trajectories (depending on initial increases or decreases of biomass or numbers and subsequent rates of return to pre-cyclone conditions (Fig. 2.9; Zimmerman *et al.* 1996).

2.10 Floods

Floods, like fire and drought, can be found in all types of terrain and climates, and, like cyclones, impact approximately 15% of all land masses. Rivers of any size can flood their banks during the seasonal spring melting of accumulated winter snow and ice. Where soils are still frozen or vegetation dormant, there is little absorption of the melt water and floodplains become quickly inundated. In other areas, the onset of the rainy season can be a time of heavy flooding. In some cases, flooding is aggravated by dried out

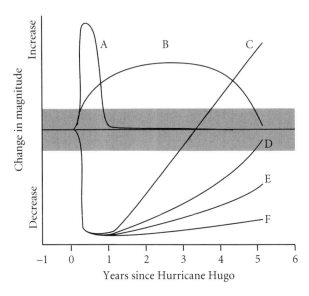

Figure 2.9   Idealized 5-yr trajectories of responses of different components of wet subtropical forest in the Luquillo Experimental Forest, Puerto Rico, to disturbance caused by Hurricane Hugo (1989). Curve A: transient increase (represented by responses of forest floor biomass and nitrate concentrations in streams). Curve B: slow increase then drop (net primary productivity, shrimp abundance, some snail and adult frog densities). Curve C: sharp decrease then rise above pre-hurricane levels (aboveground potassium and magnesium pools, bat and some terrestrial snail densities). Curve D: sharp decrease then return to prior levels (tree biomass and density, some bat densities). Curve E: sharp decrease and steady but partial recovery (fine litterfall). Curve F: sharp decrease and little recovery (fine root biomass and walking stick [stick insect] density). Values in the shaded bar are not distinguishable from pre-hurricane values. (From Zimmerman *et al.* (1996) with permission from Wiley-Blackwell.)

soils that develop a hydrophobic surface during the dry season and are not good absorbers of the renewed rains. Other river floods can be triggered by sudden melting of glacial ice or from natural or human-made dams bursting. Floods also occur on shorelines of lakes and oceans from wind-driven waves (e.g. during a cyclone) or tsunamis (see Chapter 3). More gradual flooding is associated with a regional rise in water levels from upstream conditions, dam construction, or global warming. Humans both benefit and suffer from floods. We depend on waterways for drinking, bathing, irrigation, transportation, fertilizer for crops, waste disposal, and energy sources (e.g. for milling or electric power). However, this dependence, in addition to our propensity to live and grow crops in fertile floodplains, makes us vulnerable to flooding. Many people die in floods every year. Humans have, in many cases, lowered flood frequency but increased the extent and severity of floods that do eventually occur on rivers where water flow has been curtailed by dams and levees (Reice 2003).

River floods occur when there is more discharge than can be accommodated by the river channel or when the channel fills with sediments, reducing channel volume. Estimated flood frequency is used as the basis for engineering projects (e.g. roads or dams that must survive the discharge produced by a 100-yr flood). Such estimates of frequency are only statistical averages. A 100-yr flood has a 1% chance of occurring in a given year but a greater than 50% chance of occurring in a 100-yr period (Abbot 1996). Of course, two floods of this frequency and intensity may occur in consecutive years. Floods can have significant downstream effects at long distances from where the rains occur. For example, rains on the western slopes of the Rocky Mountains in the central part of the U.S. control water levels along the entire Colorado River to the Gulf of California in northern Mexico (Christensen *et al.* 2004). Floodplains can be created or altered by glacial retreat, volcanoes, and landslides that provide sediments, cyclones that provide precipitation, and insect outbreaks that reduce vegetation that protects against erosion.

The degree of damage (severity) that a flood causes is often evaluated in human lives, property destruction, and loss of productivity of agricultural fields. The impact on natural environments is less frequently discussed (Hughes 1997). The loss of soil in the upper regions of a river is partially offset by the deposition of soil on floodplains in the lower regions. The fertility of flood deposits can vary, depending on soil fertility upstream, but they are generally at least as fertile as the surrounding uplands. High fertility, plus favorable water tables, leads to high rates of primary productivity along floodplains, often dominated by plants that can survive some inundation. Plants and animals use river corridors for dispersal. Plant succession is often reset by major flood events because new sediments are deposited and less flood-tolerant species lose their dominance. River floodplain conditions (e.g. soil stability and fertility; biodiversity) reflect the ecological status of their watersheds. Human alterations of the timing and volume of water flow (e.g. by building dams) have altered the floodplain environment in sometimes unpredictable ways (Nilsson and Berggren 2000).

2.11 Droughts

Droughts are defined as deviations from the normal range of precipitation in a given season or climate. When droughts are abrupt and disruptive of the biomass, they can be considered a disturbance, even in a generally dry habitat. When precipitation levels decline slowly over time, droughts appear gradually and are a part of a long-term climatic pattern; then they are considered a stress. Droughts occur in all climatic zones and affect about 30% of the earth's terrestrial surface. In a warm, wet climate, a drought may occur when there are several weeks without rain. In a warm, dry climate, months without rain are needed before a drought occurs. In climates with highly seasonal precipitation, the definition of a drought will vary by season. Droughts that last for several growing seasons can lead to shifts in vegetation (e.g. tree to grassland or grassland to desert). In hot deserts, spatial variability in precipitation tends to increase as mean annual precipitation decreases, so readings from rain gauges on a 1-km grid, for example, will differ greatly in rainfall values (Evenari 1985). Droughts that last for years can only be considered abrupt (and therefore a disturbance) if one considers time frames of centuries. Such time spans are relevant for the long-lived shrubs found in deserts such as the Mojave Desert in western North America (Vasek 1980). Droughts that damage or kill organisms within weeks are clearly disruptive and affect community composition and diversity. Humans have long been affected by droughts that destroy crops, promote insect outbreaks (e.g. locusts, family Acrididae), and sometimes force people to move to wetter regions in order to survive. Droughts are mostly natural features but their effects can be accentuated by human activities such as excessive plowing or overgrazing of unstable soils.

Large-scale droughts often occur when there are latitudinal shifts in moisture-bearing winds. The Dust Bowl in the U.S. in the 1930s was due to a descending high-pressure system that sent dry, hot air to the surface of central North America (Worster 1979). The dust storms that resulted were worsened by the replacement of drought-tolerant grasses in the prairies by species that were not as drought tolerant. The failure of the grasses, plus the plowing used to prepare the ground, contributed to the severe dust storms that plagued the area for several years (Abbott 1996). Many people migrated to California. The southern edge of the Sahara Desert in Africa (the Sahel) is also a dynamic ecotone between desert and semi-arid land that is subject to drought when the intertropical convergence zone shifts south. Severe droughts affected the region between the late 1960s and early 1980s and resulted in widespread famine and death (Tucker *et al.* 1991; Batterbury and Warren 2001). Longer and more severe droughts have occurred during the last 3000 years (Shanahan *et al.* 2009), but during these recent droughts the vegetation in the region shifted from productive grasslands to a less productive, desert-like landscape to the detriment of local organisms (Ellis and Galvin 1994; MacMahon 1999). Large-scale drought can also occur when a cyclone removes foliage, leading to reductions in regional evapotranspiration. After Hurricane Hugo in Puerto Rico (1989), defoliation resulted in an increase in temperature and albedo (reflectivity) in low-elevation rainforests. These changes displaced cloud layers to higher elevations in the rainforest in the Luquillo Mountains until the trees regained their foliage 3 months later (Scatena and Larsen 1991). Local droughts are frequent occurrences in many ecosystems and can impact the abundance, distribution, and phenology of organisms. For example, large dipterocarp trees on ridges in a rain forest in Borneo were particularly susceptible to drought-caused mortality, leading to an uneven forest canopy (Leighton and Wirawan 1986).

Both plants and animals have, over evolutionary time, developed myriad, long-term adaptations to droughts (Alpert 2000; Schwimmer and Haim 2009) but rapid responses are needed for organisms to survive the sudden onset of a drought. Examples of how plants acclimatize to droughts include: postponement of germination, growth, or reproduction (e.g. desert annuals); reduction of metabolic activity (e.g. mosses); or wilting to decrease evaporative surfaces (Chaves *et al.* 2003; Hietz 2010). Animals may flee drought, reduce their activities (estivation), reduce reproduction rates, or eat less food (Wiggins *et al.* 1980; Gibbons *et al.* 1983, Ismail *et al.* 2011). Soil erosion and subsequent dispersal by wind during a drought can be an important source of nutrients for drought-tolerant plants (Belnap and Gillette 1998).

2.12 Fires

Fires, like volcanoes and earthquakes, engender a visceral human reaction. We delight in a cozy setting by a fireplace but are in awe of large forest fires. Candles may have a ceremonial purpose, and many cultures cremate their dead. Once humans tamed fire they expanded into colder climates, burned forests to plant crops, and burned grasslands to renew soil fertility. Today, we worry about forests where we have controlled fires for so long that fuel build-up promotes large, hot fires (Agee and Skinner 2005). Meanwhile, human population densities have increased in fire-prone forests and prairies, making our lives and our properties increasingly vulnerable.

Fires regularly affect more than 50% of the terrestrial surface of the earth (Willig and Walker 1999), and are important determinants of species composition and ecosystem processes in grasslands, boreal forests, and Mediterranean-type shrublands. Grasslands dominate where annual droughts alternate with enough precipitation for growth, and fires are a common occurrence in such grasslands because grasses grow in dense swards and dry out above ground on a regular basis (Zedler 2007). Grasses regrow quickly from buried, undamaged rhizomes but shrubs and trees are often destroyed. Grassland biomass recovers within several years after a fire but, because of accumulated leaf litter, becomes highly flammable again. Humans have long used fire as a tool to maintain grasslands and clear forests. Today, overgrazing has led to some former grasslands reverting to shrublands (Archer *et al.* 1988) and introduced grasses, which burn at higher temperatures than native grasses (Rossiter *et al.* 2003).

Fire is a key disturbance throughout the boreal forests of Alaska, Canada, and Russia, with fire return intervals of between 40 and 200 yr and 5–15 million ha burn every year. Most large fires in boreal forests are started by lightning. Burn severity reflects existing vegetation and moisture levels with crown fires (typical in North America), understory fires (typical in Russia), and some that consume both plus the organic matter in the soil. Two dominant trees in North America (*Picea mariana* and *Pinus banksiana*) are serotinous, meaning that their cones do not release their seeds until they are heated. Seeds of serotinous trees also germinate best in the exposed mineral soil after a fire when the dominant understory and moss layers have been removed (Greene *et al.* 1999). Fire intensity, fire intervals, and climatic conditions can affect the balance of these two tree species and more open, lichen–heath communities (Lavoie and Sirois 1998). Climate change may markedly increase the length of the fire season, fire frequency, and fire severity (Weber and Flannigan 1997). Understanding the mechanisms behind the effects of fire on plants (e.g. combustion, heat transfer, burn area, and shape and nature of unburned patches) will aid in understanding future changes in the effects of fire on boreal forests (Gutsell and Johnson 2007).

Mediterranean-type shrublands are found in California, central Chile, southwestern Australia, and the Cape Region of South Africa, in addition to the Mediterranean Basin. These regions have cool, wet winters and hot, dry summers. The dominant sclerophyllous shrubs and trees burn during the summer months at intervals of one or more decades (except in Chilean shrublands where fire is rare). Centuries of human influence have deforested and urbanized several of these regions and humans have introduced invasive species native to other Mediterranean-type shrublands (Rundel 1999). These introductions can cause or be the result of altered fire regimes (Brooks *et al.* 2004).

Fires interact with other disturbances in several ways (del Moral and Walker 2007). Volcanoes can trigger fires and, in turn, fires can trigger erosion (e.g. landslides) when plant cover on slopes is destroyed. Extensive tracts of partially damaged trees can provide opportunities for insect outbreaks. Droughts promote the likelihood of fire by drying potential fuel but over time reduce the amount of plant growth available for fire. Grazing can also reduce fuel loads and thereby lower fire frequency. Finally, alterations of natural fire frequencies can promote invasions or enhance the abundance of non-native species.

Fire severity, or a fire's effect on the biota, is highly variable, depending on the intensity (temperature) and duration of the fire and the flammability of the biota (Bond and van Wilgen 1996). Slow, hot fires can burn all vegetation and even the organic matter of the soil, but most fires spread rapidly, burning only some of the vegetation, either in the canopy or understory. Fast-moving fires result in substantial plant survival. Trees and shrubs sprout back from protected buds, grasses grow from rhizomes, and herbs germinate from surviving seeds. Fires can be lethal to animals that do not flee or burrow deep enough. Yet many animal populations depend on the new herbaceous growth, insects in decomposing wood, and other resources that fires initiate. Grazers

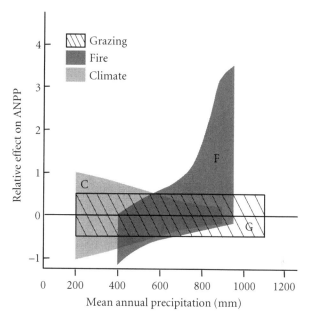

Figure 2.10 A conceptual model of the relative effects of grazing, fire, and climate on annual net primary productivity (ANPP). (From Oesterheld *et al.* (1999) with permission from Elsevier.)

in grasslands can reduce fire frequency by consuming between 15 and 94% of annual aboveground net primary productivity, thus reducing the amount of fuel available to a fire (McNaughton 1985). Grazers, fire, and climate variability likely interact and vary in importance across precipitation gradients (Fig. 2.10). At low levels of precipitation, variability in rainfall has a relatively strong influence (positive or negative) on primary productivity (more rain, more productivity). At high levels of precipitation, the influences of climate are replaced by the positive effects of fire (e.g. enhanced nutrient cycling). Grazing effects purportedly remain constant across a wide precipitation gradient (Oesterheld *et al.* 1999) because they are dwarfed by climate variability at low precipitation and fire effects at high precipitation. Plants that are heavily grazed at higher precipitation likely respond with compensatory growth. The relationship between grazing and fire is further complicated by the effects of grazing on productivity (Oesterheld and Semmartin in press).

2.13 Conclusions

Terrestrial habitats have been shaped historically and are modified daily by natural disturbances. These disturbances occur at all spatial scales from microsites to whole biomes and influence many biotic processes from competition to evolution. Despite the wide range of disturbances that affect terrestrial habitats, each can be compared by frequency, severity, and extent; intensity is more specific to each disturbance type (see Table 1.2). Another commonality is the influence of terrestrial disturbances on the structure of soils, which has important implications for the recovery of organisms (see Chapter 5). Disturbances do not act alone but instead cause and are caused by other disturbances, combining to form a disturbance regime at any given location. Organisms respond to the net effect of the interacting disturbances.

Humans have intentionally initiated or enhanced terrestrial disturbances to grow crops. Successful agriculture has involved cultivating fertile soils (often on volcanic slopes or floodplains), responding to droughts and desertification (by irrigating and controlling advancing dunes), stabilizing slopes to avoid erosion, managing the effects of fire, and coping with outbreaks of insect herbivores. This process of learning to manage disturbance has led to the development of principles of soil conservation, crop genetics, and plant physiology. Animal husbandry has led to similar developments in genetics and physiology. Broader effects of disturbances on environmental processes have led to an understanding of aspects of nutrient cycling and plant succession. Clearly, humans have a close level of interaction with natural disturbances. The intensification of natural disturbances by anthropogenic ones is the topic of Chapter 4.

3 Aquatic habitats

3.1 Introduction

The study of natural disturbances in aquatic habitats has been slower to develop than for terrestrial disturbances (Chapter 2), in part because of their reduced visibility, the fluid and rapidly changing environments of rivers and lakes, and the perceived vastness of the ocean. However, awareness has grown rapidly in recent decades as rivers become polluted and are dammed to become reservoirs, lakes are drained and filled in, and oceans become better known. The rapid deterioration of marine fish populations, polluted, anoxic dead zones along many coastlines, and the destruction of coral reefs by human activities (Chapter 4) now make headlines, although the damage continues unabated. With a growing interest in aquatic environments has come an appreciation for the natural disturbances that preceded humans. Although natural disturbances of aquatic habitats are now closely linked with anthropogenic ones, they still provide the framework for understanding all aquatic disturbance regimes. This chapter covers those natural aquatic disturbances least affected by human activities. Following the same structure as Chapter 2, I will discuss aquatic disturbances in order of increasing extent, and consider the relevance to humans, disturbance characteristics, and the nature of recovery. Riverine, lacustrine, and marine examples will be discussed together or separately, depending on the presence or absence of analogous disturbances.

Aquatic habitats are disrupted by many of the same natural disturbances that affect terrestrial habitats, including tectonic activity (volcanoes and earthquakes), erosion, floods, droughts, and storms (where waves replace winds as the destructive force). These abiotic disturbances interact with biotic ones such as herbivory, predation, diseases, and mass mortality events to produce disturbance regimes as complex as those found on land. Typical limiting resources include stable surfaces for colonization (e.g. intertidal habitats and riverbeds), light and oxygen (e.g. benthic habitats), P (e.g. lakes), and iron (e.g. open oceans). Responses to disturbances by aquatic organisms can sometimes be more rapid than by terrestrial organisms, in part because aquatic organisms are often (though not always) more readily dispersed. Nutrients can also readily mix, in part because of currents and water turnover times of days (rivers), to years or centuries (lakes), to thousands of years (oceans; Molles 2010). Benthic surfaces and coral reefs, however, can provide a degree of substrate stability. Covering over two-thirds of the earth's surface, it is timely that we focus more on disturbances that affect aquatic biomes.

3.2 Carcasses

The carcasses of whales and other large animals, plus logs, kelp, and other debris that fall to the ocean floor, can constitute a localized disturbance if they physically disrupt benthic communities or if the nutrient pulse they provide results in the rapid demise of the local organisms due to competition from organisms adapted to high-nutrient conditions. Lake fish can create similar disturbances on a smaller scale. Carcasses that wash ashore have analogous influences on terrestrial nutrient cycles (Fig. 3.1). The disturbance created by a carcass can be characterized by its intensity (force of impact), severity (damage to benthic communities), and extent (size of the carcass). Large carcasses are not widespread, but one estimate suggests that there might be one grey whale carcass every 300 km^2 (Smith *et al.* 1989). The physical state of the carcass when it falls, local conditions (e.g. temperature), and the availability of decomposer organisms will all influence the duration of the disturbance. Colonization of new disturbances is typically slower in the benthic regions than in surface waters due to colder temperatures, high pressure, and low-nutrient conditions in the **benthic zone** (Jones *et al.* 1998). However, a number of organisms thrive at depth and biodiversity can be relatively high, given the harsh conditions (Grassle and Morse-Porteous 1987). Grenadier fish and some benthic invertebrates specialize in feeding on unpredictable food sources such as carcasses within minutes of their arrival (Wilson and Smith 1984). The monitoring of one carcass that was experimentally sunk indicated that whale flesh was consumed within months by fish, sharks, and crabs, followed by colonization by **polychaetes** (Smith *et al.* 1998). Sulfur-based chemosynthetic communities then develop and can persist for at least 5–6 yr (Bennett *et al.* 1994), suggesting that carcasses may serve as alternative habitats for fauna of deep-sea hydrothermal vents (see Section 3.3.2; Van Dover 2000). Such detritus-based food webs are analogous to heterotrophic succession on carcasses in terrestrial habitats (see Section 8.5.3).

Figure 3.1   A fish skeleton on a beach of Kasatochi Island in the Aleutian Islands, Alaska (U.S.).

3.3 Tectonic activity

3.3.1 Volcanoes

Volcanoes and earthquakes comprise the vast majority of submarine tectonic activities. Submarine volcanoes are less visible and have been less intrusive into human activities than aboveground volcanoes. However, about three-quarters of all magma output comes from submarine volcanoes. Submarine volcanoes can also eject lava into the air, sometimes with lethal effects. The eruption of the Kolumbo Volcano in the Aegean Sea killed people on the nearby island of Santorini in 1649–50 when its caldera briefly reached sea level (Vougioukalakis *et al.* 1995). There are thousands of seamounts or mountains that rise at least 1000 m above the ocean floor but do not reach the surface. Many of these are eroded volcanoes that used to reach the surface, while others are young, building volcanoes. One of the latter is Loihi, which is the newest Hawaiian volcano. Based on its past rate of growth, scientists estimate that it will emerge above sea level in about 100 000 years. Volcanic activity is largely along continental plate boundaries that can be colliding or spreading apart, but can also occur over hotspots formed by upwelling magma not associated with plate boundaries (e.g. Hawaii) or even by combinations of plate boundary dynamics and hotspots (Iceland, Canary

Islands; Walker and Bellingham 2011). The dynamics of individual eruptions are largely unpredictable. Lava flows quickly form crusts under water and new lava flows into the crusted area create pillow lava. The emergence of new volcanoes above sea level (Anak Krakatau, Indonesia, 1929; Surtsey, Iceland, 1963) has sparked interest among both marine and terrestrial geologists (explosion dynamics, chemical composition, and weathering patterns of rocks) and biologists (primary succession on new land and marine surfaces). Eruptions of island volcanoes such as Kasatochi (Alaska, 2008; DeGange *et al.* 2010) have advanced the integration of marine and terrestrial research. Aerosols, beach drift, mammals, and seabirds move from sea to land, while sediments, seabirds, and plants move from land to sea (Fig. 3.2).

The responses of marine organisms to volcanoes have been best studied in shallow, coastal communities (see Fig. 1.5). On Surtsey, algal colonization is most abundant in the top 5 m of the water column and is dominated by annual species that reinvade each year. Species numbers present during each survey peaked within a decade of the eruption and have since remained

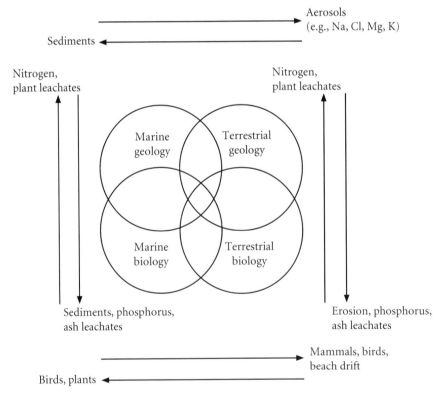

Figure 3.2   The flow of energy and matter among plants and animals in both marine and terrestrial habitats on Kasatochi Island, Alaska. (From DeGange *et al.* (2010) with permission from the Regents of the University of Colorado.)

around 50. Active abrasion by sand and erosion of basaltic cliffs by waves up to 17 m high appear to deter more extensive colonization to the same levels of cover and species richness in algal communities found on nearby, older islands (Jónsson and Gunnarsson 2000; Gunnarsson and Hauksson 2009). Benthic fauna are also slow to colonize the unstable surfaces around Surtsey (Hauksson 2000), although amphipods and isopods more closely resemble populations on undisturbed neighboring islands (Galan 2000). A lava flow (on nearby Heimaey) that is a decade younger than Surtsey has more species and more algal cover, perhaps because of more sheltered habitats and shorter dispersal distances from mature algal communities (Gunnarsson 2000). The earliest colonists following the recent eruption of Kasatochi were diatoms, much like what occurred following the emergence of Surtsey. A few amphipods have also colonized the coarse sandy substrate and a few surviving pockets of the once dominant kelp forests may expand when erosion clears the loose sediments and exposes hard rocks (Jewett *et al.* 2010). On both Surtsey and Kasatochi, unstable surfaces are slowing both terrestrial and marine colonization that can be quite rapid on more stable, rocky substrates (Bertness *et al.* 2004). On Kasatochi, the rapid erosion of newly deposited sediments (about 10^4 m^3 km^{-2} yr^{-1}, a rate typical of other recently active volcanoes) is likely to persist for several years until the older, more stable coastline is exposed (Waythomas *et al.* 2010). During that time, stable, marine surfaces are unlikely to be extensive.

3.3.2 Hydrothermal vents

Deep-sea hydrothermal vents are marine phenomena that were only discovered in the late 1970s (Van Dover 2000), even though their existence was postulated by the theory of plate tectonics generally accepted almost two decades earlier (Vine and Matthews 1963). Superheated gases rich in sulfur, hydrogen, methane, iron, manganese, and many other elements (Elderfield and Schultz 1996) react with cold ocean water and precipitate to form complex structures shaped like columns or mounds (Fig. 3.3). These vents are commonly found on spreading ridges (on the canyon floor or walls) but also anywhere there is volcanism (e.g. on active seamounts such as Loihi; Karl *et al.* 1988). Vents can therefore be considered as secondary disturbances of regional volcanic activity—at least to the degree that they disrupt existing organisms. The unique communities of organisms that are found around these vents use chemosynthesis (largely sulfide reduction) instead of photosynthesis as the basis for their productivity. These communities must continually adjust to disruptions in the rate and direction of the gas plumes they use as an energy source (Mullineaux *et al.* 2009) and typically survive for only a few years after the vents stop releasing heated gases. The mounds cause further disturbances in their surroundings as they gradually erode away. The succession of organisms that colonize these vents is therefore rapid but not reliably directional, depending on many abiotic

Figure 3.3   Large sulfide mound associated with a submarine thermal vent. (From Robigou *et al.* (1993) with permission from the American Geophysical Union.)

(e.g. temperature, fluid chemistry) and biotic (e.g. dispersal abilities, colonization order, competitive exclusion) variables (Sarrazin *et al.* 1997). Vestimentiferan tubeworms (e.g. *Tevnia* or *Riftia*) are common features of vent communities, likely due to their ability to rapidly colonize new vent surfaces and grow quickly to reproductive maturity (Van Dover 2000). Bivalves commonly replace tubeworms over time on the most stable vents (Mullineaux *et al.* 2000). Existing communities can be disturbed when new vents appear. Given the sporadic movements of subsurface lava flows and the resulting spatial and temporal heterogeneity of vents, a mosaic of communities of various successional stages is therefore the norm (Shank *et al.* 1998).

Shallow hydrothermal vents are also disruptive to marine organisms, particularly when they are explosive or have larger than usual emissions due to volcanoes or earthquakes (Dando *et al.* 1995). For example, seagrass beds may have reduced growth due to the high temperatures and reduced light from turbidity caused by the release of volcanic gases (Vizzini *et al.* 2010). These conditions offset any potential benefit from the higher carbon dioxide concentrations that usually characterize shallow vent activity. Both photosynthetic and chemosynthetic organisms are associated with shallow hydrothermal vents, although there are fewer vent-obligate taxa than at the deeper vents (Tarasov *et al.* 2005). Chemosynthetic organisms are also found at both shallow and deep methane, brine, and hydrocarbon seeps (Van Dover 2000).

3.3.3 Earthquakes

Most earthquakes are caused by movements of the earth's crust, both at subduction zones and spreading ridges, when the friction that has built up between plates is suddenly released as energy. Slowly spreading ridges (< 60 mm yr^{-1}) are more conducive to earthquake initiation than fast-spreading ones (Van Dover 2000). Earthquakes can also be caused by volcanoes (often along those same crustal boundaries), massive landslides, or floods. Humans can also cause earthquakes by flooding reservoirs or other large construction projects. Earthquakes, in turn, can trigger landslides and tsunamis (tidal waves). In 1960, the most intense earthquake ever recorded (9.5 moment magnitude; see Section 2.4) occurred off the central Chilean coast in the eastern Pacific Ocean. Earthquakes just offshore from Chile (2010; 8.8 magnitude), and Japan (2011; 9.0 magnitude) generated tsunamis that crossed the Pacific Ocean. A 9.4 magnitude earthquake occurred in the Indian Ocean in 2004 that generated tsunamis that killed about 230 000 people in Indonesia and nearby countries. During that earthquake, coral reefs also sustained damage from seismic uplift (Searle 2006). A 1995 earthquake off the Chilean coast raised coral reefs by up to 80 cm, and the death of coralline algae by desiccation was used to calibrate uplift (Ortlieb *et al.* 1996). Seagrass beds (Short and Wyllie-Echeverria 1996) and intertidal invertebrates (Haven 1971) were damaged by vertical shoreline displacement from an Alaskan earthquake in 1964.

3.4 Erosion and deposition

Erosion and deposition occur at many scales and have many causes in aquatic environments. In areas of rugged topography, large, submarine landslides are not uncommon. Some of the largest include events that occurred several hundred thousand years ago around Hawaii and the Canary Islands and involved volumes of up to 5000 km^3. The submarine landslides surrounding the Hawaiian Islands are caused by volcanic activity and slumping of material down the long slopes of the volcanic mountains. The landslides occupy a greater area than the terrestrial parts of the islands (Walker 1994). Several landslides that occurred 8000 years ago moved about 5580 km^3 of sediments in the North Sea. In Japan, a landslide that occurred 6000 years ago moved an area of land about 1 million km^2 in extent into the ocean and created many small islands (Matsushima Bay; Strom 2008). An earthquake in 1929 in the western Atlantic Ocean generated a current that transported more than 100 km^3 of sediments down the Laurentian Fan to the Sohm Abyssal Plain (Piper *et al.* 1988). The resulting "gravel waves" from the 1929 landslide now support chemosynthetic clam beds, presumably generated by the exposure of reservoirs of reduced, sulfurous fluids (Van Dover 2000).

Deposition in rivers occurs when water currents slow down due to obstructions in the river bed, a reduction in slope, or a reduction in water volume. The maximum size of particles carried downstream depends on the force of the current. Massive glacial outbursts (del Moral and Walker 2007), with flows of up to 500 000 m^3 sec^{-1}, can transport ice chunks the size of houses. Many alpine streams tumble boulders the size of cars or larger (Fig. 3.4). Gravels and sands are moved by most flood events. Silts and clays are

Figure 3.4   Severe erosion in a tributary of the Kokatahi River, South Island, New Zealand. The region receives 9000–10 000 mm of rain each year. Note the man leaning against a rock that is the size of a large house; the rock had been dislocated from a nearby hillside since our previous visit to this tributary.

suspended in the water column until precipitated—often where rivers end in broad, flat estuaries at the coast. Sediment deposits can be washed further downstream during the next storm or remain for long enough to form sedimentary rocks. Sand particles that reach the ocean often wash up as beaches and form coastal dunes (Fig. 3.5). Biotic communities in river drainages and estuaries are temporary, as species comprising such communities depend on intervals between major flood events to grow and reproduce.

Marine and freshwater sediments can harm aquatic and terrestrial (floodplain) organisms through abrasion, suffocation, and physical burial (Sousa 1985). Sediments can be deposited quickly (e.g. landslides) or gradually (e.g. steady bank erosion), year-round or seasonally (e.g. during periods of low wave energy and velocity; Robles 1982). Local spatial and temporal variations in topography or currents can alter deposition patterns (Littler *et al.* 1983). Encrusting life forms are more likely to be smothered than upright forms (Connell and Keough 1985). Corals, for example, are impacted directly by sedimentation or indirectly through loss of effectiveness of their photosynthetic algal symbionts. Sediments that are rich in nutrients can trigger **eutrophication**, a process where algal blooms respond to nutrient loading, then die and decompose. The bacterial populations that decompose the algae respire, using up the dissolved oxygen in the vicinity; the result is **anoxia**

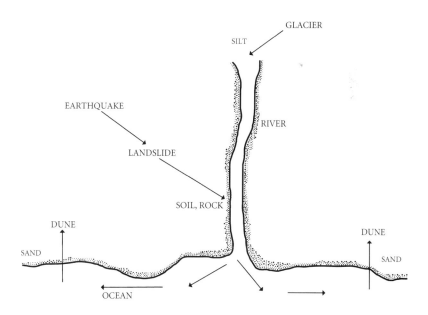

Figure 3.5   Sediments from glaciers (silt) and earthquake-triggered landslides (soil and rock) are washed down rivers to the ocean where they are furthered weathered, sorted, and then deposited as silt and sand on beaches and form dunes. (From Walker and Bellingham (2011), with permission from Cambridge University Press.)

that leads to the death of fish and other organisms. Recovery and reinvasion of disturbed aquatic habitats can be rapid. Limpets along the Central and South American Pacific Coast recovered quickly from landslides (Garrity and Levings 1985). Such rapid recovery on unstable sediments in aquatic habitats occurs through colonization by plants with short life spans and rapid growth rates. Initial growth can quickly reduce the instability, allowing more plant species to establish and animals to return. Eventually, the complexity (e.g. food webs, kelp forest structure) and services (e.g. clear water) of aquatic ecosystems are also restored, provided that severe disturbances are infrequent.

3.5 Floods and droughts

Rivers vary in length from 0.1 km (D River, Oregon, U.S.) to over 6000 km (the Nile, Amazon, and Yangtze) but each has a watershed, a characteristic geology, channel morphology, and flood regime. The rate of geological uplift in the watershed determines the erosional characteristics of the river (del Moral and Walker 2007). Entrenched rivers can cut well-incised channels, continuously eroding successive layers of a variety of uplifted rock types (e.g. the Grand Canyon, U.S; Fig. 3.6); fan out across huge outwash plains in

Figure 3.6   Erosion patterns of the Green River near its confluence with the Colorado River, Canyonlands National Park, Utah (U.S.).

Figure 3.7   A 100 × 30 km sand plain (Skeiðarárssandur) created in southern Iceland from cata-strophic floods that are triggered by a volcano that erupts from under a nearby glacier. Note constructed berms to direct water flow in the closest channel.

glacial rubble (Fig. 3.7; Box 3.1; Skeiðarársandur, Iceland); carve a series of underground caverns in limestone (Río de Camuy, Puerto Rico); and create spectacular, narrow slot canyons in sandstone (Blue Mountains, Australia). Water quality also reflects geological (and biological) parameters. Suspended silts cloud glacial melt waters (Fig. 3.8), tannins darken forest streams, and granite stream beds leave crystal clear water. Aquatic fluvial habitats vary as widely as the geology, morphology, and chemistry of river channels but generally include the water column and a **hyporheic zone** beneath the river bed where ground water mixes with surface water.

Flood frequency reflects rainfall patterns and the absorptive capacity of the floodplain. Floods can reset horizontal and vertical contours, erode banks, scour the hyporheic zone, and rearrange sediments. Disturbances in low-order, headwater regions of rivers with generally well-incised and narrow floodplains typically involve erosion of steep banks and rearrangement of rocks and coarse sediments (Fig. 3.9; Nakamura and Inahara 2007). In higher-order rivers that are braided, channel migration is a prominent dis-turbance and floods distribute sediments across a wider floodplain. Sorting of different particle sizes from boulders to fine silts depends on fluctuating channel morphology and obstructions such as islands or channel debris, so

Figure 3.8   Flood waters coursing over Dettifoss, one of Iceland's largest waterfalls.

Box 3.1 Catastrophic floods result from volcanoes melting glaciers

Skeiðarársandur is a 100 × 30 km outwash or sand plain (see Fig. 3.7) created by flood waters from Iceland's largest glacier, Vatnajökull. When Grímsvötn Volcano (and its nearest neighbors) erupts from beneath the glacier, a lake is formed that eventually fills sufficiently to lift the ice and drain out from underneath the glacier. Such releases are relatively sudden and can produce very damaging floods called glacial outburst floods or jökulhlaups, notably in 1996, 1998, 2004, and 2010. In 1996, following a 5.0 magnitude earthquake, the subsequent eruption created a 3.5 km long and 350 m wide canyon in the surface of the glacier. One month later the subglacial lake lifted the ice and the hot melt water quickly emerged from the glacier, spilling out over the sand plain. Floodwaters reached a speed of 50 000 m³ sec⁻¹, carrying multi-ton blocks of ice as they destroyed protective berms and smashed several sturdy bridges on a highway that crosses the sand plain, before reaching the ocean 30 km away. Such floods have occurred in the past and keep the sand plain devoid of human settlements, although there is enough vegetation to support about 300 grazing sheep. A few scientists study plant succession in the kettleholes (depressions left by melting ice) and are examining the effects of sheep. Tourists also cross the sand plain to reach scenic Skaftafell National Park at the foot of Iceland's tallest mountain, Hvannadalshnúkur (2119 m). In such environments, humans are clearly at the mercy of powerful and unpredictable natural forces.

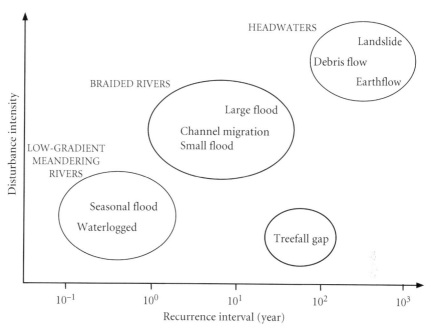

Figure 3.9   Dominant disturbance regimes in riparian zones arrange by recurrence interval and level of physical disturbance of geomorphic surfaces (intensity). (From Nakamura and Inahara (2007) with permission from Elsevier.)

a mosaic of habitats is often created. Sudden outbursts from subglacial lakes, the rupture of temporary dams caused by landslides or lava flows, or flash floods from particularly intense rainfall result in the substantial realignment of river channels. The highest-order rivers near the coast experience predictable seasonal floods and often feature natural levees and backwater swamps (Nakamura and Inahara 2007).

Droughts occur in rivers when periods of low flow are caused by decreased inputs from precipitation and groundwater. Droughts can occur in the headwaters or elsewhere along a river channel and involve partial or complete disruption of surface flow (Lake 2003). Droughts can be seasonal or stochastic, ephemeral or relatively permanent. Direct effects of a drought include loss of water, loss of habitat, and loss of connectivity among aquatic habitats. Indirect effects include deterioration of water quality (reduced N, P, and carbon inputs), conversion to primarily autotrophic rather than heterotrophic production, and alterations in how species interact (Lake 2003). When droughts persist, species able to resist or recover from drought through physical, morphological, physiological, or behavioral refugia replace drought-intolerant species (Humphries and Baldwin 2003). However, restricted movement within catchments and crowding in favorable microhabitats can negatively affect fish populations (Matthews and Marsh-Matthews 2003). Recovery from temporary drought can be rapid (e.g. shrimp populations in Puerto Rican rivers; Covich *et al.* 2006), but prolonged drought can alter population levels and even evolutionary development of species. Human alterations of riverine ecosystems add new threats to wildlife. For example, hydroelectric dams on the Columbia River (U.S.) and increased use of water by farmers have reduced habitat and water levels, respectively, for spawning salmon (Harden 1997; Dauble *et al.* 2003). River droughts may become more frequent and severe in arid ecosystems with climate change.

Lakes also vary in geologic setting, morphology, and flood regime. Sometimes lakes have precipitous origins (e.g. following dam construction; see Section 4.10) or a sudden demise. When the huge Lake Bonneville (Utah, U.S.; 52 000 km²) emptied abruptly about 15 000 yr ago, the flood waters reached depths of more than 100 m and left many strange depositional and erosional features, including melon gravel deposits (rounded basaltic boulders averaging 1 m in diameter), gigantic potholes up to 100 m deep, and polished surfaces (Jarrett and Malde 1987). Most lakes have more gradual origins and permanent water supplies that reflect geological uplift and erosion and regional precipitation patterns. Water levels in lakes can suddenly rise or fall and this can constitute a disturbance to aquatic organisms adapted to particular water levels (Hayashi and van der Kamp 2007). Water levels are controlled by the balance between inputs (precipitation, rivers, runoff, snow drift, ground water) and outputs (evapotranspiration, streams, ground water; Fig 3.10). Small ponds that depend on seasonal inputs may dry out every year. Lakes with emergent or bordering vegetation increase transpiration losses but can act as catchments for wind-blown snow in temperate climates (Fig. 3.11). Groundwater inputs tend to be greatest where there are highly permeable substrates such as sands around a lake and low where impermeable clays dominate (Hayashi and van der Kamp 2007). In arid climates, high runoff from occasional rains can rapidly fill lakes, while evaporation usually controls long-term water levels. Most watersheds are impacted by human land use that generally increases variability in lake levels.

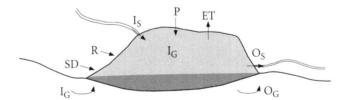

Figure 3.10   The sources and sinks of lake water, including precipitation (P), evapotranspiration (ET), stream inflow (I_S), stream outflow (O_S), diffuse runoff (R), snow drift (SD), groundwater inflow (I_G), and groundwater outflow (O_G). (From Hayashi and van der Kamp (2007), with permission from Elsevier.)

Figure 3.11   Lakes ringed by forests such as Lake Groton, Vermont (U.S.) trap wind-blown snow and nutrients. The shoreline, although calm in this photo, is annually scoured by the breakup of ice.

3.6 Storms

All organisms are adapted to some environmental fluctuations, but storms are widespread phenomena that affect both aquatic and terrestrial habitats and frequently lead to loss of biomass. Storms can impact aquatic biota at the scale of local reef damage to the realignment of large currents such as the Gulf Stream by a cyclone (Davis *et al.* 1991). In this section, the focus will be on wind-driven storms that affect marine habitats.

3.6.1 Tsunamis and other waves

Ninety per cent of tsunamis are generated from submarine earthquakes yet less than 1% of submarine earthquakes trigger tsunamis (Bryant 2001). Five per cent of tsunamis are triggered by volcanoes; other triggers include meteorites, landslides, and floods (e.g. torrential rains, glacial outbursts, bursting of dams). Lakes can also have tsunami waves, typically from volcanic activity (e.g. Lake Taupo in New Zealand). Waves generated by massive underwater landslides (see Section 3.4) can reach 30 m in height (coastal Europe, 8000 yr ago), 365 m (Hawaii, several hundred thousand years ago), and even

524 m (southeastern Alaska, 1958) above sea level. Little is known about the underwater damage caused by such enormous waves. One obvious effect is the churning up of coastal sediments and sometimes their deposition on land. Breakage of coral reefs is common (Marris 2005; Wilkinson *et al.* 2006) and all sedentary organisms such as corals, shellfish, burrowing animals, and soil organisms can be affected, particularly in shallow water (Bryant 2001) or when growing on unconsolidated sediments (Baird *et al.* 2005). When not buried in sediments, the recovery of marine organisms can be rapid. However, damage to reefs from over-fishing may reduce recovery rates, even when initial damage is not high, as found for the 2004 tsunami in Indonesia (Foster *et al.* 2006).

Waves are a permanent feature of most coastlines and organisms adapt to their impacts. Productive kelp forests thrive on wave-swept, rocky shorelines in cooler water, while coral reefs thrive on wave-swept shorelines where the water is warmer. The waves provide aeration and nutrients. However, unusually large waves associated with storms can be considered disturbances when they move sediments, tear out seagrass beds or kelp, break corals, and alter shoreline topography (Sousa 1979b). A series of studies of rocky intertidal zones, many of them experimental, have contributed to our understanding of the role of disturbance (e.g. waves, driftwood logs, desiccation, grazing) in shaping competition hierarchies (of barnacles, algae, and mussels), species diversity, predation (e.g. by whelks), and temporal and spatial patterns (Dayton 1971; Sousa 1979b; Paine and Levin 1981; Underwood 1999, 2000). Wave action turns over boulders that are then recolonized, creating a patchwork of sessile organisms. One study found that tide pools were disturbed by waves, excessive heat, or other disturbances on average every 1.6 yr (Dethier 1984). Ice scouring can impact succession on rocky shorelines in both predictable (e.g. nearness to melting glacier) and unpredictable (e.g. large glacial calving events) ways (Pugh and Davenport 1997). Sandy shorelines are also impacted by waves that alter the distribution and recolonization of aquatic organisms (Savidge and Taghon 1988; Peterson 1991), but such environments are harder than rocky intertidal habitats to experimentally manipulate because of their instability (Schoeman *et al.* 2000).

3.6.2 Cyclones

Most cyclones are circular, tropical storms that are generated by air rising from above warm oceans. Temperate cyclones are oval-shaped and are due to the convergence of warm and cold air masses. Cyclone intensity is measured by wind speed (> 118 km hr^{-1}) and barometric pressure (the lowest ever recorded being 87 kPa; Japan, 1979). Cyclone is a standard name for storms that are also called hurricanes (Atlantic and central Pacific Oceans) and typhoons (northwestern Pacific Ocean). Centers of tropical cyclone activity include the Caribbean Sea, off the western coast of northern Mexico, in the western Pacific Ocean, and in the Indian Ocean. The effects of cyclones on marine habitats include the generation of large, often destructive storms (most notably the two storms that thwarted the Mongols in their attempts to invade Japan in 1274 and 1281). These storms can generate heavy rains that trigger terrestrial erosion and increase sediment loads in nearby water bodies; dilute ocean water; transport sediments along sea beds and shorelines; disrupt wildlife; and damage coral reefs. The effects of one of the most intense, recent cyclones (Hurricane Hugo) have been widely examined, in part due to good pre-disturbance data (Bénito-Espinal and Bénito-Espinal 1991; Finkl and Pilkey 1991). Dilution of the ocean from heavy rains and terrestrial runoff during Hurricane Hugo upset recruitment of fish and crustaceans dependent on highly saline estuaries (Knott and Martore 1991) and potentially altered currents such as the Gulf Stream, with widespread climatic consequences (Davis *et al.* 1991). Sediment transport during cyclonic storms can rearrange coastal morphology and either erode shorelines or build them up from sediment deposition (Nelson 1991). Eroded beaches often recover with normal wave action in the months following a cyclone (Sexton and Hayes 1991). Wildlife disruptions can occur when habitats are altered. For example, after a cyclone in 1971 dugongs in northeastern Australia increased their movements in search of alternative food after damage to their usual seagrass habitats (Heinsohn and Spain 1974). Damage to reefs can be very patchy (Done 1992) and depends on the depth of the reef, the orientation of the reef relative to the cyclone, and the degree of damage by previous cyclones (Hubbard *et al.* 1991; Hughes and Connell 1999). Cyclones can also act as dispersal agents for coral colonies (Massel and Done 1993). Where large volumes of sediment are removed, there may be a long-term renewal of reef habitat; abrasion and erosion can hinder reef development (Harcombe and Carter 2004). Some corals are very resilient (Halford *et al.* 2004), while others such as large, branched corals can take a decade or more to recover and algae may dominate during that time (Adjeroud *et al.* 2009). Corals are generally less resilient to chronic or anthropogenic disturbances than they are to cyclones (Connell *et al.* 1997; Bellwood *et al.* 2004; Hughes *et al.* 2005).

3.7 Additional disturbances

Herbivory and other forms of predation, disease, and algal blooms are often considered as forms of biological disturbance (Sousa 1985) and are nearly as ubiquitous as waves. Herbivory, predation, and disease, for example, are common natural disturbances in seagrass beds (Short and Wyllie-Echeverria 1996). Both herbivores and predators are affected by the prevailing disturbance regime. Rocky intertidal zones can contain a high degree of spatial heterogeneity due to a combination of wave action and shifting sediments. Within that disturbance regime, herbivory and preda-tion create different patch sizes (Sousa 1985), depending in part on whether the predators prefer the patch (e.g. algal beds) or the matrix (e.g. mussel beds). Sometimes grazers prefer the relative protection of mussel beds and only forage a short distance into algal patches, leaving more algae grow-ing in larger patches. Similarly, some predators such as the snail *Thais* are more active in protected than in open habitats (Menge 1978). Herbivores alter algal productivity depending on grazing intensity but generally facili-tate the flow of energy from highly productive algal turfs to higher trophic levels (Carpenter 1986).

A special case of herbivory combined with structural damage is **bioerosion** (Glynn 1997). Numerous species, including bacteria, fungi, worms, sponges, mollusks, and fish graze algae growing on coral reefs and often burrow into the calcareous skeletons of coral reefs, partly for protection from predation. Their assorted activities can result in coral reefs becoming riddled with holes and cavities that are then susceptible to breakage during storms (Fig. 3.12). When bioerosion rates exceed coral building, the reef can be eroded. The buildup of bioeroded sediments can contribute to the burial of coral reefs. Predation of live coral by seastars (*Acanthaster*) and snails (*Drupella*) makes coral reefs more susceptible to bioerosion. Similarly, loss of coral to diseases, decreased salinity, or increased organic matter all favor a proliferation of bioeroders that are primarily filter feeders compared with reef builders that are autotrophic (Hallock 1988; Glynn 1997). Over-fishing can lead to domi-nance by sea urchins (*Diadema*), which are much more effective bioeroders of coral reefs than the few fish that feed on mostly dead coral or protuber-ances but not the flat matrix of the reef itself (Bellwood and Choat 1990) (see Chapter 4).

Aquatic diseases are naturally occurring phenomena; however, many dis-eases are becoming increasingly prevalent and have been linked to anthro-pogenic factors such as increases in ocean temperature (Rosenberg and Ben-Haim 2002; Eakin *et al.* 2010) and nutrient enrichment (Bruno *et al.* 2003a). Pollutants may or may not be causes of cancerous tumors in fish

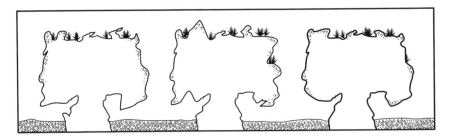

Figure 3.12   Coral reefs are frequently pockmarked by bioeroders. (From Glynn (1997) with permis-sion from Springer.)

(Mix 1986). Bacterial infections appear to play a role in coral bleaching (Kushmaro *et al.* 1997) and, along with other coral diseases (e.g. black band disease, sea-fan disease, and coral-white plague), are becoming more virulent with increasing water temperatures (Rosenberg and Ben-Haim 2002). Several water-borne fungal and viral diseases also contribute to declines in the populations of amphibians (Daszak *et al.* 1999). Little is known about the role of aquatic diseases in wild populations of fishes because of the difficulty of tracking diseased fishes (Bakke and Harris 1998).

Algal blooms occur in freshwater and marine ecosystems when there is an increase in nutrient availability, sometimes associated with eutrophication or upwelling of cold, nutrient-rich water (Pitcher *et al.* 2010). Red tides are a type of harmful bloom because the dinoflagellates that comprise them secrete toxins that directly kill fish and indirectly kill dolphins and manatees through bioaccumulation in fish and seagrasses (Flewelling *et al.* 2005). Algal blooms also occur when nutrients are added to marine ecosystems and herbivores decline either due to disease or over-fishing (see Chapter 4).

3.8 Conclusions

Aquatic habitats are dynamic and continually influenced by a variable natural disturbance regime. This variability has produced numerous types of habitats and promoted the evolution of high levels of aquatic biodiversity. Much of this diversity arises from the spatial heterogeneity provided by disturbances of different extent. Widespread disturbances of aquatic habitats include cyclones and storms; regional disturbances include volcanoes and earthquakes; local disturbances include hydrothermal vents and predation. Microsite variability (e.g. from mussel beds or bioerosion) contributes to the dynamics of herbivory and predation. Floods regularly scour river beds and deposit sediments on floodplains and lake edges. Waves move sediments around, break up coral, and alter the topography of the shoreline. Aquatic disturbances also represent the full range of disturbance intensity and severity. Lava provides surfaces where no biological legacy remains. Deltas, sand bars, and river bank sediments can vary from nearly sterile to quite nutrient rich, but, until stabilized, do not support sedentary organisms. Whale carcasses provide a rich source of nutrients for heterotrophs and are rapidly decomposed. Aquatic disturbances also vary by frequency, from rare events (e.g. volcanoes, earthquakes) to more frequent events (e.g. rogue waves, local predation). However, separating natural aquatic disturbances from anthropogenic ones is an increasingly futile exercise.

Aquatic habitats, like terrestrial ones, are all altered to some extent by human activities. For example, anthropogenic disturbances are now arguably more common than natural ones in seagrass beds (Short and Wyllie-Echeverria

1996) and coral reefs: Connell *et al.* 1997, Foster *et al.* 2006). Natural patterns of resilience are being tested because the effects of anthropogenic disturbances are like a press, pushing aquatic communities past thresholds of recovery. Following most natural disturbances, aquatic organisms recover quite well, but faced with accumulated pressures from humans, the future of many aquatic communities is bleak. Anthropogenic disturbances in both terrestrial and aquatic environments are the subject of Chapter 4.

4 Anthropogenic habitats

4.1 Introduction

Humans have evolved within the limits and opportunities presented by the natural disturbance regime but have also incrementally added anthropogenic disturbances. Some of these anthropogenic disturbances are novel, with no natural counterparts, while others resemble natural disturbances (see Table 1.1). People in pre-agricultural societies could respond to volcanoes, cyclones, floods, droughts, or fires by moving away, at least until the vegetation recovered and game animals returned. The major manipulation of the environment by early humans was probably the setting of fires to drive animals toward traps and cliffs or to improve forage for them. Human manipulations of the environment began in earnest with the development of sedentary agricultural communities that were more dependent on a stable environment. Extensive irrigation systems were developed, particularly in China, the Middle East, and South America. Small settlements grew into major urban centers as agriculture fueled increased health and reproduction. Transportation networks were built to bring resources to the cities and deliver products to markets. As marine navigation improved, fishing became concentrated in productive regions of the oceans such as the Great Banks in the North Atlantic Ocean. Exploration and the colonization of Africa, North and South America, and many Pacific Islands by Europeans between 900 and 1900 inevitably led to resource extraction (e.g. phosphate mining), deforestation, and introduction of non-native crops, weeds, and animals (Crosby 1986; Diamond 1997; Walker and Bellingham 2011).

Industrialization and the drive for energy brought new anthropogenic disturbances from coal mines to dams and canals and, most recently, oil wells. In the last several centuries, forestry has become important as a management tool for dwindling forest resources. Agricultural activities have intensified with the development of chemical fertilizers and reliance on pesticides and herbicides. Globalization has led to the exploitation of formerly remote regions. Extensive areas are still being deforested, both for wood products and for land to raise cattle and grow soybeans, coffee, bananas, sugar cane, and palms for oil production (Pimentel 1993; Terborgh 1999). Harvesting of natural fish populations is reaching its limits after a free-for-all extracting the ocean's resources to feed surging human populations (Myers and Worm 2003; Clover 2004). As a result, aquaculture is expanding, but with its own resource problems (Birt *et al.* 2009). Finally, mass consumerism of all things plastic and metal has led to an enormous amount of waste that is not easily biodegradable. Human impacts on biotic resources have accelerated exponentially, mirroring a similar rise in human population.

Humans compound and intensify natural disturbances (Fig. 4.1), in addition to contributing a novel set of disturbances (del Moral and Walker 2007). The human population is arguably at or near the ecological carrying capacity of the earth (Cohen 1995). Humans have expanded into and altered most environments (McKibben 1989; Flannery 2005); this expansion increases our exposure to disturbances. As we deforest the Amazon Basin, mine coal from China, or plow up the grasslands of Asia, we are reducing the resource capital for ourselves and future generations. Once rain forests have been fragmented or cut down they are difficult to re-establish. An open pit mine may never support the same biological communities that existed before the mining began. Soils blown away or washed to the sea are not easy to recover. Our activities, augmented by our population growth, will expand until we have reached the limits to our resources (unless wise resource management is instituted before this happens). Our activities exacerbate previous disturbances on land such as dunes, landslides, insect outbreaks, cyclones, floods, droughts, and fires (Chapter 2). For example, ocean warming is likely to lead to more intense cyclones (Emanuel 2005), levees and channelization promote more severe floods (Reice 2003), and deforestation expands and extends droughts (Laurance and Williamson 2001). Each disturbance impacts more people, threatening more lives and property than when there

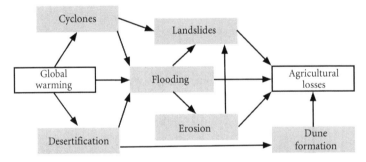

Figure 4.1   Anthropogenic disturbances (clear boxes) can alter interactions among natural disturbances (shaded boxes). In this example, global warming increases hurricane intensity, thereby increasing landslide and flood frequencies and agricultural losses. (From del Moral and Walker (2007) with permission from Cambridge University Press.)

were fewer humans on earth. Uniquely human disturbances such as commercial agriculture and fishing disrupt natural food webs and reduce diversity, a situation that increases the vulnerability of our food sources to disease and extinction. Human impacts on natural resources are not all negative (e.g. we increase soil productivity, establish nature reserves, and conserve endangered populations) but our net effect is a complex alteration of disturbance regimes around the world.

This chapter addresses the biological implications of 12 widespread anthropogenic disturbances (Table 4.1) discussed in ascending order of spatial extent for terrestrial (military activities, mining, urbanization, transportation, silviculture, and agriculture) and aquatic (benthic structures, shoreline structures, sedimentation, oil spills, recreation, and fishing) environments. Other, less extensive or indirect, anthropogenic disturbances are also mentioned for terrestrial (radiation releases, landfills, recreation, and tourism) and aquatic (dam failures, toxic spills, and garbage dumps) habitats. Within each topic, I discuss, as appropriate, the relevance to humans, the physical characteristics of the disturbance, the biotic conditions of the disturbed habitat, and various dynamics that result. This organization follows the format used in Chapters 2 and 3. Many anthropogenic disturbances could not be covered, including ones where little is known about their variable biological impacts (e.g. industrial accidents) and ones that are localized (e.g. power line right-of-ways). Invasive species as agents of disturbance are covered in Chapter 6. Many forms of land and water pollution were not covered,

Table 4.1 Anthropogenic disturbances and their major habitat alterations. Within each category, disturbances are presented in approximate order of increasing spatial extent in that habitat (approximate percentage of habitat affected in parentheses). See Table 2.1 for natural disturbances that are discussed in Chapter 2 (terrestrial) and Chapter 3 (aquatic).

Habitat	Name	Habitat alterations
Terrestrial	Military activities (1)	Damage and contaminate biota; protect from other development pressures
	Mining (1)	Destroys and contaminate biota; alters topography
	Urbanization (5)	Destroys biota; favors human-linked species
	Transportation (5)	Destroys biota; fragments habitat; alters microclimate
	Forestry (10)	Damages forests; creates gaps; promotes fire, regeneration
	Agriculture (45)	Alters topography; promotes crops, weeds
Aquatic	Novel benthic structures (1)	Damage and create benthic habitats; promote biodiversity
	Novel shoreline structures (1)	Damage biota; create habitats; alter composition
	Sedimentation (5)	Reduces water clarity; buries biota; creates habitats
	Oil spills (5)	Destroy and contaminate biota; disrupt food webs
	Recreation (20)	Reduces heterogeneity; damages biota; promotes eutrophication; promotes conservation in reserves
	Fishing (40)	Reduces populations; disrupts food webs

in part due to their diffuse nature and gradual onset. Air pollution is also beyond the scope of this book. However, oil spills are a current concern and have relatively abrupt impacts that have been reasonably well studied. The remaining chapters in this book cover general characteristics common to many disturbances and various management approaches.

4.2 Military activities

Military activities are found in virtually every country and can occupy from 1% (U.S.) to 40% (Vietnam during American occupancy) of the land (Demarais *et al.* 1999). They create severe and often long-lasting damage to natural environments (Rogers 2002) both in peace time and in war. Intriguingly, there is a positive feedback loop between environmental damage (loss of resources) and the occurrence of many military conflicts (Westing 1988). Soils are compacted or churned and low-lying vegetation is crushed from tanks, trucks, troop movements, and trenches (Becher 1985). Soil disruption can result in erosion and alteration of the local hydrology (Demarais *et al.* 1999). River fauna can change as banks erode and silt accumulates (Quist *et al.* 2003). Disturbances that impact soil fauna and recovery include oil and gasoline spills and fires (e.g. Iraq), chemical defoliants and napalm (e.g. Vietnam; Westing 1971), and radioactive contaminants (e.g. Japan). Mass graves or even unburied corpses are another gruesome aftermath of war. Removal of vegetation to clear spaces for roads, increase visibility, feed displaced populations, or provide fuel wood are other side-effects of conflicts. In Japan during World War II, 15% of the forests were felled for the war effort, resulting in long-term soil erosion and an increase in infestations of pine bark beetle on dead trees (Tsutsui 2003). Military training activities can create many of the same disturbances. Bombing ranges are particularly vulnerable to damage to plants and soils, in addition to remnant unexploded and exploded shells remaining on the surface. The islands of Culebra and Vieques in Puerto Rico and Kahoolawe in Hawaii are examples of bombing ranges that are now being restored (Walker and Bellingham 2011). Military infrastructure can also have a large environmental impact. Buildings, roads, fuel leaks, radioactive contamination, and noise pollution are a few of the side-effects of military activities. There is also a positive aspect of military activities. Land fenced off but not actively used for military training can have higher species diversity, lower percentage of invasive species, less grazing, and support rare species not able to cope in the surrounding, unfenced areas (Lathrop 1983, Demarais *et al.* 1999; Beavers and Burgan 2002). Where secondary succession to low-diversity woodlands is undesirable (e.g. in parts of Europe), military activities can promote a mosaic of habitats that support a diverse flora and fauna. However, where regular military exercises are practiced, the accumulated effects on habitat are negative.

In addition to direct disruption of soils (compaction and erosion), plants (crushing), and animals (noise pollution, habitat loss), military activities can have long-term detrimental effects on productivity, populations of endemic species, and habitat fragmentation. Species diversity can increase, but these increases are usually due to the destruction of native organisms and subsequent success of invasive species adapted to early successional environments (Wilson 1988).

4.3 Mining

Extracting or mining products from the earth is a widespread phenomenon that covers about 1% of the earth's terrestrial surface and occurs in most countries of the world (Mining Annual Review 1995). Early human cultures mined flint, gold, and obsidian. During the Bronze and Iron Ages, humans mined copper and zinc; clay, limestone, and lead were important to the Romans. With the onset of industrialization, demand for additional iron, copper, and zinc led to the rapid expansion of mines. Today, coal mining from both underground and surface mines is the most widespread extractive process. Other mined fuels include shale oil, peat, and uranium. Mining is used to extract construction materials such as granite, marble, slate, gravel, and sand and other products such as gypsum, potash, and rock salt. Minerals such as aluminum, zinc, lead, and nickel are mined for use in a variety of industries. Technological advances now allow the mining of many of the 17 rare earth elements (Spedding and Daane 1961) for use in manufacturing television cathodes, lasers, camera lenses, personal computers, cell phones, batteries, and magnets. In addition to these products, humans are now dependent on the products of mining for energy, construction, and transportation (Cooke 1999).

Surface mines are very intense disturbances where all plants and soils are removed before miners dig huge open pits, but even when mines are underground, tailings are usually deposited on the surface (Fig. 4.2). Typical mine tailings consist of unused topsoil (overburden), unprocessed, non-ore-bearing rocks, and processed rock, gravel, or sand piles (Majer 1989; Munshower 1993). Mine tailings are often extremely acidic (coal mine tailings; Bradshaw and Chadwick 1980) or alkaline (soda ash wastes; Ash *et al.* 1994). Uranium mines leave radioactive tailings. Bauxite mines produce a ton of alkaline mud for every ton of bauxite ore. Mine wastes also include areas of subsidence and various types of waste ponds that can contain toxic residual chemicals from mining and processing ores such as arsenic, cadmium, cyanide, copper, lead, mercury, silver, and sulfur. These ponds are typically only several hectares in size but can reach several thousand hectares (Williamson *et al.* 1982). The spatial extent of early mines was largely local, but modern mining has expanded to cover much larger surfaces. Some open pit mines and many mine tailings cover areas of more than 1 km^2 (Williamson *et al.* 1982). Secondary disturbances from mining include the infrastructure of transportation networks (roads, railroads, docks, airports) needed to access and distribute the product and processing plants and mills to refine the product. The spatial extent of mines can also be extended across the landscape when tailings are dumped in rivers or the sea or when mine wastes enter groundwater (Cooke 1999). Toxic rivers are a common feature of Appalachia in the eastern U.S. where coal mining has removed whole tops of mountains, inevitably dumping tailings into adjacent watersheds. Sometimes slurry ponds break through their walls and accidentally enter rivers, causing much damage downstream (Reece 2006).

Mining is a severe form of disturbance that typically disrupts animal populations and destroys plant and soil communities (Ali 2003, 2009). Biological legacies may remain, particularly where operations are limited, mining is mostly underground, or where caution is taken to avoid piling tailings around waterways, in habitat protected for wildlife, or in visually unappealing locations. Mines, like military activities, can provide a variety of disturbed habitats that increase local biodiversity (Prach *et al.* 2011). However, most modern mine operations completely denude the local area. Natural recovery of pre-mine ecosystems is rare because the topsoil is removed or buried and mine pits often remain (Lottermoser 2003). Alternative aquatic and terrestrial habitats do eventually develop in all but the largest mines.

Figure 4.2   Mine wastes in Yorkshire, England.

Acceleration of succession is often required of mining companies today. Yet the obstacles to successful revegetation are severe and removing or ameliorating toxic substrates and adding organic materials is often required (see Chapter 9; Cooke 1999; Walker and del Moral 2003).

4.4 Urbanization

Across the entire globe, half the human population lives in cities with more than 100 000 people. In some countries a large majority live in cities (Australia, Egypt, Venezuela, England) but in others rural living still predominates (Kenya, Ethiopia, Nepal, Guyana). Extensive urbanization is a recent phenomenon, beginning in Europe with the Industrial Revolution. London was the first city in the world to record over 5 million inhabitants (in 1891). Today there are 46 cities of that size or larger (Fig. 4.3). The acceleration of technological and agricultural advances has allowed efficient transport and distribution of goods and the subsequent centralization of jobs. In recent decades, medical advances have increased longevity and, combined with still high birth rates in many countries, the number of people in the world grew from 2.5 to 6.8 billion between 1950 and 2010. As a majority of people are born in cities, the proportion of the population that is urban continues to grow (Wu 2008). Recent growth in tourism has also led to urbanization in places like Hawaii (Sheldon *et al.* 2005) and the Canary Islands (León and González 1995). Cities now cover approximately 5% of the terrestrial land mass and land covered by urban centers continues to expand. For example, Cairo (at least until its revolution in early 2011) had been planning two new cities to ease congestion and house 5 million of its current 20 million inhabitants. As cities increase in size, their ecological footprint grows because cities are concentrated centers of consumption and producers of waste. Urban Seattle, for example, uses resources that if produced entirely from crops would use land equivalent to most of the state of Washington; densely populated Holland uses resources that would occupy cropland the size of France (Wackernagel and Rees 1996). Urbanization is not only a striking demographic phenomenon of the last 200 years but also a major ecological disturbance.

Cities are built from relatively impermeable materials on bulldozed and excavated land, thereby destroying most biota and many ecosystem processes. Most cities are built where people have naturally congregated for centuries because there is nearby food (fertile, flat land for agriculture) and transportation (river floodplains, seashores). Potential or active agricultural land is thereby often lost to urban growth. Very few builders preserve any of the original biota or hydrology of an area being urbanized, although newly planned cities have a better opportunity to save remnant vegetation than older, established cities (Carreiro 2008). Urban ecosystems (the subject of

Figure 4.3   Tokyo, Japan, with about 12 million inhabitants, is an example of a densely settled urban area.

urban ecology; Bornkamm *et al.* 1982, Sukopp *et al.* 1990; Carreiro *et al.* 2008) therefore rarely result from intentional landscaping but more often from the haphazard colonization by plants and animals wherever they can establish. These ecosystems, scarce as they are compared with the hectares of impermeable surfaces, offer vital ecosystems services to urban dwellers including absorption of carbon dioxide and other pollutants, production of oxygen, **microclimate** amelioration, noise reduction, rainwater retention, and aesthetic benefits (Chen and Jim 2008).

The most common ecosystems in cities include areas that are not completely covered with buildings or pavement such as parks, vegetated road medians or edges, graveled or dirt pathways, vacant lots, stream banks, cemeteries, lawns, other landscaping, and dump sites. These habitats are subject to subsequent disturbances that include compaction, flooding (particularly when floodplain hydrology has been altered), soil erosion, direct wind damage, or indirect wind damage (from falling glass or building parts), and deposition of toxic materials (when land is used for dumps or vehicle maintenance). Abandonment of an urban ecosystem can lead to plant succession. Sometimes the more fertile patches become community gardens or urban forests. Animals such as insects and rodents can have rapid population increases (outbreaks) due to fluctuations in resources (rats in a dump, mosquitoes in pooled water from tires, swimming pools, or water fountains). Urban fauna include rats and mosquitoes but also coyotes, pigeons, mice, cockroaches, and other resilient species that utilize the wastes from humans. Sometimes rare species such as the peregrine falcon (*Falco*) benefit from plentiful nest sites and readily available food in cities. Many of these animals live in the patchy urban ecosystems described above. Urbanization destroys natural ecosystems but is now a dominant feature of the landscape with its own novel ecosystems (Pickett *et al.* 2001). Understanding and expanding urban natural areas (Grimm *et al.* 2000) can not only improve the quality of urban living but also ideally reduces energy use (e.g. for heating and cooling) and energy imports (e.g. through the use of gardens and green buildings; Oberndorfer *et al.* 2007).

4.5 Transportation

Approximately 5% of the earth's terrestrial surface is covered with roads and related surfaces dedicated to transportation such as airport runways, railroad beds, subway lines, bridges, and parking lots (Walker and Willig 1999). New roads are being built all the time, including a major effort to connect Beijing to Bangkok (Fig. 4.4). There are 8 million km of roads in North America alone (Forman *et al.* 2003), and in the U.S. they have a combined impact on about one-fifth of the land (Forman 2000). Road construction can require the creation of rock or gravel quarries, staging areas for equipment, and even the clearing of campsites for road crews. In urban areas, the coverage by roads can be as high as 35% (Gilbert 1989) and continues to expand despite the widely ignored Jevon's paradox (Giampietro 1999) that states that increased efficiency in use of a resource leads to further increases rather than a reduction in use (e.g. more roads mean more, not less, road traffic). Many residents of densely populated cities rely more on trains and buses than on cars, but the suburban streets and rural highways of developed countries are still often accessed mostly by cars. Where finances and infrastructure allow, the seductive freedom of personal vehicles is too tempting to resist, even where fuel prices are high. Transportation infrastructure is clearly a type of disturbance that must be addressed.

Road construction varies in intensity from simply cutting vegetation and compressing or grading a dirt surface to the removal of all biota and organic soil and the addition of crushed gravel that is then compressed and covered with asphalt or concrete. Regrading or repaving continues as long as a community can afford repairs but when roads are abandoned they undergo succession (Walker and del Moral 2003). Road succession (time to recovery of pre-disturbance vegetation) can be slow in arid (Bolling and Walker 2000) or cold (Auerbach *et al.* 1997) climates but surprisingly fast in the wet tropics (Heyne 2000), provided the abandoned roads are subject

Figure 4.4   Construction on the ambitious Beijing to Bangkok highway in southern China near the Laos border. Note the retention walls that are planted to help reduce erosion.

to organic inputs from nearby natural areas. Despite growing interest in mitigating the effects of roads, ecological effects are still poorly documented (Spellerberg 1998).

The effect of each road on the abiotic and biotic components will vary depending on the ecosystems it bisects (Coffin 2007). Factors that influence the effects of roads include climate (temperature and precipitation), geology (parent rock, soil formation, slope), and characteristics of the road (width, presence and type of pavement, maintenance type and frequency) (Lugo and Gucinski 2000). Roads make microclimates drier and funnel winds through forest canopies, leading to treefalls along the road. Vehicles can kill animals directly, and indirectly they affect animal populations up to several kilometers away from a road by changes in light, temperature, dust, invading organisms, and other fragmentation effects (Forman *et al.* 2003). Roads can also have some favorable impacts on local ecosystems such as the road edge by loosening the soil, favoring some colonizing ant species, exposing buried seed banks that can then germinate, providing travel corridors for wildlife, and creating nutrient-rich runoff to plants at the road's edge.

Roads interact with other disturbances in many ways. Earthquakes and landslides disrupt roads, and road construction, in turn, often destabilizes slopes and triggers landslides (Larsen and Parks 1997). When roads alter hydrological conditions, changes in floods and debris flows occur (Jones *et al.* 2000). Roads are also frequently damaged by floods. Roads provide access for loggers, settlers, and invasive species of plants and animals into previously inaccessible locations. The fragmentation that occurs when a road bisects an intact ecosystem has a spatial effect on that ecosystem far beyond the actual road surface (Reed *et al.* 1996). An estimated 20% of the U.S. has its ecology affected by roads (Forman and Alexander 1998).

4.6 Forestry

Managed forests cover about 10% of the earth's terrestrial surface and are a critical part of the economy of many nations. Widespread removal of trees is as old a human activity as agriculture and the first step in slash-and-burn agriculture (see Section 4.7). Historically, fire was frequently used to reduce tree growth and promote forage for animals that were hunted, such as buffalo in North America and kangaroos in Australia. Trees were also used by early human societies for construction of buildings, boats, bridges, weapons, and tools. In areas with dense populations (e.g. eastern China, western Europe), native forest stands were reduced to small remnants during the last millennium and partially replaced with managed forests. In the last century, deforestation has accelerated and impacted most forested regions of the world, particularly tropical rainforests (see Chapter 9; Laurance and Luizão 2007). Hopeful signs include Brazil's willingness to try to stop tropical deforestation within its borders (Nepstad *et al.* 2009), but for that to happen alternatives to clearcutting for the beef and soy industries must be found. Modern deforestation is caused by the harvest of an array of wood products (lumber, plywood, furniture, tools, firewood, and charcoal). Forests are still cleared for agriculture (e.g. palm oil in Indonesia, soy in Brazil, beef cattle in Costa Rica) or for the establishment of tree plantations (e.g. *Pinus* in New Zealand, *Tectona* in India), including for rubber production (Fig. 4.5). Sometimes forests are subjected to **coppicing**, that is, the continual trimming of trees that resprout (e.g. *Eucalyptus* in the Andes, *Quercus* in Spain).

Anthropogenic tree removal can resemble natural disruptions of forests (Shiels *et al.* 2010; see Chapter 2). Selective logging via horses or helicopters is the least disruptive method, resembling loss of leaves, branches, and some trunks as occurs during natural treefalls or low-intensity cyclones (McEvoy 2004). Under such regimes, losses of biodiversity and disruptions of ecosystem function are minimal and regeneration of the forest can be rapid. However, selective logging with heavy machinery can damage up to 80%

Figure 4.5   A rubber tree plantation in Yunnan Province, southern China.

of a forest because of increased erosion and creation of trails, debris piles, and isolated forest patches (Nussbaum *et al.* 1995). The effects of clear-cutting (Fig. 4.6) are more extensive; it can cause damage that resembles high-intensity cyclones, fires, or insect outbreaks with widespread mortality of trees, loss of once-shaded understory species from **photoinhibition** (Fernández and Fetcher 1991; Lovelock *et al.* 1994), and a reduced capacity for forest regeneration. The speed and degree of recovery from logging is inversely proportional to the severity of the damage (Hartshorn and Whitmore 1999).

Management and restoration of forests have become more essential as total forest cover declines (see Chapter 9). Sometimes forests are planted to protect crops from desiccation, to protect soils from water or wind erosion, or to reclaim deserts or quarries (Hartshorn and Whitmore 1999; Haeupler 2008). Trees planted in cities or industrial zones need to be tolerant of pollution (Lee and Cho 2008). Trees planted to restore ecosystem function and successional dynamics need to be short-lived pioneer species that do not resist invasion. Direct planting of late successional trees is sometimes possible but most habitats in need of restoration are too unstable, nutrient-poor, or have light conditions that are not appropriate for late successional trees to thrive.

Figure 4.6   Recently logged pine plantations near Nelson, South Island, New Zealand. (Photograph by Peter Bellingham.)

4.7 Agriculture

The historical development of agriculture is an essential part of human history (Sauer 1952; Rindos 1987), and even today, in the time of mechanized and centralized farming with fewer farmers than in the past, nearly half (45%) of earth's terrestrial surface is devoted to agriculture. Two-thirds of that area is in pasture, the other third in crops. Pasturing of animals gradually evolved from simple tethering of animals in natural meadows or forest openings where grass was abundant, to small, fenced compounds near villages. Further advances included seasonal migrations with animals to higher elevation summer pastures (e.g. Switzerland), and the current rapid rotation of milk cows in fertile grasslands (e.g. New Zealand). Modern pastures are now fertilized, mowed (for hay or to prevent tree invasion), reseeded with high-nutrient grasses, and protected from excessive soil erosion. Natural grasslands are imperiled by the introduction of modern pasture grasses (D'Antonio and Vitousek 1992).

Crop-based agriculture began as a local, subsistence effort where enough food was grown to avoid starvation. Techniques gradually improved with additions of soil amendments (e.g. fireplace ashes, dead fish, seaweed, night soil), crop rotations (*in situ* over sequential seasons or by moving to a new location), crop preservation (drying, pickling), terracing (Fig. 4.7), irrigation, and protective stone walls. Later innovations included the plow (White 1962) and other implements drawn by draft animals, a wider selection of potential crop species, and additional techniques for preserving the products (canning, freezing). The plow started the trend of severe destruction to native prairies and other grasslands by disrupting deeper soil layers and exposing them to nutrient depletion, wind, and water erosion. Modern agriculture has intensified the ecological impact of raising crops. Genetically modified crops grow in chemically fertilized soils that are irrigated, tilled, and harvested mechanically with big, soil-compacting tractors (Raeburn 1996). Plows reach down to the subsoil and watering and fertilizer regimes are often determined by computers. Runoff from fertilized fields contributes to the eutrophication of nearby bodies of water (see Section 3.4). Agricultural products are now shipped around the world, sometimes traveling thousands of kilometers from

Figure 4.7   The Dragon's Backbone rice terraces, Yangshuo, Guangxi Province, China. These terraces have reduced erosion and resulted in the efficient use of soil and water resources for centuries.

producer to distributor to packaging plant to consumer—often in refrigerated containers. The environmental costs of a farm are thus magnified many-fold compared to the days when one got vegetables fresh from the local garden and ate only what was in season. Modern approaches that integrate ecological principles with agricultural practices are clearly needed (Altieri 1995).

The recent intensification of agriculture has led to larger fields, more monocultures, salinization of soils from excessive irrigation, export-oriented production for small countries, and the loss of critical microhabitats such as hedgerows that housed birds, insect pollinators, invertebrates, and natural mouse catchers such as foxes. Biodiversity plummets when vast monocultures are maintained (Perfecto *et al.* 1997). In Japan, the breakdown of traditional cooperatives that maintained rural rice fields and terraced slopes led to slope erosion, downstream silting, and damage to coastal fisheries (Walker and Bellingham 2011). In Japan and elsewhere, farmland is further damaged by conversion to industrial and urban development. Erosion of fertile soil from agricultural areas is so extensive (see Chapter 2) that it threatens food security for the future. Although detailed erosion rates are not well established (Trimble and Crosson 2000), globally soil is being lost 10 to 40 times faster that it is renewed (Pimentel 2006). Erosion reduces soil productivity and microbial diversity. Chemical fertilizers do not offset these losses, and eventually soils are abandoned, leading to expansion of croplands into pastures and forest clearings (Pimentel and Harvey 1999). One cause of increased erosion is that soils are often stripped bare of crop residues by farmers in search of fuel for cooking and heating (Wen 1993). Another reason is the steady conversion of marginal lands to agriculture, particularly erosion-prone slopes (e.g. in the Philippines, Jamaica, and the Himalayas) or to forests that lack the rich, stable topsoil of grasslands. Methods to reduce severe erosion are known but not always implemented due to cost, politics, and inertia. Reforestation of slopes, less plowing, maintenance of vegetative cover, cover crops of N-fixing plants, and crop rotations have been successful in reducing erosion and improving soil quality. Abandonment of former farmland sometimes results in invasion by non-native species (see Chapter 6) that may create novel ecosystems with altered fire regimes, erosion rates, plant succession, nutrient retention, and many other unknown characteristics. Finally, the slow food movement (eating locally grown products; Nosi and Zanni 2004) attempted to steer us from our dependence on agribusiness and transportation costs. However, for economies dependent on cash and export crops, transitions to such movements are likely to be difficult and may not produce enough food. Feeding 3 billion more people by the year 2050 presents a huge challenge for humanity that will not happen by ignoring the environmental damage caused by our current system of agriculture (Tilman *et al.* 2002).

4.8 Other terrestrial disturbances

Additional impacts that humans have on terrestrial environments include radiation releases, landfills, recreation, and tourism. Radiation releases can be associated with intentional activities such as the testing of nuclear bombs (e.g. the nuclear test site at Nevada, U.S.) or for scientific experiments to see how the biota responds (Luquillo Forest, Puerto Rico). The Puerto Rico study involved a 93-day, controlled release of gamma radiation in 1965 from radioactive cesium into a rainforest. Most plants died within a 25 m radius, but pioneer plants began colonizing the bare site within a few years (Odum and Pigeon 1970). Some organisms (e.g. palm trees) were more damaged than others (e.g. ferns) but the forest has since recovered, as demonstrated by over 50 yr of study (Robinson 1997). Radiation can also be released by accident from nuclear power plants (e.g. Three Mile Island, U.S.). Perhaps the best studied biological response to a nuclear power plant accident has been the response to the 1986 meltdown of a reactor in Chernobyl, Ukraine, that contaminated over 300 000 km^2 in Belarus, Russia, and Ukraine with several types of radioactive nucleotides. Studies found high initial contamination and death of some plants and animals, but subsequent increases in biodiversity in a nature reserve from which humans were excluded (Baker and Chesser 2000; Flanary *et al.* 2008). The recent radiation releases from nuclear reactors in northeastern Japan (March 2011) are already affecting water and food supplies, with high levels in local produce and detectable levels as far away as the U.S. With continued reliance on nuclear power, radiation releases are going to be an on-going disturbance.

The production of rubbish and waste by humans has led to a variety of disturbances to terrestrial ecosystems. Landfills are the most obvious and widespread, because every community has one. Early human landfills or **middens** were fertile patches where edible plants grew and got used in an early form of agriculture (Sauer 1952). As cities grew, rivers and shallow seas became dumping grounds (see Chapter 3). When the landfills of modern cities reached their capacity, new ones were created on the periphery. Landfills potentially create a variety of disturbances, including burial of natural areas, ground water and soil contamination, and release of methane gas. They can also harbor disease-carrying rodents. Old landfills are sometimes covered with soil, revegetated, de-gassed, and used for suburban or industrial development, recreational facilities, or wild lands (Gilman *et al.* 1985; Robinson and Handel 1993; Rebele and Lehmann 2002). The long-term success of revegetation efforts, however, will require better mixing of soil cover and deeper horizons (Parsons *et al.* 1998). Modern recycling efforts strive to reduce the volume of paper, plastic, glass, and metal products; all but paper are slow to decompose. Nevertheless, most landfills remain as long-term monuments to our centralized, consumer society and present a loosely compacted mixture of toxic and benign materials for colonizing species. Industrial wastes vary widely from fine-textured material dredged from river-bottoms (fertile) or sand (infertile) to coarse-textured logging wastes (fertile) or metal parts (infertile). Grasses and rats are often successful colonists of these heterogeneous environments, but landfills also attract ferns, shrubs, trees, frogs, snakes, lizards, foxes, coyotes, and bears.

Recreational activities on land and water (see Section 4.13) also cause a variety of disturbances. Terrestrial examples include golf courses, off-road vehicles, and hunting. Golf courses present a fertile but homogenous habitat favored by wild animals such as geese, rabbits, and bighorn sheep. Off-road use of motorized vehicles, dune buggies, and bicycles can trigger soil compaction or erosion, damage vegetation, and disrupt animal behaviors. Sport hunting of native animals in locations with a long human history can sometimes help reduce exploding populations of herbivores whose predators have been removed by humans over time (Gordon *et al.* 2004). More often, there is little net effect on population reduction. Importation of non-native herbivores for sport hunting (particularly into countries with no analogous native herbivores) can have negative effects on the environment because the non-native herbivores can severely damage native vegetation and compete with native herbivores (see Section 6.4; Petrides 1975), especially on islands (Walker and Bellingham 2011). Native species that are hunted can also cause environmental degradation, especially when their numbers are kept artificially high.

Tourism can promote the preservation of natural environments (see Chapter 3), but its net environmental effect is usually detrimental (León and González 1995). This impact is not surprising, given that tourism is the world's largest industry, with an infrastructure worth almost US$3 trillion (Walker and Bellingham 2011). Damage includes soil compaction on trails, fires, vehicular traffic, and construction of the copious infrastructure to house, feed, and entertain tourists and house tourism employees (e.g. airports, trains, boats, roads, hotels, visitor centers, tramways, and trails). Wildlife is always a little less wild around tourist centers, as animal populations either adapt to the presence of tourists or leave. Even ecotourism destinations open up relatively pristine habitats to potential colonization by invasive species and alteration or loss of soils, plants, and animals (Buckley 2004). Nonetheless, ecotourism represents an important first step in addressing the myriad environmental impacts of tourism.

4.9 Novel benthic structures

Humans have accidentally and intentionally altered the floor of the world's oceans and larger lakes. Shipwrecks provide a disturbance to benthic communities when they first occur and can subsequently have both negative and positive effects. Nearly 100 000 wrecks have been located and more occur every year (Pickford 2006). Although limited in spatial extent, shipwrecks have important ecological consequences that in some ways resemble those of carcasses (see Section 3.2). The intensity of a shipwreck depends on the force and extent of impact. The severity depends on the type of surface where it sinks. More damage occurs in shallow water where biodiversity is generally higher than in deeper water. Initial damage can be minimal compared with any long-term effects of fuel or other chemical leaks as the ship breaks down from settling and corrosion (Leewis 1991).

The positive effects of shipwrecks on benthic biodiversity can be substantial. Shipwrecks provide a variety of habitats for colonization by barnacles, sponges, and bryozoans (S. Walker *et al.* 2007) and the fish that feed on them. Similar effects are intentionally created when artificial reefs are constructed by sinking ships or the dumping of vehicles, tires, cement blocks, or other material into the water. None of these materials is inert, so long-term disintegration has to be weighed against short-term benefits.

4.10 Novel shoreline structures

Dams, canals, dikes, breakwaters, docks, jetties, and piers are novel anthropogenic alterations of shoreline habitats. Dams are usually constructed in rivers to provide flood control, water for irrigation or drinking, electricity, and lakes for recreation (Fig. 4.8, Box 4.1). Canals channel river water for boat traffic and irrigation and often employ locks. Dikes are coastal dams to protect land that is near or below sea level from inundation. Breakwaters protect harbors for boats and docks. Jetties are often built to maintain sand on beaches or protect river outflows, while piers are structures built to provide boats with a place to load and unload cargo. Each of these structures involves both short-term damage to existing shoreline habitats and long-term consequences that are poorly understood from an ecological perspective. Early human manipulation of rivers and coastlines revolved around fish traps and fish ponds (Farber 1997). Construction of extensive irrigation canals for agriculture began about 5500 yr ago with the Sumerians and was also used by the Egyptians, Greeks, and Romans. The Chinese also constructed elaborate canals, including the remarkable 1776 km long Grand Canal (Hangzhou to Beijing) between 2500 and 1500 yr ago (Needham 1986). Canal building became more widespread with the arrival of mechanized digging tools and improved lock technology in the last few centuries. Similar advances have been made recently in the construction of dams and other shoreline structures.

The disruption of shorelines by novel structures alters water flow and temperature and, especially in the case of dams, begins a process of sediment

(a)

(b)

Figure 4.8   Manipulation of the Nile River in Egypt: (a) the Aswan Dam; (b) floodplain agriculture made possible by irrigation canals.

Box 4.1 The pros and cons of damming the Nile River

The Nile River is the reason that Egyptian civilization developed and remained so stable for over 4000 yr. When the Sahara Desert formed with the drying of savannas in northern Africa over 10 000 yr ago, humans migrated to the Nile River valley where they had a steady source of water. Complex irrigation schemes gradually developed to harness the annual flooding. Egyptians were taxed based on "nilometers" that indicated the height of the flood waters and therefore the likely degree of agricultural production (Fig. 4.8b). The fortunes of everyone, including the pharaohs, depended on the seasonal floods. Without the floods, people starved and, when the droughts persisted, kingdoms met with social upheaval. To meet the demands of a growing population, Egypt built the original Aswan Dam from 1898–1902. This dam increased the amount of cultivatable land and supplied Egypt's electrical needs, but was supplanted in the 1950s by the High Dam 6 km upstream of the original dam. The High Dam (Fig. 4.8a) created Lake Nasser, the world's largest artificial lake, flooded many Nubian villages and ancient ruins (although some such as Abu Simbel were moved piece by piece to higher ground and then reassembled), and further increased Egypt's cultivatable land and hydroelectric power supply. However, there are drawbacks to the High Dam. The fertility of the Nile Valley depends on the annual deposition of silt brought from upriver, but this silt now accumulates in and will eventually fill Lake Nasser. Year-round irrigation has led to the spread of the bilharzia parasite and extensive use of chemical fertilizers without an annual flushing of the soils has increased soil salinity. These problems are not unique to Egypt or the Nile but are in sharp focus because of Egypt's total dependence on the Nile for water, irrigation, hydropower (that they now have) and fertile silt (that they now do not have).

---

accumulation against the upstream edge of the structure. Both sediment accumulation and increased erosion from altered water flow (as can occur where a breakwater or jetty meets the shoreline) can result in continuous revision and reconstruction of shoreline structures. Dams create reservoirs that reduce water flow, cool water temperatures, reduce suspension of sediments in the water column, and reduce oxygenation of river water. Reservoirs often provide unstable shorelines because water removal for downstream agriculture, urban use, or electricity generation can abruptly change water levels by several meters. Reservoirs also evaporate large quantities of water (Vörösmarty and Sahagian 2000). At Tsukuba Science City in Japan, models of rivers, dikes, dams, spillways, and reservoirs are flooded to identify flood responses, minimize sedimentation in reservoirs, control riverbed erosion, and learn how to adjust water temperature and oxygen content for fish (Clark 1982; Fukami *et al.* 2009).

Biotic damage can be substantial when large structures are introduced to shoreline habitats. The construction of dikes, breakwaters, jetties, and piers can involve dredging, blasting, drilling, infilling, and other activities. Some of these produce noise, others increase local sediment deposition, and the immediate area is completely altered by the human-made structure. Organisms living in the coral reefs, mudflats, rocky intertidal zones, seagrass beds, mangrove swamps, or other shoreline habitats are clearly affected, but new opportunities for colonization are provided by the added surfaces. Human activities in the area (e.g. motorboat docking, swimming, dredging) will further disturb the new colonists. Dams offer the most dramatic alteration of aquatic habitats, often converting a river into a reservoir with such different physical conditions that entirely new flora and fauna are likely to establish. These changes favor a different fauna (e.g. bass, trout) from that found in flowing, turbid, and warmer conditions of undammed rivers (Ligon *et al.* 1995; Poff and Allan 1995; Gehrke *et al.* 2002). Drowned riparian communities are a feature of many reservoirs from initial flooding and also from the rise and fall (drowning and subsequent exposure) of water levels. One benefit they provide is potential fish habitat for reservoir-adapted species (Barwick *et al.* 2004). Novel shoreline features are pervasive and biotic communities have adapted to utilize them.

4.11 Sedimentation

Human activities have altered many natural patterns of deposition of sediments (see Section 3.4) into lakes and the ocean. Dams trap river sediments and fill reservoirs until they are no longer useful for water storage. Removal of a dam results in rapid downstream movement of large amounts of sediments. Deforestation (for wood products, agriculture, urbanization, or roads) increases soil erosion into rivers. Burning of vegetation often has the same effect, at least until vegetative regrowth again stabilizes the soil. Artificial river channelization (deepening, straightening, and sealing drainages) further exacerbates the transport of sediments downstream that would otherwise be trapped by vegetation, meanders, and shallow water. Sedimentation can damage or destroy plants and animals but can also favor a new suite of species adapted to high sediment loads. Humans are affected by the loss of native riparian communities and native fish, as well as various ecosystem services that rivers and estuaries normally provide. The biggest loss is the 10 million ha of croplands that are lost annually due to soil erosion, because humans get 99.7% of their food from terrestrial ecosystems (Pimentel 2006).

4.12 Oil spills

Natural seeps on the ocean floor emit about 600 000 Mg of oil per year, although that value may vary ten-fold (Wilson *et al.* 1974). Most of that comes from the Pacific Ocean, where some of the resulting slicks are large enough to be visible from space (MacDonald *et al.* 1993). The Caspian Sea is also an area of high seepage (Guliev and Feizullayev 1996). Extraction of oil by humans in recent decades has likely reduced natural oil seepage (Hornafius *et al.* 1999). Oil spills related to human drilling and shipping activities do not usually surpass the natural background rates of oil seepage in a given year, but when very large spills were included (Gulf of Mexico in 1979, 752 000 Mg; Arabian Gulf in 1991, 122 400 000 Mg), the annual mean between 1967 and 1994 reached 4 606 000 Mg (Fig. 4.9; Paine *et al.* 1996). The 3-month-long Gulf of Mexico spill in 2010 resulted in a release of 598 400 Mg of oil (Crone and Tolstoy 2010). This amount doubled the total from natural seeps in the Gulf of Mexico during that time (Camilli *et al.* 2010). It is rare that more than 10% of the released oil is cleaned up in an oil spill (Teal and Howarth 1984).

The physical effects of an oil spill include, in increasing order of approximate duration in the environment: floating oil, submerged oil plumes, dispersed oil mixed with chemical dispersants, benthic oil films, and debris on the sea floor such as broken machinery, drill bits, or drilling mud and rocks (Grant and Briggs 2002). Coastal environments are also affected. High levels of hydrocarbons were found in marsh sediments 33 yr after a spill in Massachusetts, U.S. (Reddy *et al.* 2002). There is some controversy about how long oil persists in various environments and much to learn about its

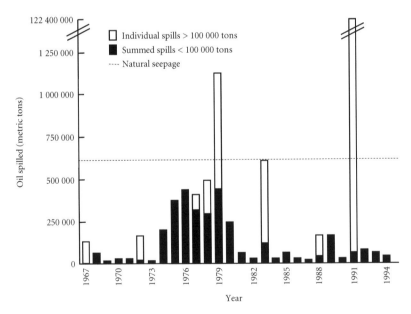

Figure 4.9   Output of oil from natural seeps (mean per year equals 600 000 Mg; Wilson *et al.* 1974) and human-caused spills from 1967 to 1994. (Modified from Paine *et al.* (1996), with permission from Annual Reviews.)

environmental effects (Teal and Howarth 1984). Immediate effects include the death of wildlife that uses the ocean surface. These deaths occur from hypothermia, smothering, drowning, and ingestion of toxic hydrocarbons. Planktonic communities are also impaired. Algae and invertebrates in benthic and shoreline habitats are killed due to chemical toxins and smothering. Following the immediate death of organisms ("acute mortality"), longer-term consequences include decreased growth and reproduction, disease, and disruption of food webs (Loya and Rinkevich 1980; Peterson *et al.* 2003). Oil persists in the water column for several months but in sediments for years to decades, so spills that reach benthic zones will have different impacts than surface spills. Some organisms may survive oil spills if they can get away or if they were dormant when the spill occurred (e.g. the littoral alga *Fucus* in the 1977 Tsesis spill in the Baltic Sea; Teal and Howarth 1984). Mobility is also important for recolonizing benthic areas affected by oil spills and sometimes invasive species dominate the recovery process. Remediation (see Chapter 9) of oil spills includes the use of chemical dispersants (Lessard and DeMarco 2000), fertilization to accelerate bioremediation (Bragg *et al.* 1994), burning of surface oil, and physical removal from water and land surfaces. Sometimes mechanical removal on beaches using high-pressure hoses causes substantial secondary harm to organisms (Peterson *et al.* 2003). Long-term monitoring will be needed to best assess on-going and future damage (Paine *et al.* 1996).

4.13 Recreation

The increase in aquatic tourism in the last 50 yr has led to a variety of disturbances. Physical damage to shoreline habitats includes trampling by surfers, swimmers, and boaters; inundation by sand imported for beaches (Fig. 4.10); and breakage of coral reefs by boat anchors, divers, and snorkelers. Discarded or broken fish nets, hooks, poles, buoys, and other equipment pollute aquatic environments and can create hazards for wildlife. Recreational fishing can severely deplete populations of both freshwater and ocean fish. Tourist boats impact wildlife such as whales (Scheidat *et al.* 2004), seals (Boren *et al.* 2002), and dolphins (Constantine *et al.* 2004). Oil and gasoline leaks or spills from tourist boats and organic wastes from recreational fishermen and tourist boats pollute aquatic ecosystems (Fig. 4.11). Although many of these disturbances are limited in spatial extent and temporal duration, their combined effect has grown with the rise of aquatic tourism.

The ecological consequences of aquatic tourism include reductions in spatial complexity and biodiversity, and nutrient imbalances. Trampling and breakage result in coral reefs losing their vertical structure when corals (especially emergent forms such as staghorn corals) are damaged. Dam-

Figure 4.10   Waikiki Beach, Honolulu, Hawaii is a popular tourist destination, but sand is regularly imported from offshore or other nearby islands to maintain the beach.

Figure 4.11   Sightseeing along the Li River near Guilin, Guangxi Province, China, is popular with tourists.

aged corals are then prone to bleaching (ejecting their photosynthetic algal symbiont) and susceptible to diseases. Local extinction of game fish and competitive dominance by invasive fish, algae, and mollusks can reduce biodiversity (Gray 1997). Eutrophication from tourism and other human activities can lead to nutrient imbalances, acidification, and disruption of the life cycles of aquatic species (Camargo and Alonso 2006). Benefits from tourism accrue when marine reserves are established, fishing is prohibited, and income from tourists is funneled to the protection of aquatic ecosystems. Conservation of the Florida manatees (*Trichechus*) had the additional benefit of reducing the need to dredge canals because the manatees eat the invasive hydra plant (*Hydrilla*; Solomon *et al.* 2004). Ecotourism, however, is only a temporary amelioration of the stresses imposed by an increasing tourist clientele.

4.14 Fishing

Commercial fishing is perhaps humanity's major impact on marine ecosystems. Early human settlements were often along coastlines and overfishing sometimes occurred, particularly on islands where fishing provided the main source of protein. Polynesian migrations across the Pacific Ocean could have been driven in part by declines or alterations in fish populations, including ciguatera poisoning (Rongo *et al.* 2009). Fish catches became more efficient with the development of modern spear guns and rotenone or dynamite to kill reef fish (Jennings and Polunin 1996). The pursuit of certain desirable products such as whale oil for lanterns and seal skins for hats and coats led to the globalization of the harvest of marine products. In recent decades, refrigeration, sonar fish finders, trawls that rake the sea floor, and satellite navigational systems have all contributed to a surge in fish harvests. Global fish harvests peaked in the 1970s and have declined since then as many fish populations were drastically reduced (Brander 2007).

The development of aquaculture has been promoted in the last few decades as a ready supply of protein for a hungry human population. Detrimental effects of aquaculture increase with its intensity. Aquaculture operations that grow shell or fin fish in natural settings with minimal inputs are not big polluters. Intensive aquaculture that grows fin fish in cages requires substantial inputs of feed and produces copious waste, particularly when carnivorous fish are fed wild-caught fish (Naylor *et al.* 2000). Wastes that pollute the surrounding environment include nutrients from feces and wasted food, anti-foulants to deter encrusting algal growth, and chemicals from medicines such as antibiotics, anesthetics, vaccines, and parasiticides (Read and Fernandes 2003). Aquaculture and stocking of game fish, particularly in fresh water, can lead to invasions by non-native species.

When fish are taken in the open ocean, their removal can disrupt food webs by depleting food or prey items, altering nutrient cycles, and altering the reproductive success of the targeted species. Shoreline and reef fishing directly damage breeding and feeding grounds for fish populations. Benthic trawling disrupts habitat used by flounders and other bottom-dwellers as well as destroying benthic communities. With an average global fish consumption of 16 kg per person per year (much higher in some island countries; Walker and Bellingham 2011), the disruption of aquatic ecosystems continues. As populations of species after species drop, fishermen are simply switching to other, more available species that were previously considered less delectable. Decreases in whales, sharks, fish, turtles, oysters, and corals, for example, have been accompanied by increases in other organisms, including sea urchins, sponges, jelly fish, worms, and phytoplankton (Daskalov 2002). When fishing removes jelly fish predators, jelly fish are competitive and their populations often expand rapidly. Jelly fish compete with fisherman because they eat fish and fish larvae, clog fish nets, and compete for zooplankton (Purcell *et al.* 2007). This fundamental shift to more disturbance-adapted species has led some to suggest that in the near future we will have a diet featuring sponges, jelly fish, and seaweed. Humans have, in just several generations, severely disrupted most aquatic ecosystems in their pursuit of food.

4.15 Other aquatic disturbances

Humans disturb aquatic habitats in a multitude of additional ways. Relatively localized disturbances include heated water, saline ponds, and mine wastes. Broader concerns include aquifer pollution, dam failures, toxic spills, and ocean dumping. Power plants often release heated water into streams. Saline ponds are built along shorelines of calm inland or estuarine water bodies to evaporate salt. Liquid mine wastes that pool in slurry ponds are a human-made aquatic habitat. These pools often have to be covered with nets to avoid killing migrating birds that try to land on them and sometimes leak into nearby drainages. Aquifers are depleted for crop irrigation and urban use and industries often pollute them (e.g. Puerto Rican pharmaceutical companies; Hunter and Arbona 1995).

Dam failures are sudden and unfortunate events, unexpected because local populations have come to rely on the structural integrity of a dam. Dams can fail from poor placement, construction, or maintenance; inadequate spillways; gradual deterioration from age; earthquake tremors; or some upstream event (e.g. landslide or heavy rainfall) breaking the dam by sending a wall of water over it (Box 4.2, Fig. 4.12). On average, three large dams (at least 15 m tall) failed each year during the 20th century (Clark 1982). The

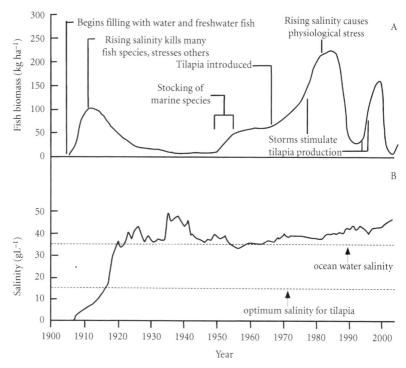

Figure 4.12 Historical changes in (A) fish biomass and (B) salinity in the Salton Sea in southern California, U.S., 1907–2004. From Hurlbert *et al.* (2007) with permission from Taylor and Francis.

Box 4.2 The Salton Sea ecosystem

The Salton Sea (Fig. 4.12) was created by accident when irrigation canals broke that were designed to funnel Colorado River water to the Imperial Valley in southern California. For 16 months in 1905 and 1906, the Colorado River flowed into the Salton Sink, creating a freshwater lake 1348 km² in size. Fish populations initially surged as species adapted to the murky water conditions and a relatively predator-free environment, but then died as salinity levels increased (Hurlbert *et al.* 2007). Marine species were introduced in the 1950s, several decades after salinity levels had become comparable to those of ocean water. Tilapia were introduced in the 1960s and have been part of several fish population explosions and crashes in the last few decades. Populations of piscivorous birds have followed parallel fluctuations. Evaporation has decreased the size of the Salton Sea to about 980 km². Plans have begun to try to maintain one portion of the lake through periodic infusions of river water, sealing off the rest for its likely fate as a return to a salt flat. Maintenance of populations of the many migratory birds that have come to depend on the Salton Sea has been one critical argument for preserving it. Who knew that a simple historical accident could lead to such a complex and evolving ecosystem?

flash floods scour the floodplain and deposit organic debris and sediments downstream. Such floods provide moist, exposed sediments that are optimal for germination or vegetative growth of native riparian species. However, most of these plants (e.g. *Populus* trees) disperse their seeds to coincide with normal spring floods (Mahoney and Rood 1998). More opportunistic species such as the invasive *Tamarix* shrub may then out-compete the natives. Native aquatic organisms are frequently better adapted to spring floods and warm, sediment-rich water than invasive fish that prefer the colder, less silty water conditions that are promoted by dams.

When dams that were holding back toxic mine wastes break, the damage downstream is severe. In 2010, 184 million gallons of strongly alkaline sludge from an aluminum factory spilled in Hungary, eventually reaching the Danube River and killing at least seven people. Also in 2010, 2.4 million gallons of acidic copper wastes spilled in Fugian Province in China, entering the Ting River. These are just a few of the many industrial pollutants that enter aquatic ecosystems each year and affect both terrestrial and aquatic organisms (Korte and Coulston 2000; Pastor *et al.* 2001).

Ocean dumping creates major disruptions for aquatic ecosystems. Direct dumping of sewage sludge has been a common practice around the world. The largest operation has been off the New Jersey coast where dumping of wastes from New York City occurred until 1992. Other aquatic wastes are those that float. Huge gyres in both the Atlantic and Pacific Oceans contain floating debris that stretches for thousands of square kilometers. One sample from the North Pacific Central Gyre contained six time more plastic material than plankton by mass (Moore *et al.* 2001). Plastic does not readily degrade (Weisman 2007) and is often mistaken for food by fish, but when it lodges in their digestive tract they can die. Some of the most lethal pieces of ocean trash are "ghost nets" left over from fishing activities and plastic bags that turtles confuse with their typical food, jelly fish. Ocean dumping has been associated with contaminated food fish, fish diseases, viruses that affect humans, radioactive contamination, genetic alterations, shifts in the composition of benthic fauna, and death of marine wildlife (Kennish 1991, 1997; Wurgler and Kramers 1992; Nielsen *et al.* 1997).

4.16 Conclusions

Anthropogenic disturbances are now an integral part of any terrestrial or aquatic disturbance regime. Initially, many human activities mimicked natural disturbances (e.g. fires to guide game animals to traps, floods to fill irrigation canals). With increasing population densities, disturbances in human-dominated ecosystems became more frequent, intense, and severe. Human landscapes now dominate half the land surface and we make use of more than half of all accessible fresh water (Vitousek *et al.* 1997b). Nutrient cycles have been altered, particularly by artificial N-fixation in the manufacture of fertilizers. These fertilizers help crop growth but also cause eutrophication in nearby rivers. Sediments and nutrients then drain into the oceans and alter shoreline and reef habitats. Increases in emissions of carbon dioxide from industry, urbanization, and transportation are altering climates and acidifying the oceans. The future of disturbance ecology will be dominated by exploring measures to mitigate these sometimes novel disturbances and trying to preserve diminishing pockets of biodiversity.

Mitigation efforts (see Chapter 9) can begin by exploring the parallels between natural and anthropogenic disturbances. Studies of succession have examined responses of biota to natural disturbances for over a century (Walker and del Moral 2003) and some general principles are emerging (see Chapter 8). When biotic responses to anthropogenic disturbances parallel responses to natural ones, restoration efforts will be most straightforward. However, anthropogenic disturbances are often novel in extent, frequency, and severity and can accelerate losses of native ecosystems not accustomed to them. As the dominant species, we have a growing population of individuals that each wants not only minimal food and shelter but comforts that have traditionally come from resource consumption. This consumption drives our immense impact on our environment. Present and future anthropogenic ecosystems or "anthromes" are characterized by new species mixtures and respond differently than natural ecosystems to both natural and anthropogenic disturbances.

5 Ecosystem processes

5.1 Introduction

Disturbances disrupt ecosystem processes, thereby affecting the distribution and abundance of biota. Disturbances in terrestrial and aquatic habitats alter light and temperature regimes, carbon dioxide and nutrient fluxes, and productivity. Disturbance of one of these ecosystem processes is likely to affect others as most biogeochemical cycles (e.g. of carbon, N, P) are coupled—not only with each other, but also to landscape and anthropogenic factors (Chapin *et al.* 2011). For example, disturbances alter light through structural changes in an ecosystem and temperatures through air and water currents and removal of structure. Carbon dioxide fluxes and concentrations are altered in many ways, including by disruptions of air and water currents, soil microbial dynamics, nutrient inputs, and productivity. Nutrient dynamics are also influenced by all of the above parameters. Removing biomass, adding organic matter, or altering microbial decomposition through changes in light or temperature can each affect nutrients. Productivity represents the integrated biotic response to the availability of light, temperature, carbon dioxide, and nutrients and is affected both directly (mortality) and indirectly (growth) by disturbances that alter their availability. This chapter addresses how organisms respond to alterations of ecosystem processes in the air, soil, and water. Specifically, I address how disturbances alter light, air temperature, and carbon dioxide levels, then explore the effects of disturbance on soil nutrients (N, P), soil pH, and soil organisms. Next, I address nutrient fluxes, enrichment, and acidification in aquatic environments. I also cover transfers of energy and matter across interfaces of aerial, terrestrial, and aqueous parts of an ecosystem (Silver *et al.* 1996a,b). I end with an examination of the responses of productivity to disturbance.

5.2 Light

Terrestrial disturbances that remove a canopy increase light levels reaching the understory vegetation and the soil. Initially, such increases can cause damage to once-shaded plants (e.g. photoinhibition; Long *et al.* 1994), raise soil temperatures, increase vulnerability to drought, and accelerate rates of mineralization of nutrients and organic matter. Full sunlight and high red–far red ratios favor the germination and growth of early successional species (Fenner 1985; Walker and del Moral 2003). The subsequent development of layers of canopy returns the understory to a shadier environment (Tilman 1988) characterized by low red–far red ratios and high heterogeneity in the quality and penetration of light (Fig. 5.1; Clark 1990). **Leaf area index**, leaf litter depth, and the height of the region of energy exchange all increase as the canopy develops (Bazzaz 1996). Variations in leaf morphology, canopy thickness, solar inputs, and seasonal changes contribute to the high heterogeneity in light levels, especially for understory plants (Fernández and Fetcher 1991). Understory light levels therefore generally decline with time following a disturbance. For example, in a forest in Jamaica, light levels returned to pre-disturbance levels 33 months after a cyclone (Bellingham *et al.* 1996).

Figure 5.1   Sunflecks in a redwood forest (*Sequoia*) in coastal Oregon (U.S.).

However, spatial heterogeneity of light levels can increase as a forest canopy returns (Nicotra *et al.* 1999), in part due to patches of sunlight (sunflecks).

Various disturbances can disrupt normal patterns of the attenuation of light with depth in aquatic ecosystems. Sewage (Pastorok and Bilyard 1985) and suspended sediments (Dollar and Grigg 1981) can negatively affect aquatic organisms through light reduction. For example, continual inputs of tannins in drainage water from organic soils can permanently diminish the penetration of light and allow aquatic organisms adapted to low light levels to grow nearer the surface, although lowered salinity was perhaps more important than reduced light levels in the distribution of a black coral in New Zealand (*Antipathes*; Kregting and Gibbs 2006). Disruptions of physical structures (e.g. coral reefs) or emergent vegetation (e.g. cattails) can initially increase light levels in the water column, resulting in changes in photosynthesis and photoinhibition (Valiela *et al.* 1997). Experimental manipulations indicate how the quality of light changes during aquatic succession. Diatoms characteristic of early primary succession (on ceramic tiles in the Mediterranean Sea) were more susceptible to ultraviolet radiation in early than in late succession (Santas *et al.* 1997). Light in aquatic systems parallels patterns on land and generally decreases with time following a disturbance.

5.3 Air temperature

The major determinant of air and soil temperature is solar radiation, in conjunction with water content and biotic variables. Air temperature is most responsive to radiation, while soil temperature is less responsive. Following a disturbance that creates open ground, temperature fluctuations at the soil surface will be much greater than those where the vegetation cover is denser. Large fluctuations occur in boreal soils, for example, when the insulating layers of canopy and moss are removed and soils experience much greater diurnal fluctuations (Bonan and Shugart 1989; O'Connell *et al.* 2003). Plants that colonize surfaces with large temperature fluctuations maintain adequate leaf temperatures for photosynthesis by remaining short, developing protective hairs, growing in protected microsites, and developing high transpiration rates (Bazzaz 1996).

Aquatic temperatures are fairly stable within ranges determined by latitude, season, and currents. Lakes at higher latitudes have a period of equal temperature along a depth gradient that allows vertical mixing to occur. Disruptions in these predictable cycles (fluctuations) that can increase temperatures include lowered water levels (e.g. reservoir releases), inputs of warmer flood waters, and climate change. Disturbances that can lower temperatures include invasion by surface plants, inputs of cooler flood waters, or additions of sediments that reduce solar heating. Lake temperatures affect aquatic biota in many ways. In cooler lakes, increased consumption by zooplankton can result in a clearer water column (Scheffer *et al.* 2001). Warmer water tends to promote algal growth and can stabilize water columns, reducing nutrient upwelling and fish productivity (O'Reilly *et al.* 2003). Past temperature changes are sometimes estimated using diatom remains in lake sediments, although these remains are more robustly linked to changes in lake salinity and ice cover (Smol and Cumming 2000).

5.4 Carbon dioxide

Carbon dioxide, produced by plant, animal, and microbial respiration and consumed by photosynthesis, can vary in concentration on a vertical profile from the top of a forest canopy into the soil. Moist, warm, organically rich soils often have the highest concentrations of carbon dioxide along that vertical profile, particularly in late successional stages with little air movement at the ground surface (Schwartz and Bazzaz 1973). Daytime concentrations of carbon dioxide are depleted in the canopy by both photosynthesis and the circulation of air. Carbon dioxide produced by the respiration of plant roots and soil organisms remains mostly within the soil or the air near the soil surface, and is a principal source for understory seedlings (Medina *et al.* 1991), although significant amounts also leave the soil and enter the atmosphere (Kutsch *et al.* 2009). Disturbances that reduce the forest canopy (or other vegetation cover) allow mixing of surface gases and canopy concentrations of carbon dioxide may be lower in early than in late successional environments. However, small seedlings in early succession still have two conditions that favor growth: high light and high carbon dioxide levels. Concentrations of carbon dioxide within the soil are decreased by disturbances that remove organic matter and associated roots. Microbial populations that grow in disturbed soils increase mineralization of the remaining organic matter, leading to further losses of carbon dioxide. Soil movement or intentional manipulations (e.g. plowing) can increase soil aeration, temperature, and sometimes fertility (indirectly through increased mineralization or directly through the addition of fertilizer), and can offset these losses of carbon dioxide by promoting plant growth (Bazzaz 1996). However, less intense forms of tillage generally involve lower releases of carbon dioxide than more intense tillage of agricultural soils (Al-Kaisi and Yin 2005). There is widespread concern that global climate change may increase net release of carbon dioxide to the atmosphere by enhancing decomposition rates, leading to a positive feedback on climate change (Cox *et al.* 2000). However, there is much uncertainty about the interactions of climate change and soil processes because of the many confounding variables (e.g. drought or invasive species; Davidson and Janssens 2006; Bardgett and Wardle 2010).

Carbon dioxide in aquatic systems comes from the weathering of rocks and sediments, diffusion from the atmosphere, and respiration. There is an approximate equilibrium between levels of dissolved carbon dioxide in marine ecosystems and the atmosphere. However, many lakes contain more carbon dioxide than the atmosphere and are therefore net sources of carbon dioxide to the atmosphere—at a rate of approximately half that of the carbon transported by rivers as organic matter to the ocean (Cole *et al.* 1994). Carbon dioxide levels are also closely tied to pH. For example, the respiration of bacteria and fishes in poorly mixed water can increase carbon dioxide levels, reduce pH, and inhibit **nitrification**, while aeration generally releases carbon dioxide and raises pH. Carbon uptake through photosynthesis by **autotrophs** in rivers and lakes can be limited by shade from shoreline or aquatic plants and generally declines with depth as light levels attenuate. In oceans, where primary production is by phytoplankton, the attenuation of light with depth also limits photosynthesis. Nutrients can also limit photosynthesis in both marine and freshwater ecosystems. Disturbances that disrupt aeration, primary or secondary productivity, light, or nutrients can therefore each influence the concentration of carbon dioxide in and subsequent release from aquatic ecosystems. Aeration can come from wind storms or turbulence in moving water and can reduce dissolved carbon dioxide by increasing photosynthesis if nutrients are not limiting. Higher carbon dioxide levels are generally found in eutrophic lakes due to nutrient inputs from agriculture or peatlands in the catchment (Huttenen *et al.* 2003). Recent increases in atmospheric carbon dioxide levels have increased the amount of carbon dioxide in shallow water, likely leading to changes in species distributions (e.g. seagrasses and phytoplankton; Short and Neckles 1998). Marine animals most sensitive to acidification from increased carbon dioxide levels appear to include squid, but the long-term consequences for fish reproduction could be important (Pörtner *et al.* 2004). Previous mass extinctions of marine fauna are likely to have been linked to ocean acidification. Suggested reduction of atmospheric carbon dioxide levels by disposal in deep ocean water may lead to the accumulation of ammonia, inhibition of nitrification, and possible eutrophication if the plumes of carbon dioxide reach shallow water (Huesemann *et al.* 2003).

5.5 Soil nutrients

5.5.1 Overview

Soils are the source of most nutrients for terrestrial organisms, and disturbances affect soil structure (see Section 2.2) and nutrients in a wide variety of ways (Fig. 5.2). Severe disturbances (Walker 1999b) destroy most organisms and remove or cover up most soils (e.g. volcanoes, landslides). Such conditions drastically reduce available nutrients and initiate the process of primary succession (see Chapter 8). Natural disturbances of intermediate severity might partially remove (e.g. blown down trees), burn (e.g. fires), inundate (e.g. floods), or desiccate (e.g. droughts) soils, with variable effects on nutrients. For example, following Hurricane Hugo in Puerto Rico, litter deposition from the storm immobilized soil N (Zimmerman *et al.* 1995). Even less severe disturbances such as soil compaction from vehicle tracks (Bolling and Walker 2000) or insect outbreaks (Schowalter and Lowman 1999) can affect the availability of soil nutrients, as when the periodic emergence of cicadas alters N availability (Yang 2004).

Intermediate to minimal losses of biomass and nutrients are associated with secondary succession (see Chapter 8). Anthropogenic disturbances also range in order of decreasing damage to soils from severe (e.g. mining or pavement), to moderate (oil spills or military activities), to mild (forestry). Many disturbances (e.g. landslides, floods, dunes, erosion) transport soils to new locations, creating substrates that have lost their typical horizontal layers and that vary in fertility. Agriculture purposefully manipulates the

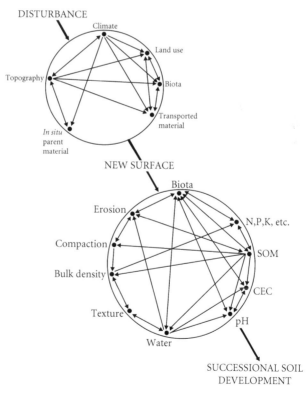

Figure 5.2   Interactions of factors that affect soil development. (From Walker and del Moral (2003), with permission from Cambridge University Press.)

stability, texture, and fertility of soils for the production of food (Ghersa and Leon 1999). Soil fertility therefore varies as a function of the severity of the disturbance. Disturbance frequency, intensity, and extent are also important, as are disturbance interactions. For example, the effects of flooding (both direct on soil aeration and temperature and indirect on soil stability and vegetation composition) depend on how often the floods occur, inundation depths and duration, and the area and types of soils that are covered (Sojka 1999). Along eastern Canadian rivers, flooding scoured soils and decreased nutrients by removing organic matter (Day *et al.* 1988). Both flood severity and duration affected vegetation composition (Fig. 5.3). In another example, inundation of coastal forests by sea water during a cyclone in South Carolina (U.S.) reduced nutrient availability from leaf loss and microbial immobilization, and decreased cation exchange (Blood *et al.* 1991). Soil conditions also reflect the activities of organisms over time in response to the interplay of climate, topography, and parent material, as well as to the prevailing disturbance regime (Bardgett and Wardle 2010). Ultimately, soil nutrient losses from disturbance affect biodiversity, vegetation structure, soil biota, and most terrestrial ecosystem processes.

Most disturbances trigger at least a temporary loss of soil structure and fertility. However, there are at least five ways in which nutrients can increase following a disturbance. First, the physical disruption of soils can cause an immediate rise in nutrient availability from mineralization (which releases nutrients previously bound in organic matter). Second, weathering of parent material that has been newly exposed by a disturbance can release key nutrients such as P. Third, disturbances such as fire change complex organic compounds such as wood into readily available forms of minerals. Fourth, animal activities including insect outbreaks (with associated frass) or burrowing (soil turnover) can increase soil fertility (Kurz *et al.* 2008). Finally, disturbances that leave animal bodies behind (e.g. earthquakes, landslides, floods, fires, military activities) provide a rich and readily decomposable form of nutrients (Parmenter

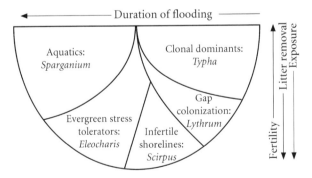

Figure 5.3   Fertility and the severity and duration of disturbance determine typical marsh vegetation. (From Day *et al.* (1988) with permission from the Ecological Society of America.)

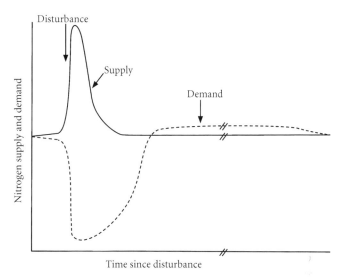

Figure 5.4   Change in supply of and demand for nitrogen during recovery following disturbance in forest ecosystems. (From Vitousek and Walker (1987) with permission from Wiley.)

and MacMahon 2009). Immediately following such nutrient pulses, the most labile forms of nutrients (especially nitrates) are often leached away by rains or dispersed by erosion (Bardgett 2005), although they can contribute to spatial heterogeneity of resources. Losses are reduced when nitrates are converted to ammonium or the most labile nutrients are taken up by rapidly growing algae, bacteria, or roots of neighboring plants.

Two temporal patterns of nutrient dynamics predominate following a disturbance. Organic matter and total N tend to increase over time while total and available P and pH tend to decline (Walker and Syers 1976; Crews *et al.* 1995; Walker and del Moral 2003). Nitrogen availability may peak in early succession and then decline as demand exceeds supply (Vitousek and Walker 1987; Fig 5.4) and N is bound in unavailable forms in litter and humus (Wardle *et al.* 2004, 2009). Nutrient dynamics are influenced by physical changes in soils during succession. These changes include stabilization of the soil surface, addition of organic matter, and development of a litter layer in early succession (Bazzaz 1996); and soil weathering, leaching, and **iron pan** formation in late succession (see Section 5.8). Nutrients are also affected by changes in both aboveground and belowground organisms during succession.

5.5.2 Nitrogen

Nitrogen is often a limiting nutrient for organisms because although atmospheric N is abundant, only a few bacteria possess the ability to transform it into forms available for plant growth (Sprent 1987). Nitrate is the most commonly absorbed form, although ammonia and organic N in the form of amino acids (Chapin 1995; Jones and Kielland 2002) can also be absorbed, particularly in infertile ecosystems (Nordin *et al.* 2004). Some plants, such as the Antarctic hair grass (*Deschampsia*), can even obtain N directly from short-chain proteins (peptides; Hill *et al.* 2011). The nitrate/ammonia ratio often declines during secondary succession as nitrates are leached out (Bazzaz 1996) and the relative importance of organic N increases. Nitrogen availability is influenced by many abiotic and biotic factors but it is generally most available in moist but well-drained, organic soils with high rates of mineralization and nitrification (conversion of ammonia to nitrite and nitrate). Nitrogen is least available in infertile soils with high C:N ratios and in late successional stages when soil lignin content, acidity, and recalcitrant organic matter content increase.

Nitrogen losses following disturbances occur through erosion of soils, removal of animals that are N-rich, removal of plants that hold the soil and retain N and other nutrients, and volatilization (e.g. from fires). Nitrogen is also lost by leaching, especially of nitrates and nitrates through the oxidation of ammonia via nitrification under aerobic conditions. **Denitrification** (the conversion of nitrite to N gas) is most prevalent in anaerobic conditions that commonly occur in waterlogged soils. Denitrification may lower plant productivity by as much as 7% (Schlesinger 1997; 2009).

Additions of available forms of N occur through rainfall, salt spray, volcanic gases, decomposition of urine, feces, dead plants, and animals, and the fixation of atmospheric N. Nitrogen fixation can be from lightning, free-living N-fixing bacteria, or N-fixing bacteria in symbiotic relationships with plants. Nitrogen-fixing bacteria can be in a loose physical association with leaf litter, lichens, and mosses (Matzek and Vitousek 2003; Lagerström *et al.* 2007; Menge and Hedin 2009) or reside inside root nodules of vascular plants including legumes and **actinorhizal** plants (23 woody genera including *Alnus, Myrica, Coriaria*; Fig. 5.5, Box 5.1; Sprent and Sprent 1990). Free-living N-fixing bacteria are often found in inhospitable environments including dry or cold regions such as Antarctica (Wynn-Williams 1993), hot desert soil crusts (Belnap 2003), and recently de-glaciated terrain (Worley 1973). The bulk of soil N is produced by the symbioses that occur in root nodules on vascular plants (Vitousek and Walker 1989; Blundon and Dale 1990).

Accumulation of total soil N following a disturbance is a robust pattern that applies across many types of disturbances (Walker and del Moral 2003) but is most pronounced after severe disturbances that initiate primary succession. Rates of accumulation during the first several centuries vary (across mines, glacial moraines, floodplains, and dunes) from 27 to 163 kg N ha^{-1} yr^{-1} and generally reach an asymptote of several thousand kg N ha^{-1} (Walker

Figure 5.5   Fire tree (*Morella*) on the edge of Halemaumau Crater, Kilauea Volcano, on the Island of Hawaii. This invasive, N-fixing tree has many ecosystem effects on the low-nutrient volcanic ecosystems. Note the dark shaded forest in the background that is also fire trees.

Box 5.1 An invasive tree alters ecosystem properties

Morella (formerly *Myrica*) *faya* or fire tree is native to the Azores, Madeira, and Canary Islands and was introduced to the Hawaiian lowlands by Portuguese settlers in the late 1800s. A few planted individuals reported in 1961 at 1000 m elevation led to the rapid invasion of nutrient-poor volcanic soils around Kilauea Volcano in the Hawaii Volcanoes National Park (Fig. 5.5), despite extensive efforts to control it. Several factors promoted the invasion. Native plants grow slowly in the open areas created by past eruptions; the roots of *Morella* are infected by a N-fixing bacterium that gives it an advantage in the volcanic soils; and *Morella* seeds are widely dispersed by another invasive, the Japanese white-eye bird (*Zosterops*)—an example of an "**invasional meltdown**" (Vitousek and Walker 1989; see Chapter 6). In some areas, *Morella* has out-competed the native trees, facilitated the spread of earthworms, and shaded out the dominant native trees, despite efforts to control it physically, chemically, and biologically. Dieback from insect damage led to understory succession dominated by graminoids (both native and non-native) instead of native shrubs (Adler *et al*. 1998). Some invasive shrubs were able to establish when *Morella* was experimentally removed by logging or girdling (Loh and Daehler 2008). *Morella* appears to be more susceptible to acidic volcanic fumes and a leafhopper insect (*Sophonia*) than some native plants (Lenz *et al*. 2006), but it will likely remain as a permanent member of the local plant community, creating novel mixtures of species composition and ecosystem conditions.

1993). At least 400 but typically 1000–1800 kg N ha^{-1} are needed to support woody vegetation (Olson 1958; Marrs *et al.* 1983). Current human interference in N cycles is generally increasing N availability. Nitrous oxides (from industrial and automotive exhausts) and N fertilizers (added to agricultural ecosystems) are important anthropogenic additions. Inputs from anthropogenic N fixation (for the manufacture of fertilizers) now exceed inputs from naturally occurring N fixation (Vitousek *et al.* 1997a; Schlesinger 2009). These inputs occur in both early and late successional environments, saturating systems that were N limited (Vitousek 1994) and potentially altering rates and trajectories of succession (Bazzaz 1996). The transformation of the global N cycle by humans comes not only from demands for N as a fertilizer, but from the combustion of fossil fuels and increasing deposition of N-rich wastes (Galloway *et al.* 2008).

5.5.3 Phosphorus

Phosphorus is another critical nutrient for plant and animal growth. Its availability to plants depends on the type of P, and ranges from highly available forms that are extractable by water or resin, to intermediate forms that are extractable by hydrogen chloride, to the least available forms where P is bound to calcium in alkaline soils or iron and aluminum in acidic soils. Organic forms of P are considered generally unavailable, although Turner (2008) argues that organic P is likely to be utilized directly by plants. Organic forms of P can constitute more than half of all soil P in wetlands and in highly weathered soils such as those that characterize Australia and parts of Africa. Phosphorus often peaks in availability immediately after a disturbance as easily extractable forms are removed from newly exposed rocks. With time, the availability of P declines as the available forms of P are leached or become bound to other elements or to organic molecules. This decline in the availability of P occurs over thousands of years in a variety of ecosystems (Peltzer *et al.* 2010; Vitousek *et al.* 2010), including New Zealand and Australian dunes (Walker and Syers 1976; Walker *et al.* 1981) and Hawaiian volcanoes (Crews *et al.* 1995). In addition to declines in availability, losses of P occur from both erosion and leaching. For example, although new volcanic ash deposits contained substantial levels of P in a 2008 eruption in Alaska, much of that P was susceptible to loss from both erosion and leaching (Wang *et al.* 2010).

Additions of soil P occur primarily from exposure of rocks (including new volcanic surfaces) but also from bird droppings, dust, and agricultural fertilizers. Bird droppings add substantial P (and N) to volcanoes (Sobey and Kenworthy 1979; Clarkson and Clarkson 1995; Magnússon *et al.* 2009), rocky shorelines (Skaggs 1995), and many island ecosystems (Fukami *et al.* 2006). For example, additions of P from bird droppings helped maintain higher soil P levels and grassland vegetation on the Aleutian Islands (Alaska) without foxes compared with lower soil P levels and shrub vegetation on islands where foxes had been introduced and reduced bird populations (Maron *et al.* 2006). The disturbance regime characterized by soils with high nutrient content and high acidity, as well as the presence of densely packed nest sites of many coastal birds can, however, have detrimental effects on vegetation and soil organisms (see Section 5.7.3; Mulder and Keall 2001; Bancroft *et al.* 2005).

5.5.4 Soil pH

Soil pH (hydrogen ion concentration) affects nutrient availability in several ways. Phosphorus is most available at intermediate pH levels but becomes fixed to aluminum or iron oxides at low pH and fixed as calcium phosphate at high pH. Ammonium uptake by plants lowers pH but nitrate uptake increases pH. The cation exchange capacity represents the ability of soils to supply and store cations (e.g. calcium, magnesium, potassium, and sodium). These cations are held to the negatively charged surfaces of soil clay particles but, at low pH, excess hydrogen ions displace the cations that can then be leached from the soil.

Soil pH generally decreases over time following severe disturbances as plants produce organic acids, although volcanic deposits and some mine wastes that start with an acidic pH do not usually decline further. Post-disturbance substrates that are initially alkaline (pH 7–9), including glacial moraines, floodplains, and dunes, all showed a decline in pH over time (Walker and del Moral 2003). The gradual increases in acidity and organic matter and loss of readily available nutrients favor soil fungi over soil bacteria, which dominate early successional stages (Bardgett and Walker 2004) when nutrients are more readily available. Decreases in soil pH are also correlated with changes in vascular species. On Alaskan floodplains, for example, declines in pH were accompanied by replacement of graminoids and deciduous shrubs and trees by evergreen shrubs and trees (Davidson 1993). The more acidic litter of evergreens reinforces the trend of increasing acidification of the soils.

5.5.5 Organisms

Perhaps the most critical of the many ecosystem functions that soil organisms regulate are the decomposition and mineralization of organic matter, recycling of nutrients, and the formation of soils in disturbed habitats. Rates of decomposition and soil formation vary but are highest in mesic climates with high net primary productivity and lowest in arid or cold climates with low productivity (González and Seastedt 2001; Walker and del Moral 2003). Severe disturbances will delay these processes because of the removal of most soil and soil organisms and possible microclimatic effects (e.g. increased aridity). Where there are biological legacies (surviving organisms), decomposition and soil formation proceed more quickly. In general, the soil fauna responds negatively to disturbances and large-bodied animals are most susceptible. Progress is being made in evaluating the responses of certain traits (e.g. abundance, diversity, species composition) in different ecosystems (Cole *et al.* 2008; Barbercheck *et al.* 2009) and different disturbance regimes (Wardle *et al.* 1995a).

Soil animals range from burrowing mammals (e.g. gophers, mice, rats, voles) to a diverse array of invertebrates (Bardgett 2005) and they have multiple roles as colonizers of disturbed habitats (Schowalter and Lowman 1999). Macrofauna (> 2 mm body width) include beetles, earthworms, ants, and termites (Fig. 5.6). These organisms break down litter, aerate soils with their tunnels, and fertilize the soil with their feces and dead bodies. Beetles are often the dominant predator among soil fauna (e.g. the green carab beetle, *Calasoma*) but they can also be herbivorous (e.g. larvae of bark beetles, family Curculionidae, that eat plant roots) or detritivorous (e.g. dung beetles, family Scarabaeidae; Hanski and Cambefort 1991). Some beetles tolerate extremely arid conditions such as the Namib Desert (Lovegrove 1993). Earthworms dominate the soil fauna on non-acidic soils and have long been recognized as important contributors to the development of soil structure, water infiltration, soil organic matter, and decomposition (Edwards 2004). Earthworms are not common in arid or acidic soils. Ants are widespread among soil types and can increase soil nutrients around their ant mounds, as found on sandy soils in Texas (U.S.; McGinley *et al.* 1994). Termites alter soil structure through their mounds, increase decomposition through their diet of organic matter and roots, and influence nutrient availability, in particular when their gut bacteria fix N (Abe *et al.* 2000). Dung beetles, earthworms, ants, and termites are so influential in soil nutrient dynamics that they are sometimes considered as soil ecosystem engineers (Bardgett and Wardle 2010).

Mesofauna (0.1–2.0 mm) include **enchytraeids**, mites, and springtails (Collembola). Enchytraeids have functions similar to earthworms, feed mostly on fungi, and are dominant components of the soil fauna in many acidic soils such as those found in the boreal forest and tundra (Didden 1993). In agricultural ecosystems, enchytraeids can benefit from the periodic disruption of populations of macrofauna by plowing (Wardle 1995). Mites, like many soil organisms, are sensitive to desiccation and can be slow to colonize disturbances (Siepel 1996a; Bardgett 2005). Mites have several critical roles in the soil food web (Fig. 5.7), feeding both as primary consumers on mycorrhizae and fungi and also on nematodes, collembolans, and other mites. Oribatid mites stimulated microbial growth and minimized nutrient leaching in experimental freezing and heating experiments involving forest litter (Maraun *et al.* 1998). There is also a wide range of tolerance of disturbance among different groups of mites. Oribatid mites in temperate beech

Figure 5.6   Termite mound in northeastern Australia.

forests that are parthenogenic and have opportunistic feeding habits were most tolerant of disruptions of the litter layer (e.g. from earthworms) and springtails were less tolerant than mites (Maraun *et al.* 2003). Mesofauna are important decomposers but their responses to disturbance are still poorly understood.

The most numerous organisms in the soil are the microflora (bacteria and fungi) and microfauna (nematodes, protozoa, and rotifers) that are all < 0.01 mm in size (Swift *et al.* 1979). The microflora produce enzymes that break down organic matter (Bardgett 2005) and are remarkably diverse (there are thousands of genetically distinct types per gram of soil), abundant (millions of bacteria per gram of soil), and important (present at levels of hundreds to thousands of kg ha^{-1}). Bacteria and fungi are omnipresent in soil pore space, although they may occupy only 0.5% of that space (Kilham 1994). Microflora are the principal decomposers of plant matter (Fig. 5.7), they influence nutrient uptake, and are responsible for many other soil processes including N fixation, mineralization, nitrification, and denitrification. Their populations are governed by carbon and nutrient availability, temperature,

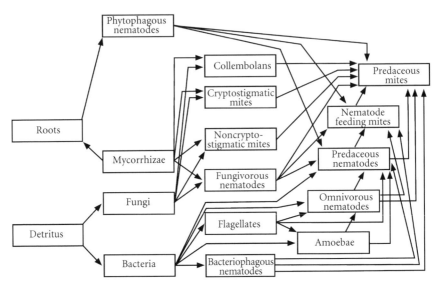

Figure 5.7 Structure of the soil food web. (Modified from de Ruiter *et al.* (1995), with permission from the American Association for the Advancement of Science.)

moisture, pH, and time since disturbance, as well as by food web dynamics (e.g. predator abundance). Populations of soil microflora are most abundant in moist, warm, and fertile soils and in the **rhizosphere** (next to plant roots). Bacteria are most abundant where mineralization and nutrient cycling rates are high but fungi are dominant by biomass and become more dominant as succession proceeds (Bardgett *et al.* 2005). Binkley *et al.* (1997) found that total fungal biomass on an Alaskan floodplain was about 100 times that of bacterial biomass. Fungi are the sole decomposers of high-lignin compounds such as wood and most readily decompose complex organic acids found in more mature soils such as on late successional glacial moraines in Alaska (Bardgett 2000; Bardgett and Walker 2004) and New Zealand (Wardle and Ghani 1995). The soil microflora clearly have a substantial influence on nutrient availability, plant productivity, and soil formation.

Nematodes are the most abundant multi-cellular animal in soils (there are millions of individuals per square meter) and feed on plant roots, bacteria, fungi, and other soil organisms, thereby occupying several links in the soil food web (Bardgett 2005; Fig. 5.7). Nematodes are also used as indicators of soil health and status following disturbance because their composition correlates with N cycling and decomposition (Moore and DeRuiter 1993; Neher 2001).

Mycorrhizal fungi provide a critical link between plant roots and the soil matrix (Fig. 5.7) and greatly increase nutrient and water uptake of plants by expanding their contact with the soil. Mycorrhizae are present on most species of plants, but early plant colonists in primary succession (following severe disturbances) are sometimes non-mycorrhizal (Allen and Allen 1990; Ernst *et al.* 1984) or less dependent on mycorrhizae (Hobbie *et al.* 1999) than plants in later successional stages. Mycorrhizal infection was inversely correlated with mobility of alpine plants invading areas disturbed by gophers (Onipchenko and Zobel 2000). In other examples of severe disturbances, mycorrhizae can be critical to plant establishment, including on glacial moraines (Jumpponen *et al.* 1998), floodplains (Krasny *et al.* 1984), and mine tailings (Allen and Allen 1980; Stahl *et al.* 1988). Mycorrhizae can survive disturbances that do not remove or destroy the soil but are also able to disperse as spores or fragments by water, wind, or animals. Sometimes dispersal is from local sources. On Mount St Helens (Washington, U.S.), gophers were instrumental in bringing mycorrhizae to surface volcanic materials from buried soil layers (Allen *et al.* 1992).

Populations of soil organisms are disrupted whenever soil structure is disturbed such as by flooding or erosion by water or wind. Flooding of soils can initially release N and P, increase bacterial activity, and accelerate decomposition of plant litter, although prolonged flooding can result in the dominance of anaerobic bacteria (Baldwin and Mitchell 2000). Subsequent drying can reduce the availability of N and P. Prolonged drought can kill many soil organisms, although others have the ability to withstand periods of harsh conditions by slowing down their metabolism. Short life cycles and high rates of reproduction generally accompany a return to favorable conditions. When most of the soil community is destroyed, re-invasion occurs through wind, water, and biotic vectors (Allen *et al.* 1999). Soil organisms as a whole are highly resilient to disturbance and clearly have a substantial influence on nutrient availability, plant productivity, and soil formation.

5.6 Aquatic nutrients

5.6.1 Overview

Nutrient distributions in aquatic ecosystems are determined by the same factors that govern terrestrial nutrients: decomposition and production. For terrestrial ecosystems, primary productivity is driven by aboveground photosynthesis and decomposition is spatially segregated on the surface and below ground. Roots and shoots provide the cambial structures to transfer nutrients upward and carbohydrates downward. In aquatic ecosystems, the same spatial segregation between the sites of decomposition and production occur but, except in shallow bodies of water or wetlands, there is no vertical connection through plant structures. Instead, nutrients and carbohydrates are cycled by water currents. Vertical mixing is therefore important. This occurs via seasonal turnover in lakes, nutrient spiraling in rivers, and various currents that are driven by wind and Coriolis forces from the spin of the earth in oceans and along shorelines (Smith and Smith 2009). The effects of disturbance on nutrients in aquatic ecosystems are therefore measured as disruptions of these various forms of nutrient cycling.

Natural disturbances of aquatic ecosystems (see Chapter 3) can add nutrients or reduce nutrient availability through the addition of carbon-rich substances, leading to microbial immobilization of limiting nutrients. Disturbances can also directly or indirectly remove nutrients. Additions in the ocean include whale carcasses, fallout from fish farms, or other nutrient pulses that fall to the sea floor. Estuaries receive nutrients from flood deposition. Most coastlines obtain nutrients from on-shore currents that bring organic matter from the sea floor. Increases in this input from storms could be considered as a disturbance when they alter food webs to favor fast-growing species that out-compete the original flora or fauna. Nutrient removal can occur directly through loss from burial, from thermal damage by volcanic activity or hydrothermal vents, and from erosion of nutrient-rich sediments and organic matter. Indirect losses can come from reduction of productivity due to silt or tannins that reduce photosynthesis or from anthropogenic industrial or recreational activities, including the over-exploitation of resources from fishing and the creation of dead zones from extreme eutrophication (Turner and Rabalais 1994). Without inputs from plant or animal growth, nutrient sources decline. Biotic responses to nutrient enrichment include outbreaks of aquatic diseases or algal blooms. Nitrogen and P are the most studied and probably most limiting of nutrients. Their interactions with disturbances are so intertwined that in the following section I first discuss the principal sources of each, then how they are involved in two pervasive disturbances: nutrient enrichment and acidification.

5.6.2 Nitrogen, phosphorus, enrichment, and acidification

Major inputs of N to aquatic ecosystems include runoff from land, atmospheric deposition, and fixation by planktonic and benthic organisms. The relative contribution of each depends on many variables, including proximity to land, proximity to point sources of atmospheric pollution, the amount and N content of runoff, currents, and the presence and abundance of N-fixing organisms. Lakes, estuaries, and relatively enclosed ocean basins receive most of their N from runoff. Atmospheric deposition of N is primarily due to anthropogenic activities and results in the acidification of lakes and rivers (Rabalais 2002). Reductions in sulfur dioxide emissions from power plants have left the uncontrolled nitrous oxide emissions as the main source of acidification of freshwater systems in Europe and North America (Camargo and Alonso 2006). Cyanobacteria are responsible for most planktonic fixation in aquatic ecosystems and values are generally low, although in some eutrophic lakes planktonic fixation can be high (Howarth *et al.*

1988). Benthic bacteria, including both heterotrophic and cyanobacteria, fix moderate amounts of N. The highest values of N fixation come from cyanobacterial mats but these are of limited extent. Contributions of N fixers to total N budgets can vary from less than 1% to 50% of total N inputs in freshwater wetlands with few other N inputs. Nitrogen can be dissolved in inorganic or organic forms that are differentially utilized by phytoplankton and higher plants (Berman and Bronk 2003), and in particulate matter. Nitrogen limitation is most pronounced in **pelagic** habitats, followed by hard marine benthic and stream benthic habitats. Soft marine benthic, lake benthic, and lake pelagic habitats are least limited by N (Fig. 5.8; Elser *et al.* 2007).

Phosphorus enters aquatic ecosystems primarily from runoff that picks up phosphates from erosion, mining, and fertilizers. Today, levels of P are approximately three times the pre-industrial background rates (Bennett *et al.* 2001) because inputs of P to the land from fertilizers exceed farm product outputs and mining has increased many-fold from negligible levels before the Industrial Revolution. There can be a substantial lag time between the application of fertilizers to land and the accumulation of high levels of P in aquatic ecosystems, but rates of accumulation continue to increase, particularly in developing countries. Phosphorus typically accumulates in benthic sediments but can also be recirculated, particularly in rivers, when those sediments are disturbed by vertical mixing or animal activities. Where

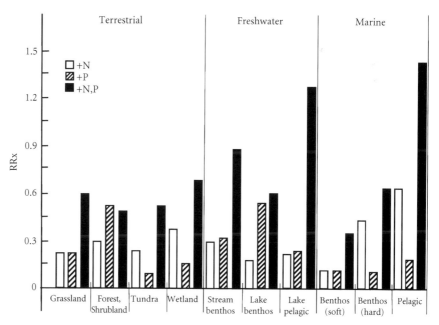

Figure 5.8 Relative responses (RR_x) to N, P, and N + P fertilization by terrestrial, freshwater, and marine autotrophs in various ecosystems. (From Elser *et al.* (2007), with permission from Wiley-Blackwell.)

there is little mixing, such as in protected bays, P can become limited, but as long as N-fixers thrive, N levels will remain stable (Smith and Atkinson 1984). Productivity in marine environments was traditionally considered to be limited by N while freshwater ecosystems were limited by P (Smith 1984). Instead, it appears that both elements limit growth in most environments. Benthic lake habitats are particularly responsive to experimental addition of P, but all studied habitats responded most to a combination of N and P addition (Fig. 5.8; Elser *et al.* 2007).

The principal disturbance that involves aquatic N and P levels is nutrient enrichment and it has many cascading influences on nutrients and other aspects of aquatic ecosystems. Although background levels of nutrients are now hard to estimate given the pervasive human influence, one study suggested that rivers in the U.S. now contain 6.4 times the background levels of N and 2.0 times those of P (R. A. Smith *et al.* 2003). Atmospheric deposition of N from industrial and vehicle emissions and runoff from fertilizers and sewage are the main anthropogenic sources that now dwarf natural sources. Industrial N fixation and mining of phosphate rocks provide the N and P essential to most fertilizers. Treated sewage has lower levels of nutrients than untreated wastes but many sources still go untreated. Wastes arriving from diffuse origins (non-point sources) are more difficult than point sources to identify and treat (Carpenter *et al.* 1998). The proportion of untreated sewage entering aquatic ecosystems is a function of the wealth of a country. While developed nations are more likely to treat their sewage, developing nations tend to have fewer point sources of industrial wastes entering their waterways and more non-point sources from domestic and local businesses.

Additions of nutrients (particularly N and P) to low-nutrient, **oligotrophic** ecosystems lead initially to enhanced productivity in most organisms (Fig. 5.9), but continued increases can result in a loss of productivity at all trophic levels (Rabalais 2002). Eutrophication results when over-enrichment promotes the excessive growth of algae and other aquatic plants (Fig. 5.10). Algal blooms result, and these can be dominated by toxin-producing species. Decomposition of the abundant plant growth results in oxygen depletion and anaerobic conditions develop (Fig. 5.11). The toxins, loss of food, reduced levels of dissolved oxygen, lower light penetration, and often higher turbidity all harm most fish and invertebrates. Their deaths release further amounts of P that can accelerate eutrophication (Correll 1998). Those species that thrive at slightly higher nutrient levels usually succumb if levels continue to increase. Eutrophication occurs in both freshwater (Fig. 5.12) and marine environments. Dead zones around fertile deltas such as the Mississippi Delta (U.S.) now cover thousands of square kilometers. Biodiversity is reduced and coral reefs and other marine organisms can be killed (Anderson and Garrison 1997). The recovery of marine dead zones and eutrophic lakes can take years because of the long-term and substantial

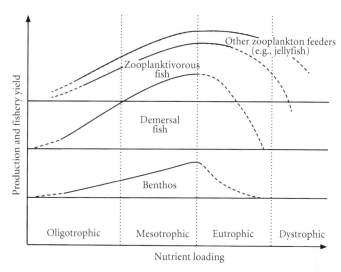

Figure 5.9   Increasing productivity of fisheries as nutrients increase, then a decline when high levels of organic acids cause a decrease in available nutrients in **dystrophic** water. (From Rabalais (2002) with permission from the Royal Swedish Academy of Sciences.)

build-up of nutrients in agricultural soils, even after inputs cease (Carpenter 2005). Where soil management remains unchanged, eutrophication is generally irreversible. However, there is some hope that N enrichment can be reduced. Although N and P both contribute to eutrophication, up to 75% of N inputs from agricultural runoff are removed as enriched water moves seaward (Zedler 2003). It now appears that nitrates are not only removed by plant uptake and microbial respiration (through denitrification) but that there are a number of other processes that scrub nitrates from lakes and rivers, providing encouragement to managers trying to reduce eutrophication (Burgin and Hamilton 2007).

Acidification is a spreading disturbance in aquatic ecosystems and a consequence of nutrient enrichment. Anthropogenic sources of sulfur dioxide, nitrous oxides, and ammonium have led to increased atmospheric concentrations, which are then precipitated through rain, snow, and fog. This acid rain has led to acidic conditions in lakes in eastern North American and northern Europe that are downwind of pollutant sources. The acids reduce pH levels in lakes and rivers, increase aluminum concentrations, and contribute to eutrophication both in the lakes and downstream in estuaries. The biological consequences can be severe, with reductions in biodiversity and abundance of aquatic species (Driscoll *et al.* 2003). Marine vertebrates appear to be more tolerant of nitrous oxides than are freshwater vertebrates (Camargo and Alonso 2006).

Oceans receive all of the acidified river water and about 25% of all anthropogenic carbon dioxide emissions (Hoegh-Guldberg *et al.* 2007); this has

A.

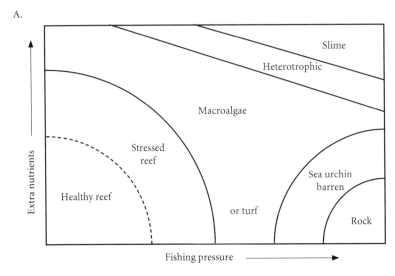

B.

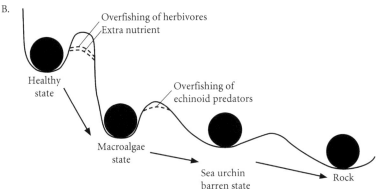

Figure 5.10   Alternate states of coral reefs from over-fishing and excess nutrients showing (a) typical organisms and (b) the influences of anthropogenic disturbances on transitions between states. (From Bellwood *et al.* (2004) with permission from Nature Publishing Group.)

led to recent acidification of marine habitats. There is substantial concern that the oceans could become substantially more acidic than they are today because carbon dioxide emissions are not declining. Such increases could severely damage marine organisms. Carbon dioxide concentrations in the atmosphere had not exceeded 300 parts per million (ppm) for at least 750 000 years but in 2011 they had reached 391 ppm. This increase has led to a drop of 0.1 pH units as the oceans have become more acidic. Increased acidity reduces the availability of carbonate for organisms that rely on it to build their shells, exoskeletons, and other structures. Corals, mollusks, coralline algae, and sea urchins are some of the organisms that are probably being affected (Kuffner *et al.* 2008). Deep-water corals may also be vulnerable

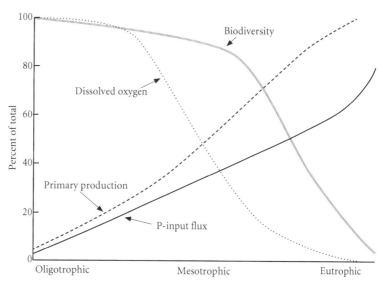

Figure 5.11 Eutrophication drivers in fresh water. (From Correll (1998) with permission from the American Society of Agronomy.)

Figure 5.12 Eutrophication in a Puerto Rican reservoir.

(Turley *et al.* 2007). At projected levels of 480 ppm carbon dioxide, coral reef formation will probably cease because corals will no longer be able to form aragonite, the principal crystalline form in calcium carbonate. High-latitude organisms may be particularly vulnerable (Orr *et al.* 2005). Coral reefs are already experiencing many other disturbances including bleaching, disease, over-fishing (see Chapter 4), and nutrient enrichments that favor algal growth over coral development. These disturbances can have synergistic effects, as when warmer ocean waters make corals more likely to bleach when exposed to acid conditions (Anthony *et al.* 2008). Consequently, the resilience of corals is declining and a tipping point may be reached where coral reef ecosystems are unable to regenerate due to the multiple onslaughts (Hoegh-Guldberg *et al.* 2007).

5.7 Interfaces

A disturbed habitat may have sharp (e.g. landslide, hydrothermal vent) or indistinct (e.g. erosion, urbanization) edges. Regardless, the habitat is embedded in a geological and biological context (see Fig. 3.2) as well as in a matrix of neighboring habitats with variable disturbance histories. Among and within habitats there are boundaries (interfaces) across which energy

TABLE 5.1 Energy and nutrient transfers across interfaces among several ecosystem parameters: atmosphere, plants, soils, water, and animals. The soils category includes soils, soil organisms, and air pockets but not soil water. The effects of disturbances on these transfers are highly variable (see text).

Interface	Transfer	Reciprocal	Transfer
Air to plants	Photosynthesis; nutrient capture by epiphylls	Plants to air	Oxygen release
Air to soils	N fixation	Soils to air	Denitrification
Air to water	Dissolved nutrients	Water to air	Volatilization
Air to animals	Oxygen uptake	Animals to air	Carbon dioxide release
Plants to soils	Carbohydrate release to rhizosphere; litterfall	Soils to plants	Nutrient uptake from sediments, microbes
Plants to water	Leached nutrients	Water to plants	Nutrient uptake from rainfall, stem flow, soil water
Plants to animals	Nutrient and energy uptake as food	Animals to plants	Nutrient release through decomposition
Soils to water	Leached nutrients	Water to soils	Dissolved nutrients; distribution of nutrients
Soils to animals	Salt uptake	Animals to soils	Nutrient addition (urine, feces, decomposition)
Water to animals	Dissolved nutrients	Animals to water	Nutrient addition (leaching, decomposition)

and nutrients are transferred (Likens and Bormann 1974). These interfaces involve transfers among the air, plant, soil, water, and animal components of an ecosystem (Table 5.1; Silver *et al.* 1996a,b). Sometimes biochemical activity is increased at these interfaces in space (hotspots) or time (hot moments) when transfers bring missing or complementary reactants to an area or a moment in time (Polis and Hurd 1996; McClain *et al.* 2003; Hocking and Reynolds 2011). These transfers both influence and are influenced by disturbance responses of local habitats in complex ways. In the following discussion, the soils category includes soils, soil organisms, and air pockets but not soil water (which is included in the water category).

5.7.1 With air

Four transfers are considered here: air to plants, air to soil, air to water, and air to animals (and the reciprocal directions when appropriate). First, plants fix atmospheric carbon dioxide through photosynthesis and some epiphylls (plants that grow on the surface of leaves) can trap atmospheric nutrients (Nadkarni and Matelson 1992). In the reciprocal transfer, plants release both oxygen to the atmosphere and some volatile substances such as **monoterpenes** (Mahmoud and Croteau 2002). The effects of disturbance on photosynthesis depend on the severity of the disturbance (loss of foliage) and the favorability of the local environment (adequate light, water, and nutrients) to support regrowth following the disturbance. Second, bacteria incorporate atmospheric N into the soil environment by the process of N fixation and release N back to the air through denitrification. These processes are also highly sensitive to disturbances that either reduce or enhance populations of responsive bacteria. Ecosystem-level effects can be measured in net N accumulation during succession, but this process is not always attributable to a given vascular species with N-fixing symbionts, even when it dominates the biomass (Vitousek and Walker 1989; Walker and del Moral 2003; Peltzer *et al.* 2009). Third, nutrients dissolve from the air into aqueous environments and volatilize back into the air. Responses of these processes to disturbance depend directly on how the disturbance alters abiotic conditions such as temperature or pH and indirectly on biotic variables such as concentrations of decaying organic matter in solution. Finally, animals take up oxygen from the air and release carbon dioxide, and these transfers depend on the effect of the disturbance on animal populations (movement away, death, or survival of the disturbance event followed by return and the rate of population growth and recovery; Willig and McGinley 1999).

5.7.2 With plants

Three transfers are considered here: plants to soils, plants to water, and plants to animals. First, plants release both energy (through litterfall and shoot and root decomposition) and nutrients (through diffusion at the root surface) into soils. These releases have a direct effect on cation concentration gradients in the rhizosphere, or the region surrounding each root. Plants are able to manipulate this environment to some degree in ways that favor the growth of microbial populations amenable to plant growth (Bardgett and Griffiths 1997), including mycorrhizal fungi when P levels are limiting (Johnson *et al.* 2003). Plants can also accumulate organisms that are detrimental to their growth. Soils, in turn, are the principal suppliers of nutrients to plants, again, largely regulated by microbial activities. Second, plants leach nutrients to water tables and also take up nutrients from rainfall, stem flow, and soil water. Leaching tends to be accelerated when a disturbance promotes decomposition of plant tissues and uptake reaches a peak during early secondary succession (Bazzaz 1979) when rapidly growing, young plants dominate.

Finally, plants transfer energy and nutrients to herbivorous animals, and animals release nutrients back to plants through urine, feces, and decomposition (Brown 1985). Plant-to-animal transfers are generally reduced immediately following a disturbance because of a reduction in both plant and animal biomass. Insect herbivory is likely to increase with successional development, but mammalian herbivory can quickly re-establish in secondary succession, and then decline (Walker and Chapin 1987). Animal-to-plant transfers are also likely to drop initially (unless there is substantial animal mortality without parallel reductions in plants—perhaps from radiation effects), then increase as animal populations re-establish. Animal–plant transfers can either retard or accelerate succession when large herbivores impact nutrient cycling (Wardle and Bardgett 2004). Animals also affect plants when they act as dispersal vectors (e.g. carrying seeds in their guts or claws) and plants can serve as attractants for animal dispersers (e.g. as perch trees; Shiels and Walker 2003).

5.7.3 With soils

Two more transfers to consider are: soils to water and soils to animals. First, soils leach nutrients to water and water distributes dissolved nutrients throughout the soil and into the groundwater. Soils intersect with water along the surface of every hydrated soil particle where water-soluble nutrients are exchanged and where pH and cation exchange capacity are determined. These two transfers can be accelerated by disturbances that do not lead to a loss of water. When soils are eroded they can transfer their nutrients into bodies of water, with the soils usually settling to the bottom or being washed to shore to form beaches and dunes (see Fig. 3.5). Excess water that does not infiltrate soil pores tends to pond on the soil surface and can induce anaerobic conditions and promote anaerobic bacteria. Second, soils provide habitat and sometimes mineral supplements for many animals, and animals enrich soils with their urine and feces or their decom-

Figure 5.13   Gannet (*Morus*) colonies impact soil fertility and plant succession on White Island, New Zealand.

posing bodies. Animals can also compact soils or erode it from excessive trampling. These transfers are most active in relatively undisturbed habitats with well-developed soils and a diverse fauna. Animals can be vectors for the introduction of spores of mycorrhizae and other soil biota. These vectors can influence recovery following disturbance. Birds that feed on aquatic resources bring a net transfer of energy from water to land when they defecate, reproduce, and die on land (Mulder and Keall 2001; Sekercioglu 2006). Seabirds (Fig. 5.13) can influence soil fertility, species composition, biomass, species interactions, and plant succession (Fukami *et al.* 2006; Wardle *et al.* 2007; Magnússon *et al.* 2009); their effects on succession are mixed, depending on their density (see Section 5.5.3).

5.7.4 With water

The remaining transfer involves water and animals (Fig. 5.14). Water provides animals with minerals and hydration, and in extremely arid situations loss of water sources can lead to the death or migration of animal populations. Animals return nutrients to water with their urine, feces, and decomposition. Salmon, for example, are key players in the addition of nutrients to streams where they spawn. They also represent a connection among marine, freshwater, and terrestrial habitats and nutrient transfers among

Figure 5.14   Endangered Hawaiian monk seals (*Monachus*) bask on the beach at Kaena Point, Oahu, Hawaii.

aquatic animals. Salmon carcasses are used by bears and by aquatic insects, with cascading consequences in both the terrestrial and aquatic habitats (Cederholm *et al.* 1999; Lessard and Merritt 2006; Hocking and Reimchen 2009). Disturbances such as dams that disrupt salmon runs also disrupt these connections.

5.8 Long-term processes

Relatively short-term ecosystem processes (lasting a few thousand years) influence and are in turn influenced by relatively longer-term processes (lasting for millions of years; Fig. 5.15). As long as litter deposition and other nutrient inputs lead to active nutrient cycling and ecosystems are periodically disrupted by disturbance and subsequent colonization events (including the arrival of non-native invasive species), ecosystems remain in a dynamic, progressive phase. When outputs exceed inputs and low nutrient availability limits growth, longer-term processes of acidification, leaching, and nutrient immobilization occur (Peltzer *et al.* 2010). Decline in the availability of P over periods of thousands of years occurs when the available

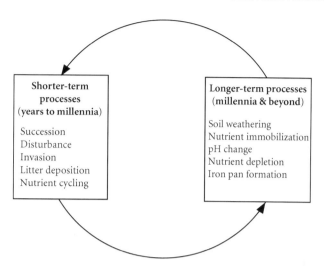

Figure 5.15 In the development of ecosystems and changing availability of nutrients, disturbance is one of several shorter-term processes that influence and are influenced by longer-term processes.

forms from parent substrates (usually apatite) are dissolved and leached or increasingly bound (occluded) to mineral surfaces or recalcitrant organic forms (Turner *et al.* 2007). Atmospheric inputs can offset long-term nutrient depletion, as in Hawaii where on the 4.1 million yr old island of Kauai, 90% of available P is derived from dust (Vitousek 2004).

Long-term aquatic processes are less well understood. Pollen records from lake sediments show terrestrial vegetation patterns over thousands of years that can lead to inferences about water quality, acidity, turbidity, and other variables. Rivers are too ephemeral to maintain long-term processes, but marine benthic habitats can represent relatively old ecosystems. With metabolic rates several times slower than those of surface organisms, due to cold temperatures and high pressures, deep-sea corals, sea stars, and other benthic inhabitants can persist for many decades, while the reefs last for millennia (van Dover 2000; Turley *et al.* 2007).

5.9 Productivity

The initial response of primary productivity, or the production of plant biomass, is inversely proportional to the severity of the disturbance (Bradbury 1999), because the most severely disturbed sites lack the necessary resources (e.g. nutrients, water, seeds, or vegetative propagules) to support growth. Productivity has reciprocal feedbacks with biodiversity (see Chapter 6; Worm and Duffy 2003), although the nature and importance of that relationship can vary greatly (Waide *et al.* 1999; Cardinale *et al.* 2006; Whittaker 2010). Spatial extent is one of the confounding variables that makes generalizations about the relationships among disturbance, productivity, and biodiversity so complex. Measurements of both terrestrial and aquatic productivity take place at microscales with leaf gas exchange and *in vitro* incubations of phytoplankton, mesoscales with radioisotope labels, eddy covariance towers, and water column samples, and macroscales with remote sensing of vegetation and chlorophyll (Geider *et al.* 2001). However, much remains unknown (e.g. belowground and deep ocean productivity and the photosynthetic efficiency of phytoplankton) so global estimates (10^{17} g carbon yr^{-1}; Field *et al.* 1998) remain approximate and responses to large-scale disturbances (e.g. global climate change) somewhat speculative. Experimental approaches such as air enrichment with carbon dioxide (Smith *et al.* 2000) have been useful in developing models of the likely responses of primary productivity to global changes. Yet fundamental issues about not only ecosystem-level but leaf-level responses to changes in carbon dioxide concentrations and the interacting effects of N limitation and temperature on photosynthesis are still unresolved (Geider *et al.* 2001).

The aboveground terrestrial responses of primary productivity to local disturbances can be readily observed. For example, opening a light gap in a forest by the removal of several trees represents a slight initial loss of primary productivity that is often offset within several years with increased rates of photosynthesis and growth of plants in or around the gap. Severe disturbances that remove more plants and soils cause longer-term reductions in primary productivity. Fires frequently increase the availability of nutrients and are followed by marked increases in primary productivity, particularly in Australia's eucalyptus forests (Christensen *et al.* 1981), Mediterranean shrublands (Rundel 1999), or grasslands (Bradbury 1999). Fire is particularly favorable to primary productivity in wet climates (Oesterheld *et al.* 1999) such as boreal forests (Peng and Apps 1999). Reductions in vegetation structure following fire can reduce fire frequencies (Pausas and Bradstock 2007) but do not necessarily reduce overall productivity (e.g. grasses can have greater productivity than the forests that they replace; Bradbury 1999). After an initial period of increasing primary productivity (progressive succession), a peak is usually reached before the maximal biomass phase and then a decline as P and decomposition decline (retrogressive succession). During **retrogression** there can be an increase in the proportion of belowground biomass (Walker *et al.* 1987), a decline in maximum rates of photosynthesis (Whitehead *et al.* 2005), and a shift to stress-tolerant species with low foliar nutrient concentrations and higher nutrient-use efficiency (Peltzer *et al.* 2010).

Secondary productivity, or the growth of animals, following disturbances generally depends on plants, either for food or habitat, except in very early primary succession where wind-blown invertebrates such as spiders and crickets can survive (but do not reproduce) without plants (Crawford *et al.* 1995; Walker and del Moral 2003). Herbivory provides a direct link between primary and secondary productivity. Insect herbivory can be a prominent part of overall production in the early stages of terrestrial succession (with relatively low primary productivity) and vertebrate herbivores congregate where disturbances are frequent and the palatability of plant resources is high (Walker and Chapin 1987; Farrell 1991; Walker and del Moral 2003). Herbivores can alter successional trajectories depending on their relative influence on transitions from early to late successional species (see Chapter 8). Herbivorous beetles on mines in Wyoming (U.S.), for example, were most abundant in areas of high plant diversity (Parmenter and MacMahon 1987).

Belowground productivity is less easily measured, but it clearly has a central role in the response of productivity to disturbance. Grasslands (Gill *et al.* 2002) and microcosms (Stampe and Daehler 2003) are habitats where belowground productivity responses can be measured, and they are often negatively correlated with aboveground responses (Wilson and Tilman 1993). Soil microbes have critical direct and indirect roles in the disturbance responses of both aboveground and belowground productivity, particularly in low-nutrient habitats. Symbiotic N-fixing bacteria and mycorrhizal fungi are responsible for up to 75% of all N and P acquired by plants (van der Heijden *et al.* 2008). Free-living microbes also provide critical benefits, including the decomposition of organic matter. Their rapid responses to the nutrient releases that often accompany disturbances allow retention of resources that might otherwise leach away. Microbes also compete with plants for nutrients, and pathogens reduce productivity in many species (Burdon *et al.* 2006). Nematodes undergo successional changes in composition and abundance following disturbances and can be used as indices of recovery in both terrestrial and aquatic habitats (Bongers 1990).

Aquatic responses to disturbance are often assessed by the balance of top-down and bottom-up influences on productivity (Paine 1980). A major corollary is the access to limiting nutrients (typically P > N in freshwater and N > Fe > P in marine ecosystems; Micheli 1999; Geider *et al.* 2001; but see Fig. 5.8). The effects of spatial distributions of disturbances and nutrients on both resident and transient organisms are particularly important in river habitats, and there is some degree of predictability in the responses of aquatic organisms to disturbances based on landscape position (Power and Dietrich 2002). Top-down and bottom-up processes are generally complicated by the many connections found in aquatic food webs. Such connections and the length of the food chain may or may not influence resistance and resilience to disturbance. In one study in 10 New Zealand rivers, web size and the number of links per species were negatively correlated to disturbance intensity (Townsend *et al.* 1998), suggesting that those features of food webs are reduced by intense disturbance, or conversely, that small, short food webs are less resistant. Damming rivers also reduces the length of the food chain and food web connectivity and thereby alters productivity because the natural flux of water levels combined with erosion and drought help to maintain their complexity (Power *et al.* 1996; Uehlinger and Naegeli 1998). In several lakes in the central U.S., manipulations of fish populations influenced the productivity of phytoplankton and zooplankton, suggesting top-down control, although abiotic factors were equally as important (Carpenter *et al.* 1987). Marine primary producers are strongly affected by both herbivory and nutrient availability, but the relative importance of these two variables depends on productivity (Burkepile and Hay 2006). Nutrient enrichment was more influential on primary productivity in highly productive habitats, while herbivores were more important in less productive habitats. This effect varied by algal groups and location. For example, in tropical regions, herbivory was more important than nutrients, suggesting that the anthropogenically driven shift from coral to algal dominance of reefs may be due more to over-fishing than to pollution (Burkepile and Hay 2006).

5.10 Conclusions

Disturbances alter not only the abiotic and biotic components of a community but also the ecosystem processes that connect them. These processes, in turn, influence the response to the disturbance. Disturbances can increase or decrease levels of light and nutrients, carbon dioxide levels, and temperatures and can also increase or decrease the spatial heterogeneity of these factors (see Chapter 7). Disturbances can impact ecosystem processes in a ripple effect across many habitats. Damming salmon rivers, for example, has effects on nutrient transfers and both aquatic and terrestrial productivity (see Section 5.7.4).

The disruption of the web of interrelated processes can lead to recovery that is fast or slow to re-form in a myriad of different configurations. Whatever trajectory unfolds following a disturbance, ecosystem processes both reflect and influence the biotic response. This relationship occurs because organisms have minimal (and maximal) levels of resources (e.g. light, nutrients) that they can tolerate and characteristic responses of productivity to the resource levels that are present. The most resilient organisms such as soil bacteria and aquatic algae respond most quickly to resource pulses and, through their influences on the availability of resources, modify the responses of larger, more conspicuous, but less resilient organisms.

Studying the transfers of resources across interfaces between habitats is a standard way to examine ecosystem processes (Pace and Groffman 1998).

Hotspots of resources frequently occur along these interfaces (McClain *et al*. 2003) because of a convergence of resources. Disturbances disrupt pre-existing flows and interfaces and some organisms rapidly take advantage of these changes, especially when they represent abrupt increases in nutrient or light levels (e.g. algal blooms in eutrophic water or tree seedlings in forest gaps). Altered light and nutrient levels and energy transfers across interfaces influence biodiversity, spatial patterns, and temporal dynamics in disturbed ecosystems, topics that are covered in the following chapters.

6 Biodiversity and invasive species

6.1 Introduction

Biodiversity is the number and relative abundance of taxa (e.g. species, genera, families) or other biological units (e.g. genes, communities, ecosystems, functional groups). Biodiversity incorporates concepts of both richness (number of species) and evenness (relative abundance of species) and indices of biodiversity combine these two concepts in various ways (Magurran 1988; Magurran and McGill 2011). When not otherwise specified, biodiversity commonly refers to richness. Species diversity (the most commonly used measure of biodiversity) reflects the integration of the disturbance regime with species interactions among and within trophic levels, and species responses to patterns of resource heterogeneity over ecological and evolutionary time. Species diversity and disturbance are related through a series of feedbacks. Disturbance alters species diversity directly by removing species and their functions or by altering their rates of reproduction or mortality. Indirect effects include altering species interactions, modifying the carrying capacity of the habitat, lowering resource abundance and availability, and reducing species vigor (Chaneton and Facelli 1991; Thrush *et al.* 2001; Hooper *et al.* 2005; Dornelas, 2010). A disturbance that removes a predator species, for example, can decrease species diversity if one of the remaining species competitively excludes other species (Paine 1966). Because each species uses the sum of all resources in its environment in a unique way (Hutchinson 1959), alteration of those resources leads to alterations in species diversity. Species diversity can, in turn, alter the characteristics of a disturbance including frequency and severity, as well as responses to disturbance, including resistance and resilience, because of the differential responses of species to a given disturbance (Hughes 2010). Biodiversity of both species and ecological functions often peaks at intermediate levels of disturbance (Grime 1973; Biswas and Mallik 2010; see Chapter 1), although this pattern is certainly not universal (Shea *et al.* 2004), particularly for multi-trophic communities (Fig. 6.1; Wooten 1998).

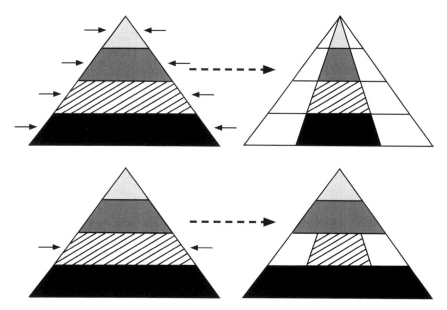

Figure 6.1   Effects of disturbance. Disturbance can have no effect on relative abundance of species because it impacts trophic levels equally (top) or it can alter relative abundance by selectively acting at only one trophic level (bottom). (From Wooten (1998), with permission from the University of Chicago Press.)

Studies of the effects of disturbance on biodiversity, and vice versa, demonstrate that the relationship between these two processes is highly variable (Fig. 6.2; Mackey and Currie 2001). In one study, species diversity affected disturbance in two-thirds of all surveyed examples, while disturbance affected species diversity even more often (Hughes *et al.* 2007). The most common effect of diversity on disturbance is to reduce the impact of a disturbance, particularly by increasing resistance to invasion (Levine *et al.* 2004). The most common effect of experimental disturbances is to decrease diversity, followed by no effect, and then (in only 18% of cases) to promote maximum diversity at intermediate levels of disturbance (Hughes *et al.* 2007). Observational studies of the effects of disturbance on biodiversity are more variable, with neutral, positive, hump-shaped, or negative effects in decreasing frequency (Fig. 6.2). The variability of the types of effects of biodiversity on disturbance might be explained by interactions among the two variables. For example, interactions can involve the effects of disturbance on recruitment (the more severe a disturbance, the higher the diversity of recruits) or competitive exclusion (the more severe a disturbance, the less competitive exclusion there is). Models of these complex interactions (Hughes *et al.* 2007) are still simplistic and do not incorporate the frequency of recruitment in a habitat,

the diversity of potential recruits in a region, the patchiness of disturbances, and many other complicating variables. Similar problems affect
the related attempts to interpret the relationship between diversity and
productivity (Whittaker 2010).

Disturbance can moderate the direct and indirect effects that biodiversity has on ecosystem functions such as productivity and nutrient cycling
(Tilman 1999; Mulder *et al*. 2001). Herbivory, for example, can differentially alter the relative abundances of plant species and thereby influence
productivity (Mulder *et al*. 1999). Herbivory can also alter the spatial and

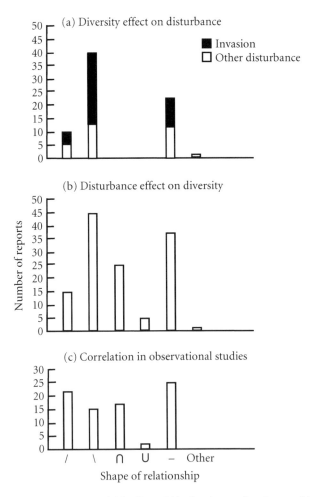

Figure 6.2   Frequency of different shapes of (a) effect of biodiversity on disturbance, (b) effect of
disturbance on diversity, and (c) correlation between disturbance and diversity. (From
Hughes *et al*. (2007) with permission from Wiley-Blackwell.)

temporal distribution of nutrients (e.g. from feces, trampling, or regulation of soil fauna; Bardgett and Wardle 2010). Such herbivore effects appear to be linked to the size of the herbivore, precipitation, and site fertility. Herbivory generally has a more negative effect on species diversity in dry, infertile sites than in wet, fertile ones (Olff and Ritchie 1998). Disturbances also influence biodiversity by altering the relative abundances of species that regulate critical ecosystem functions such as floodplain nutrient cycling (beavers; Wright *et al.* 2002), belowground productivity (grasses; Aarssen 1997), the flux of organic matter in streams (caddisflies; Cardinale and Palmer 2002), or **bioturbation** in sandflats (gastropods, crustaceans, and other macrofauna; Lohrer *et al.* 2010). Higher species richness can also reduce the disruption of ecosystem functions by disturbance (increase resilience) if surviving species are able to maintain key energy and nutrient flows across biotic and abiotic interfaces (Silver *et al.* 1996a; see Chapter 5). However, claims that increases in biodiversity cause increases in ecosystem functions such as productivity or **nutrient use efficiency** may result from experimental complications rather than biodiversity *per se* (Huston 1997).

Biodiversity is a relatively easy measurement to make and it provides a quick assessment of the complexity of a community. When species are not easily identifiable, the biodiversity of functional groups can be measured. For example, most soil bacteria are not easy to culture or identify to species level, but their diversity can be assessed at the community level by differences in phospholipid fatty acids or genetic composition (Bardgett 2005). The bacteria responsible for N fixation and N cycling or the decomposition of organic matter and organic pollutants clearly serve ecologically critical functions and are a valuable component of biodiversity (Groffman and Bohlen 1999). The responses of bacteria to disturbance appear to largely mirror those of plants and animals, with the highest levels of biodiversity in habitats that experience intermediate levels of disturbance, although many variables (soil structure, vegetative cover, land-use history) may confound this relationship (Horner-Devine *et al.* 2004).

Biodiversity has recently gained a high public profile (Sachs *et al.* 2009), particularly in conservation biology, in part because it is often considered positively correlated with the resilience of a community to disturbance (Chapin *et al.* 2000). This assumption needs to be examined further. Certainly, biologically diverse communities provide essential ecosystem services (Costanza *et al.* 1997; Hooper *et al.* 2005). For example, certain plants and animals provide community-level functions such as year-round food (e.g. tropical fig trees; Fig. 6.3), pollination, and maintenance of biodiversity through predation of species that promote competitive exclusion. In contrast, however, many species are functionally equivalent and therefore redundant to the maintenance of ecosystem services within a community. Overall, it appears that resilience is a complex topic and is not tightly linked to biodiversity, depending on the variable effects of a particular disturbance regime and the many species-specific responses to it (Brussaard *et al.* 2007). The resilience of a community can be reduced when a disturbance selectively removes the most resilient species. Diverse communities can provide functional redundancies, interacting food webs, and variable levels of sensitivity to a given disturbance (severity) that can increase community resilience. Equally likely, however, is that functional redundancy is not apparent, a disturbance can cause more damage to complex than simple food webs, and species are equally affected by a disturbance—all factors that would not necessarily lead to an increase in resilience.

In this chapter, I discuss how disturbance and biodiversity are linked in both space and time. I also address how the addition of invasive species can constitute a disturbance when there is a relatively sudden decrease in native species biomass, the reciprocal effects of biodiversity on invasive species, and the central role that humans have in altering biodiversity. Chapters 7 and 8 elaborate on the spatial and temporal dynamics of disturbance, respectively.

Figure 6.3   A strangler fig in Ulong Kulong National Park, Indonesia. Fig trees promote forest biodiversity by supplying animals with a year-round source of food. They are also important colonists of new volcanic surfaces, such as the nearby Krakatau Volcano.

6.2 Space

A landscape is a mosaic of habitats reflecting communities at various stages of recovery from the sum of all interacting disturbances in the region. Those disturbances vary in extent, frequency, intensity, and severity, with each of these characteristics having its own influence on biodiversity. Most disturbances at organismal to landscape scales increase spatial heterogeneity because few disturbances have a uniform effect (see Chapter 7). Such heterogeneity promotes biodiversity because of the number of potential habitats made available for species to colonize. However, the average rate of supply of a limiting resource may be more important than resource heterogeneity in promoting biodiversity (Stevens and Carson 2002). Colonization by species poorly adapted to local conditions (the rescue effect; Brown and Kodric-Brown 1977) can further increase biodiversity. A diverse flora or fauna reinforces a heterogeneous effect of the next disturbance. At larger, more extensive scales, (e.g. regional or continental), certain disturbances can have a homogenizing effect (e.g. boreal forest fires or pest outbreaks) and result in a synchronous, widespread loss of biodiversity (Huston 1994). The successful colonists can form extensive, low-diversity ecosystems that add a few species as they mature but then become susceptible to a recurrence of the same type of disturbance.

Biodiversity interacts variously with disturbance frequency, intensity, and severity in a spatial context. In experimental communities of marine benthic microalgae, dispersal affected species richness, but only at high disturbance frequencies (early succession; maximum richness at intermediate dispersal rates; Matthiessen *et al.* 2010). Disturbances that leave regional propagule pools untouched have the least effect on regional biodiversity. In a given habitat, as disturbance intensity increases, mortality and corresponding losses of biodiversity generally increase. For example, fewer species survive occasional extremes of cold (cold snap) or heat (fire, drought) than more normal conditions. Spatial heterogeneity in intensity leads to increased spatial variability in biodiversity. Severity can be measured as immediate mortality or decreased functioning of survivors due to partial damage or reduction of resources. Alteration in resource levels (e.g. light, nutrients; see Chapter 5) can have potentially longer-lasting effects on biodiversity than initial mortality, with increases in patchiness due to delayed mortality, competitive displacement, or accelerated senescence from prior reductions in resources.

Spatial hierarchies are useful for evaluating the effects of disturbance on biodiversity (Whittaker 1977; Pickett and White 1985). Disturbances affect biodiversity at specific microenvironments (point diversity), within a given patch or habitat (alpha diversity), and at larger scales across patches (pattern or beta diversity). Biodiversity can also be assessed within a region composed of several habitats (gamma diversity) or even across a biome (epsilon diversity). The effects of disturbance will vary depending on which spatial scale is considered but also on whether just dominant species or a larger list of ecosystem characteristics is examined. For example, both cattle grazing and flooding reduced species diversity at the scale of 1 ha (stand) in the Argentinean pampa, but at smaller scales (~5 m^2) grazing increased biodiversity and flooding decreased it (Chaneton and Facelli 1991). When the combined disturbance regime (incorporating disturbance interactions) was considered, grazing increased species richness and flooding increased species evenness (by reduction in the abundance of dominant species).

6.3 Time

Biodiversity, like other community characteristics, is altered by a disturbance and responds over time as species disperse to the disturbed site and colonize it. Replacement of communities through the successional process that ensues (see Chapter 8) often means changes in plant, animal, and microbial diversity. Biodiversity can be altered immediately by mortality from a disturbance, over longer periods by the influence of disturbance on species growth and interactions, and over evolutionary time by the survival of species that successfully cope with a given disturbance regime (Huston 1994). Disturbance frequency helps determine biodiversity by regulating which species survive. Frequent disturbances favor pioneer species with short life cycles that allow the organism to reproduce before the next disturbance. When frequent disturbances are predictable, local biodiversity can be high because competitive exclusion is uncommon. Regional diversity is not likely to be higher than local diversity under these conditions. When frequent disturbances are less predictable, biodiversity is potentially lower because there are fewer species able to survive periods of heightened competition during the longer intervals between disturbances. When disturbances are predictably infrequent, regional biodiversity can be high but local biodiversity is often lower (e.g.~a tropical rainforest). This pattern can be the result of various factors including competition, dispersal, and predation. Biodiversity not only responds to temporal aspects of disturbance but can also influence succession. For example, high species richness can impede colonization (Diemer and Schmid 2001) when all niches are filled.

There are many temporal variables that influence biodiversity during succession, including landscape dynamics, phenology, productivity responses, soil fertility, successional stage, and disturbance history. First, the physical landscape (geology, topography, soil condition) presents a mosaic of variables, each with its own temporal state. The longer-term processes of uplift and erosion and the shorter-term processes of microbial activity and nutrient flux influence species responses to disturbance. Second, the timing of the disturbance relative to the composition and phenological patterns of local organisms influences which species will colonize a newly disturbed habitat. Many organisms have evolved life cycles that reflect a dominant disturbance regime with such characteristics as dormant seeds or germination only following fire (**serotiny**; Gauthier *et al.* 1996). Third, the growth rates and productivity of the successful colonists can influence biodiversity and the response to disturbances. Biodiversity and productivity both tend to increase during the early stages of succession (Cardinale *et al.* 2004). Where there are rapid growth rates and high productivity, the relative influence of disturbance is small and short-lived. Frequent disturbances tend to increase early successional, local biodiversity because in the absence of disturbance, good competitors exclude other species and lower biodiversity (Huston 1994). Conversely, where disturbances are infrequent, growth rates are slow, and productivity is low, the impacts of disturbance on biodiversity can be relatively large because any effects are long-lasting and multiple disturbance effects are additive. Fourth, once harsh, early successional environments are ameliorated by early colonists, there is often a period of peak species richness, caused in part by overlap of species best adapted to early, middle, and late successional conditions. Species richness can decline in later stages of succession, particularly where soil fertility decreases (Wardle *et al.* 2008). Finally, historical disturbances (land use) influence

Figure 6.4   A walkway 40 m above the ground in an 80-m tall dipterocarp forest near the Laos border, Yunnan Province, China. From this walkway I was lucky to catch a glimpse of a palm civet (*Paradoxurus*) because wildlife sitings are unusual in China.

biodiversity. Fish, but not invertebrate, diversity, was affected by 50-yr-old agricultural land-use practices in surrounding watersheds in North Carolina (U.S.; Harding *et al.* 1998). Soil conditions, propagule availability, and remnant intact patches of tropical forests (Fig. 6.4) that have been altered by both natural (hurricanes) and anthropogenic (grazing, fires) disturbances modify rates of recovery and re-establishment of biodiversity (Chazdon 2003).

6.4 Invasive species

Invasive species are those species that expand their range due to their dispersal abilities. The wide distribution and mixing of species has led to speciation and to the species-rich world that we now inhabit. However, humans have accelerated the mixing many-fold through their deliberate and accidental introductions of species to habitats beyond their original range (Vitousek *et al.* 1997b). A small percentage of these non-natives spread beyond gardens or other points of introduction to become non-native invasives (henceforth "invasives"). Concern about the ecological effects of these successful invasives is now widespread (Suarez and Tsutsui 2007). Successful invaders often possess characteristics that make them good competitors so that they often outcompete native species; this displacement of natives can lead to major and often undesirable changes in biodiversity and community and ecosystem properties. There can also be a reciprocal relationship, where not only do invaders affect biodiversity but biodiversity influences the invasibility of a community. Both of these effects are discussed in this section.

6.4.1 Effects of invasive species on biodiversity

Invasive species are agents of disturbance when they cause a loss of biomass or ecosystem structure or function in their new habitat. The effect of invasive species on biodiversity is complex and depends on the functional role of the native species that are replaced or that have their abundance reduced and their niches subsequently filled by the invasive species. Invasives can play a dominant role through competitive superiority or merely contribute to the complexity of a community that is more driven by environmental factors or dispersal dynamics than by species interactions (MacDougall and Turkington 2005). Invasive species can directly affect biodiversity by increasing richness by one species (the invasive), with probable shifts in evenness as the relative abundances of native species are subsequently rearranged. There are also a number of indirect ways that invasives alter biodiversity through their influences on disturbance regimes, species interactions, and community and ecosystem properties. The net effect on biodiversity depends on how many species are lost or damaged due to the suite of effects by invasive species, a process that is difficult to document. Invasive predators and pathogens are the most likely organisms to cause extinctions (Davis 2003), particularly on islands with no effective escape terrain (Hamilton 2011). Invasive species from similar climates can also have strongly negative effects, such as *Acacia* from Australia invading South Africa (Gaertner *et al*. 2009). Overall, displacements or population reductions of native species are more likely than actual extinctions (Gurevitch and Padilla 2004).

Disturbance regimes can be altered by invasive species with many potential consequences for the biodiversity of both native and invasive species (Mack and D'Antonio 1998). Invaders can increase disturbance frequency or severity. When grasses invade arid shrublands, fire frequency and severity are often increased because of increases in the amount of fuel (D'Antonio and Vitousek 1992; Smith *et al*. 2000). When the invasive species becomes dominant, plant diversity declines due to competitive exclusion. Pest outbreaks can lead to stand dieback and loss of tree diversity (Raffa and Berryman 1983). Invaders can also decrease the frequency and severity of disturbances by stabilizing soils (e.g. invasions by grasses or extensively rooted trees along floodplains or on slopes; Versfeld and van Wilgen 1986) or by decreasing fire regimes (e.g. invasion of grasslands by shrubs; Brown and Archer 1999). Herbivores also can influence biodiversity through their interaction with invasive species (Fig. 6.5). For example, in one survey of experiments, generalist native herbivores suppressed invasive plants (provided biotic resistance to invasion), whereas invasive herbivores facilitated invasive plants and thereby invasions (Parker *et al*. 2006). Alternatively, some invasive species succeed because they have left their herbivores behind ("enemy release"; Keane and Crawley 2002). The net effects of altered disturbance regimes on biodiversity include a net loss, net gain, or no change in the proportion of invasives and natives. As an example of no net change, grubbing by feral pigs in meadows of northern California promoted colonization by annual plants that were both native and invasive (Kotanen 1995).

When species interactions are modified by invasive species, biodiversity can be affected. Alterations of food webs, particularly by removal of a keystone predator, are one way in which invasives can decrease biodiversity. Paine's (1966) removal of the sea star *Pisaster* resulted in competitive dominance by a mussel (*Mytilus*) over other benthic organisms because it was released from predation by *Pisaster*; there was then a net loss of local biodiversity. Additions of non-native predators can also reduce biodiversity; for example, the invasion of Lake Victoria by the Nile perch (*Lates*) led to the extinction of native cichlid species (Witte *et al*. 1991). Biodiversity can increase when the invasive species is part of an "invasional meltdown," whereby other nonnatives invade with it or are promoted by the presence and functions of the original invasive species (Simberloff and Von Holle 1999). Changes in biodiversity can then vary from the addition of a new suite of interacting invasives, or from alterations in species evenness in the newly formed community. A large-scale example involves the numerous European organisms

Figure 6.5   Result of herbivory on an invasive cactus (*Opuntia*) by a giraffe on Crescent Island in Lake Navasha, Kenya. The creation of a game reserve on the island has concentrated many large herbivores into a small area, thereby increasing herbivore pressure.

introduced throughout the former British colonies, many of which are now dominating many anthropogenically altered habitats (Simberloff and Von Holle 1999). Another example comes from Hawaii where a majority of the lowland flora is non-native. On the Island of Hawaii, an invasive bird, the Japanese white-eye (*Zosterops*), dispersed the invasive faya tree (*Morella*; see Fig. 5.5) that, in turn, fed feral pigs (*Sus*) that, in their turn, promoted increased litter turnover, invasive earthworms, and several invasive plant species (Vitousek and Walker 1989; Aplet 1990; Aplet *et al.* 1991). An additional, multi-trophic "invasional meltdown" occurred when the crazy ant (*Anoplolepis*) invaded Christmas Island in the Indian Ocean. The ant removed the red land crab (*Gecarcoidea*), causing a cascade of events including increased seedling cover and richness, increased populations of scale insects, increased sooty molds, and increased tree death (O'Dowd *et al.* 2003). Several years after the introduction of the term "invasional meltdown" to the ecological literature, Simberloff (2006) determined that firm evidence of two-way or multi-species facilitation of invasive species leading to increased number and impact of invasives species was rare. Most examples of this highly cited concept illustrate a one-way process rather than mutual facilitation. The lack of corroborating evidence for "invasional meltdowns" could be in part due to the lack of long-term data or the difficulty of assessing the population and community consequences of shifts in species composition. Many studies have circumstantial evidence that suggests mutual facilitation. The discussion of the relative importance of facilitation for invasions and biodiversity parallels similar discussions in the successional literature about the importance of facilitation (versus competition or neutral effects) in determining shifts in successional stages (Walker and del Moral 2003), and a renewed awareness of the important role of positive interactions in community dynamics (Bruno *et al.* 2003b; Callaway 2007).

Community and ecosystem properties are variously altered by invasive species, and these changes can lead to reduction in disturbances in the form of reduced productivity or mortality of native species. Invasive species are also affected by the processes that they modify (Table 6.1; Walker and Smith 1997). Many changes in properties affect biodiversity. For example, invasive species can increase primary productivity by exploiting a vacant niche (net species addition) or outcompeting less productive native species (net species loss). When invasive species add a limiting resource such as N (Vitousek and Walker 1989), they supply litter of higher quality, which then improves nutrient availability (Bardgett and Wardle 2010), or water availability (Busch *et al.* 1992) with possible consequences for productivity and biodiversity. Productivity and nutrient cycling can decrease when invaders sequester nutrients through additions of nutrient-poor litter (Versfeld and van Wilgen 1986); volatilization or sequestration of N, K, Ca, and Mg (D'Antonio 1990); salt addition (Busch and Smith 1995; Walker *et al.* 2006a); or alteration of water supply through reduction (Mueller-Dombois 1973) or increase (Sala *et al.* 1996) in transpiration rates. Invaders can also modify productivity by altering the distribution and abundance of root pathogens (see Section 5.9; Klironomos 2002). Finally, when invasive species alter the disturbance regime (e.g. by reducing fire frequency) they can also affect productivity and biodiversity. The responses of biodiversity to altered ecosystem properties vary depending on the effect of the changes on colonization, competition, and mortality.

The net effect of invasive species on biodiversity appears to be more benign than expected under many circumstances (Goodenough 2010; Hamilton 2011). Invasive species often reduce local biodiversity (though they do not usually cause global extinctions of species). Community structure is also frequently altered, as mammals invade and alter predator/prey dynamics, trees invade grasslands, or insects carry diseases that disproportionately affect dominant species. However, invasive species also provide (sometimes unrecognized) benefits, such as supplying critical habitat for endangered species. For example, monarch butterflies (*Danaus*) use invasive *Eucalyptus* trees in California; willow flycatcher birds (*Empidonax*) nest in invasive tamarisk (*Tamarix*) trees in Nevada and Arizona; and African tulip trees (*Spathodea*) have improved soil conditions in abandoned pastures. Functional roles such as the provision of pollen for native butterflies can be performed by invasive species (Shapiro 2002) and many agricultural crops rely on pollination by European honey bees. These benefits of structural

Table 6.1 Impacts of invasive plants on community and ecosystem attributes. (Modified from Walker and Smith (1997) with permission from Springer.)

Attribute	Positive Impact	Negative Impact
Primary productivity	Exploits vacant niche	Sequesters nutrients
	Exploits under-utilized resource	Grows more slowly than natives
	Outcompetes native species	Promotes disturbance
Nutrient cycling	Fixes N	Increases nutrient uptake
	Adds nutrient-rich litter	Sequesters nutrients
	Increases weathering rates	Accumulates salt
Water balance	Lowers leaf transpiration rate	Increases transpiration, stem flow
	Lowers leaf area	Increases run-off
	Increases infiltration and stem flow	Accesses unused water sources
Disturbance regime	Increases fire frequency	Decreases fire frequency
	Increases erosion	Controls erosion
	Decreases herbivore resistance	Increases herbivore resistance

support, habitat amelioration, or functional role occur because the invasive species grow where natives have been eliminated before invasives arrive, and the habitat has been drastically altered by humans. The increasing proportion of anthropogenic habitats and impossibility of removing all invasive species suggests that effects of invasive species on biodiversity will continue to be explored well into the future.

6.4.2 Effects of biodiversity on invasive species

The biotic resistance model suggests that species-rich communities are less vulnerable to invasion than species-poor ones because of competition for limiting resources and niches (Elton 1958; Case 1990). This view emphasizes competitive interactions between natives and non-natives and among non-natives, in contrast to the "invasional meltdown" model that suggests that facilitative interactions dominate. Support for the biotic resistance model has come from experimental seeding in oak savannas (Tilman 1997), species-poor islands (Vitousek *et al.* 1996), and grasslands (van Ruijven *et al.* 2003). In aquatic ecosystems where space is the limiting factor, species diversity can also limit invasion success (Stachowicz *et al.* 1999). The strength of biotic resistance to invasion often centers on the competitive dominance of individual native species and these may or may not be more likely to be present in species-rich than in species-poor communities. This uncertainty compounds the difficulty of separating out the effects of diversity and the influences of co-variables such as disturbance, nutrients, and productivity on invasibility (Fig. 6.6; Naeem *et al.* 2000; Wardle 2001). Less common species can also resist invasions through allelopathy and associations with mycorrhizal fungi

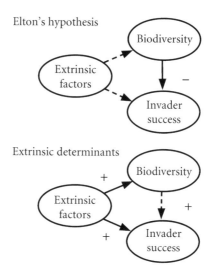

Figure 6.6   Elton's hypothesis (top panel) suggests that variation in biodiversity affects invader success because higher biodiversity promotes competition that, in turn, reduces invader success. Alternatively (bottom panel), extrinsic factors such as disturbance frequency, soil fertility, and climate can have similar effects on biodiversity and invader success, creating apparent positive effects of high biodiversity on invader success. (From Naeem *et al.* (2000) with permission from Wiley-Blackwell.)

(Lyons and Schwartz 2001). However, evidence has also accumulated to suggest that species richness can have either a positive or a neutral effect on invasibility. Forest, meadow, and prairie habitats with high species richness were readily invaded in the central U.S., with plant cover and soil fertility also promoting invasion (Stohlgren *et al.* 1999, 2008). In a Californian floodplain community, diverse communities were the most likely to be invaded (due to propagule availability that increased with diversity), even though at smaller scales, biotic resistance to invasion was present (Levine 2000). Aquatic invasions of the Great Lakes in North America faced little biotic resistance, instead invasion occurred whenever abiotic conditions were favorable and dispersal opportunities were present (Ricciardi 2001). Invasibility is frequently linked to anthropogenic disturbances that may disrupt species interactions and ecosystem functions enough to promote invasion into species-rich habitats, or alter local resource availability and therefore change competitive interactions among species, further complicating our poor understanding of what causes patterns of species richness (Palmer 1994).

6.4.3 Humans and invasive species

The world's flora and fauna are now so homogenized by human activities that studying pristine, native communities is becoming increasingly uncommon. Novel communities (Hobbs *et al.* 2006) are becoming more the norm as the pace of homogenization accelerates. This homogenization is characterized by a global reduction in biodiversity through the competitive dominance of a few species. Humans assist that process by promoting monocultures in agriculture, forestry, and landscaped habitats and by reducing the extent of high-diversity habitats such as wetlands and coral reefs (Fig. 6.7; Box 6.1). The loss of biodiversity and habitat diversity simplifies ecological communities, often making them more vulnerable to anthropogenic alterations of disturbance regimes (Hobbs and Huenneke 1992). Although diversity may not be the sole cause of ecosystem stability (Ives and Carpenter 2007), the increasing loss of species is likely to lead to destabilization and potential collapse of ecosystems and their ecosystem services that humans rely on (McCann 2000), unless those losses are compensated by addition through invasions.

Humans serve as both intentional and unintentional vectors for invasive species. Intentional introductions were initiated by explorers who brought edible species with them, releasing them on their arrival in new lands (Fig. 6.8). For example, Polynesians introduced at least 50 species of plants including yams (*Dioscorea*) and taro (*Colocasia*), as well as pigs (*Sus*), chickens (*Gallus*), dogs (*Canis*), and rats (*Rattus*) to many of the islands of the Pacific Ocean that originally had few edible species (Walker and Bellingham 2011).

Figure 6.7   A coral reef community as seen at Waikiki Aquarium, Oahu, Hawaii. (Photograph by Elizabeth Powell.)

Box 6.1 Threats to biodiversity on coral reefs

Coral reefs are under pressure from both natural and anthropogenic disturbances with consequences for ecosystem structure and function. The coral reefs of Hawaii are no exception. With millions of tourists every year, Waikiki Beach in Honolulu on the island of Oahu is the epicenter of Hawaiian tourism. Massive amounts of sand are regularly brought in from offshore and other islands to replenish the glistening white beaches, so it is a bit ironic that the beach is designated as a protected area, off limits to fishing or removal of any other aquatic organisms. Another irony is that the rich diversity of aquatic life found in the small but delightful Waikiki Aquarium (Fig. 6.7) has little resemblance to what one finds in the nearby coastal waters. Waikiki's reefs are subject to the damages that threaten reefs everywhere, including storms, diseases, increased organic matter and water temperatures, decreased salinity, invasive species, and the constant pressures of people enjoying themselves (surfers, boaters, fishermen, divers, snorkelers, tide pool waders, swimmers). The Hawaiian government has recognized the important cultural role that subsistence fishing has had in the past and has devised a rule whereby fishing is allowed on the reefs near Waikiki every other year. My stay in Honolulu overlapped the change in this regime, as my wife and I snorkeled regularly just east of Waikiki during 4 months with no fishing and then during 6 months with fishing. Such regulations might work if fishing were only for limited subsistence use, but sport fishermen are there, too. On New Year's Eve, fishermen wait for the year to begin, rushing out to catch the larger fish and octopi that have developed in the year without fishing. Many of our favorite fish, especially the larger ones, disappeared in January and even through May we contended unhappily with spear fishermen chasing the same colorful animals that we were watching. Biodiversity declined once fishing began, and we often snorkeled on the north side of the island instead. Such a tragedy of the commons (where more is taken away from shared resources that is replaced) occurs with reefs and other marine resources around the world. However, fish biodiversity on protected reefs can quickly rebound (e.g. nearby Hanauma Bay), suggesting the urgency to protect coral reef ecosystems while there are still healthy ones that can repopulate damaged reef ecosystems.

Subsequent land clearance for agriculture as well as competition and predation on native plants and animals led to the loss of biodiversity, especially on island ecosystems (Anderson 2009). Early explorers also returned to their homelands with edible and ornamental species, a policy accelerated by naturalists interested in describing and classifying the world's biodiversity. This mixing of continental and island biota continued to increase as European settlers spread around the world, taking species essential to their agriculture, animal husbandry, and landscaping tastes. For example, New Zealand forests and tussock grasslands gave way to pastures filled with European cows and sheep. Similar replacements of native vegetation occurred in Iceland, the Canary Islands, South Africa, the eastern U.S., Australia, and many other locations around the world in what has been called ecological imperialism (Crosby 1986). Today, we are still intentionally moving species around for the horticultural, pet, zoo, and aquarium trades. Agriculture, silviculture, and animal husbandry also continue to contribute to species exchanges, except where there are restrictions (e.g. to maintain genetic purity: Icelandic horses can leave Iceland but not return). Such intentional commercial introductions are frequent causes of invasions, particularly when they occur multiple times and the introduced species is capable of overcoming any local habitat obstacles to its spread (Suarez and Tsutsui 2007).

Unintentional introductions have paralleled the rise in intentional ones, beginning with rodents on ships; parasites, bacteria (e.g. cholera), and viruses (e.g. smallpox) associated with humans; and ballast water containing aquatic species. Modern distribution continues with species traveling as accidental hitchhikers on boats, trains, and airplanes. These non-native organisms come in imported potting soils, in grain for animal or human consumption, on fruit and vegetable cargo, and on tourists and business travelers (e.g. on their clothing, luggage, hair, skin). Some efforts are now made to stop such introductions through agricultural inspections at airports and shipyards, but the number and variety of vectors is difficult to monitor. Such unintentional invasives can alter ecosystem properties (Walker and

Figure 6.8   A fruit market in Bali, Indonesia representing the variety of locally grown food items available in many tropical markets.

Vitousek 1991) and outcompete native species (Bellingham *et al.* 2005a). Even modest inspection efforts appear to be worthwhile because the negative economic and environmental effects of invasive species are so large (Table 6.2; Pimentel *et al.* 2000).

Extensive development of low-diversity, anthropogenic habitats (e.g. croplands, golf courses, reservoirs) reduces native biodiversity but species adapted to these novel habitats can add a new component to local biodiversity. Such newly recruited species tend to be those that tolerate gaps, edges, and high levels of disturbance frequency. Cities provide a wide range of habitats for such species, and urban biodiversity can be substantial. Urban forests can include surviving patches of native terrestrial and riparian forest as well as planted street landscapes and wooded parks (Carreiro *et al.* 2008). In addition, there are often many scattered, low-density pockets of trees either natural or planted throughout most cities. City habitats also include aquatic habitats that can be surprisingly species-rich (Lugo 2010). Similarly, golf courses can support a relatively high biodiversity, depending on surrounding land use. For example, the biodiversity of trees, birds, bumblebees, and beetles (but not herbaceous plants) was higher on English golf

Table 6.2 Environmental and economic costs of non-native species introduced intentionally or not into the U.S. (Modified from Pimentel *et al.* (2000).)

Type of organism	Examples of introduced species	Undesirable environmental effects	Cost of environmental effect and control (×$1 million)
Plants	Tamarisk, melaleuca, purple loosestrife, salvinia	Weeds in crops, pastures, gardens, lawns, golf courses, wetlands, and lakes	34 000
Mammals	Horse, pig; mongoose, rat, cat, dog	Predators, herbivores, or competitors of native or commercial species; habitat destruction	37 000
Birds	Starling, pigeon, house sparrow	Dispersers of non-native plants and diseases; displace native birds; damage crops	1900
Reptiles and amphibians	Brown tree snake	Predators on native birds; power outages	5.6
Fishes	Carp	Competitors of native fish	1000
Arthropods	Fire ant, gypsy moth, green crab	Herbivores on trees; predators of livestock, shellfish	20 000
Mollusks	Zebra and quagga mussels, shipworm	Damage habitat for native fish; clog pipes	1300
Microbes	Dutch elm disease, influenza	Plant pathogens; livestock and human diseases	41 000

Figure 6.9   A complex landscape in Bali, Indonesia with natural forest remnants dominated by such anthropogenic disturbances as planted trees, crops, rice paddies, garden landscaping, and tourist cabins.

courses than in the surrounding farmland (Tanner and Gange 2005). While intensely manipulated anthropogenic habitats reduce global biodiversity (Gardner *et al.* 2010), they can contribute to regional or local biodiversity because they widen the range of environmental conditions (Fig. 6.9). Efforts to preserve or restore biodiversity cannot ignore such human-modified ecosystems (Tabarelli 2010).

6.5 Conclusions

Biodiversity has a central role in disturbance ecology because it is affected by disturbance regimes and can also affect the outcomes of disturbance. The complexity of interactions between biodiversity and disturbances is mediated by the characteristics of the community and of the disturbance. Net effects on biodiversity vary with species composition, disturbance frequency, and disturbance severity. The effects of disturbance on biodiversity can appear robust but be less critical than is initially apparent due to functional redundancy among species. Disturbance can modify the effects that biodiversity has on other functions such as herbivory, productivity, competitive balances among species, and efficiency of nutrient use. These multi-way interactions are difficult to experimentally verify.

Disturbances often promote biodiversity through the creation of spatially heterogeneous habitats, at least at local scales. The variable intensities of disturbances themselves (e.g. variable temperatures in a fire) help to promote this heterogeneity. Both the effects of disturbance and biodiversity can be assessed at multiple spatial scales, often with contradictory results. Disturbances also have various temporal effects on biodiversity. Disturbance return intervals are often not synchronized with life histories of organisms, so seasonal or interannual effects can vary, although frequent disturbances favor disturbance-adapted species with high resilience. Temporal responses of communities also lead to changes in biodiversity as the abiotic and biotic environments change. Biodiversity tends to increase over time following a disturbance, although declines can occur in later stages of succession.

Humans have accelerated the introduction of species to new environments, and many of these species become successful invaders in the new habitats. Non-native invasive species generally decrease native biodiversity, but documented examples of them actually causing extinctions of native species are largely limited to predators on islands. Invasive species can have even greater effects on biodiversity when they alter disturbance regimes (e.g. increase fire frequency) or food webs. Invasive species often facilitate the successful establishment of other invasives by providing vectors for their dispersal, nutrition, or other functions. Invasive species also alter productivity and nutrient cycles. The overall effect of invasive species can be positive, particularly when they stabilize, fertilize, or provide missing and critical functions (e.g. habitat for endangered species). Biodiversity can also influence the invasibility of a community, particularly by providing resistance to invasion. Humans continue to be the main vectors for the distribution of invasive species and much of the earth is now predominantly populated by introduced species.

7 Spatial patterns

7.1 Introduction

Spatial patterns, from scales incorporating organisms to biomes, are an integral part of ecology, albeit sometimes less a focus of ecological studies than temporal patterns or species interactions (Tilman and Kareiva 1997). Spatial heterogeneity provides the context (and sometimes the content) for many ecological questions regarding such variables as ambient temperatures, substrate fertility or texture, species diversity, and ecosystem patchiness across landscapes. Disturbances contribute to the number and variety of spatial patterns across all ecologically relevant scales (see Fig. 1.3). Disturbances not only alter the current spatial distribution of habitats but also influence the trajectories that ecosystems follow over time when they respond to disturbances (see Chapter 8). For example, the spatial heterogeneity of dissolved nutrients in rivers changes over successional time following floods and influences the rates of ecosystem processes (Dent and Grimm 1999).

Generalizations about the role of disturbances in determining spatial patterns across habitats and scales are difficult to make and are often unrealistic because of the many types and scales of spatial heterogeneity as well as the many types and scales of disturbance. For example, a monospecific pine forest may have lower spatial heterogeneity in aboveground structure and biodiversity than an old-growth forest but higher spatial heterogeneity of mycorrhizae in the soil, sunflecks at the forest floor, or grass species richness among colonists of a treefall gap. Similarly, nutrient availability in dune slacks in The Netherlands (Grootjans *et al.* 1998) varied at spatial scales from microsites (5 cm) to landscapes (500 m; Fig. 7.1). Limiting conclusions about spatial patterns to specific types of heterogeneity and scales is preferable for some studies, but hierarchical analyses of scales allow a broader and more realistic integration (O'Neill *et al.* 1986). This chapter examines how spatial dynamics are compared across spatial scales, how disturbances contribute to spatial heterogeneity, the role of dispersal in creating patches, spatial heterogeneity in two very heterogeneous environments (soils and

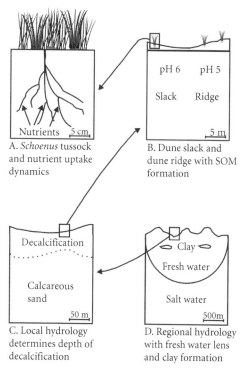

A. *Schoenus* tussock
and nutrient uptake
dynamics

B. Dune slack and
dune ridge with SOM
formation

C. Local hydrology
determines depth of
decalcification

D. Regional hydrology
with fresh water lens
and clay formation

Figure 7.1   The importance of spatial scales for studies of soil organic matter (SOM) accumulation on dunes in The Netherlands. (From Grootjans *et al.* (1998) with permission from Elsevier.)

shorelines), the dynamics of habitat patches, and the contrasting spatial patterns derived from natural and anthropogenic disturbances.

7.2 Scales

A landscape is a convenient scale at which to introduce the importance of spatial heterogeneity (Pickett and Cadenasso 1995). Sometimes described as what one sees from the top of a hill (Higuchi 1983) or out of an airplane window (Forman 1997), a landscape incorporates multiple ecosystems that can be readily examined and contrasted. Gradients (Whittaker 1967; Vitousek and Matson 1990) are one way to contrast changes in a landscape. Along an elevation gradient, for example, from a mountain slope to sea level (Fig. 7.2), one will likely encounter changes in the physical environment (e.g. air temperature and moisture or soil fertility) and changes in the biota and their characteristics (e.g. physiological or population traits). Continuing the gradient into the ocean, further changes will be observed in abiotic (e.g. water temperature, carbon dioxide and oxygen concentrations, light saturation) and biotic parameters (e.g. plant and animal populations). Once spatial patterns at a landscape scale have been identified and characterized, they can be contrasted along the gradient using any parameter of interest (e.g. fertility, biodiversity, resilience). For example, biodiversity can be measured at local (alpha diversity) or landscape (gamma diversity) scales or as the variation across a landscape gradient (beta diversity; see Chapter 6). Disturbances generally increase spatial heterogeneity at multiple spatial scales, including landscapes (Wu and Loucks 1995; Halford *et al.* 2004).

The heterogeneity that we often take for granted at landscape scales is sometimes less apparent but still present at larger (e.g. biome, global) and smaller (e.g. stand, microsite) spatial scales. The choice of scale depends on the grain or coarseness in texture of an environment (large patch sizes create a coarse grain, small patch sizes create a fine grain). Additionally, an organism of interest experiences the environment at a particular grain. A deer responds to forest patches, gaps, and edges between the two, while a pollinating insect responds to flowers at a much finer grain. Ecosystem processes can also be

Figure 7.2   Cliff habitats along the north coast of Molakai, Hawaii. (Photograph by Peter Bellingham.)

examined at various grains. Watershed erosion or nutrient fluxes (Bormann and Likens 1979) address larger patches than the microtopography of a local slope (Scatena and Lugo 1995) or soil fertility under a shrub (Garner and Steinberger 1989). Spatial patterns are often similar, even at different scales, as in the dissected leaves of a fern or the intricate angles of a snowflake (fractals; Sugihara and May 1990; Milne 1991). Some disturbances (e.g. erosion, meandering streams, and pest outbreaks) can greatly increase the sinuosity of patterns, giving such habitats a higher **fractal dimension** (Forman 1997). A much better understanding of spatial dynamics can be achieved when the influences of patterns on the scale of primary interest are examined at coarser and finer scales using hierarchical approaches.

Hierarchical approaches to spatial scales provide a context for the scale of interest (focal scale). Small units such as individual leaves are nested within increasingly larger units such as trees, forest stands, ecosystems, landscapes, and biomes. Energy and matter flow horizontally within elements of each unit and vertically across scales (Forman 1997). Habitats within the same spatial scale can be affected by similar disturbances. The next higher scale provides a stable context and constrains the consequences of a disturbance at the focal scale. The next lower scale provides the mechanistic explanations for the variety of disturbance-triggered patterns and processes at the focal scale (Fig. 7.3; Pickett *et al.* 1999). A disturbance at the focal scale may not be noticed at the higher scale (that is measured on a coarser grain) but still provide part of the structure for the higher scale (O'Neill *et al.* 1986).

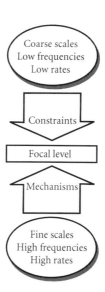

Figure 7.3   Each focal scale of interest is influenced by mechanisms at finer scales and constraints at larger scales. (From Pickett *et al.* (1999) with permission from Elsevier.)

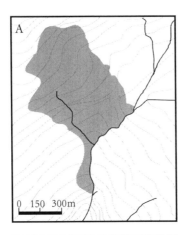

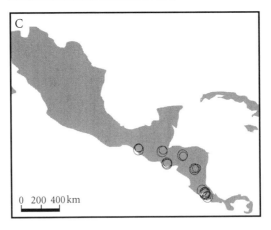

Figure 7.4   Landslide effects on the landscape. (A) A single landslide (shaded area) can be a major disruption at a local scale. (B) Landslides (dots) may affect less than 1% of the surface area of a watershed. (C) Landslides may be spatially insignificant at a regional scale (Central America). In (C), each circle represents about 100 landslides. (Adapted from Restrepo *et al.* (2009) with permission from the American Institute of Biological Sciences.)

For example, a 1000-m long landslide clearly has a local impact but may represent less than 1% of the area of a forested watershed and become a very minor component of an evaluation of landslides across Central America (Fig. 7.4; Restrepo *et al.* 2009). The largest (regional) scale constrains the influence of the watershed; the watershed's disturbance regime is explained by incorporating the sum of all smaller scale disturbances.

7.3 Disturbances create spatial heterogeneity

One of the defining characteristics of a disturbed habitat is its heterogeneity (Turner 1989; Pickett and Cadenasso 1995; Pickett *et al.* 1999). A disturbance varies in intensity, severity, frequency, and extent, and this variation results in a heterogeneous impact on the landscape. The heterogeneity applies to qualities of the resulting substrate (e.g. texture, fertility, and temperature), climate, hydrology, and newly formed (or surviving) biotic communities. Each recently disturbed habitat can be seen as a mosaic of disturbance severity, or a complex template upon which subsequent disturbances act and ecosystems respond. Later disturbances can accentuate differences in the mosaic (e.g. insects attack newly vulnerable trees that are physiologically stressed from a fire) or rearrange the mosaic (e.g. a flood deposits silt across a lava flow). Most biotic responses to the mosaic are affected by the gradients of physical conditions and resource availability. The long-term response of the biotic community (succession) begins on this heterogeneous template that is further complicated by on-going disturbances and increasingly complex species interactions. Some colonizing species decrease spatial heterogeneity (e.g. a widespread canopy dominant or a prolific grazing animal; Table 7.1)

Table 7.1 Conditions that lead to increases or decreases in spatial heterogeneity and typical habitats.

Conditions	Habitats
Increase spatial heterogeneity:	
Low evenness	Coral reef
Low competitive dominance	Alpine meadow
High rate of species invasions	Early successional weed patch
Selective, low-intensity grazing	Alpine meadow
Decrease spatial heterogeneity:	
High evenness	Salt marsh
Competitive exclusion	Shrub thicket
Low rate of species invasions	Old growth forest
Generalist, high-intensity grazing	Agricultural pasture

while others increase it (e.g. selective grazers or invasive species that form novel communities or promote disturbances such as fires). Such variable responses make predicting future patterns of spatial and temporal heterogeneity very unlikely, at least at local scales (Walker and del Moral 2003).

7.3.1 Natural disturbances

Spatial patterns are created by natural disturbances because of the variation in the characteristics of each disturbance and the complex responses of the biota to each disturbance. Different types of disturbance interact in a local disturbance regime. Disturbances, of course, operate at many spatial scales, from nutrient losses at microsites to global issues including atmospheric carbon dioxide levels. Despite the extensive influence of anthropogenic disturbances on habitats (about 90–95% of the earth's terrestrial surface), natural disturbances (particularly extensive and severe ones) continue to set the global stage, particularly with large, completely unregulated disturbances such as volcanoes and earthquakes. Natural disturbances also influence anthropogenically modified habitats, just as anthropogenic disturbances impact the few remaining natural areas. Many species adapt to a frequent disturbance regime. A few of the most common natural disturbances in terrestrial and aquatic habitats will now be discussed in the context of spatial heterogeneity.

Wind, fire, erosion, and animal activities are common terrestrial disturbances that create spatial heterogeneity (see Chapter 2), while floods, cyclones, and tectonic activities are common disturbances that create spatial heterogeneity in aquatic habitats (see Chapter 3). When high-velocity winds flow over a vegetated landscape, the resistance from the uneven vegetation creates a turbulent flow with eddies and other non-uniform patterns (Finnigan 2007). The damage caused by the wind is therefore uneven as well (Whigham *et al.* 1999). Trees and shrubs also provide variable resistance to wind damage (abrasion or loss of leaves and branches, trunk snapping or uprooting) based on stem and root morphology, position in a stand (e.g. leeward, windward, center), stabilizing root connections with other trees (Basnet *et al.* 1993), substrate stability, past damage, and landscape position of the stand (e.g. ridge, valley, slope) (Fig. 7.5; Quine and Gardiner 2007). So many factors determine damage to a forest that the resulting mosaic of damage (and response) is highly variable in space (Webb 1999).

The contributions of fire to spatial heterogeneity can be measured by variation in the fire regime and by variation in the responses of the burned plant communities. Fires are prominent features of boreal forests (Engelmark 1999), grasslands and savannas (Oesterheld *et al.* 1999), eucalyptus forests of Australia (Birk and Simpson 1980), and Mediterranean-type ecosystems (Rundel 1999), that include chaparral (California, U.S.), matorral (Chile), garrigue or maquis (Mediterranean Basin), fynbos (South Africa),

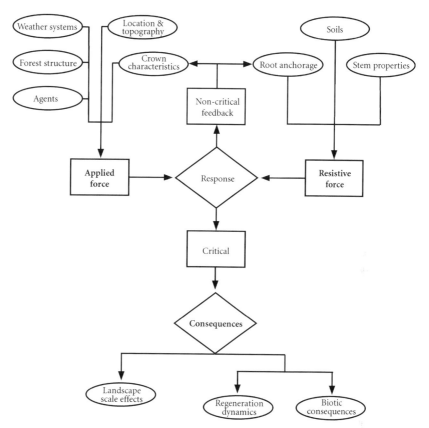

Figure 7.5   Interaction of wind and trees and the role of the landscape in determining the nature of the force, the resistance to it, and the consequences. (From Quine and Gardiner (2007) with permission from Elsevier.)

and kwongan (southwestern Australia). The fire regime varies in frequency, intensity, and severity. Variable intensity is driven by moisture, temperature, topography, and fuel quantity and flammability (Christensen 1985). Fires burning up hill move faster and are less intense than fires moving down hill. Live plant tissues are less flammable than dead ones, so when substantial amounts of dead branches and leaf litter accumulate, fires are most intense. Flammability is also inversely related to nutrient content, so low-nutrient shrubs decay slowly and increase the proportion of dead (flammable) tissues (Rundel 1981). Each of these physiological variables adds to the spatial heterogeneity. Damage from fires (severity) causes heterogeneity, with unburned patches adjacent to patches burned to varying degrees. Sometimes just the canopy or just the understory vegetation burns. Slow-burning fires can burn organic soils under some conditions. The result is the creation of numerous patches of differential damage across a burned landscape.

Erosion by water on unstabilized slopes accentuates depressions that can develop into rills, then streams with deeply incised banks, and finally even large canyons (Fig. 7.6, Box 7.1). Such dissection has a high fractal dimension, particularly as large canyons still have small, nascent rills and depressions at their headwaters. Older erosion channels that snake through a flattened landscape also add to the heterogeneity and fractal dimension. Some examples of sheet erosion may reduce spatial heterogeneity, but that process is rarely uniform across an uneven surface and erosion channels usually result. Erosion by wind can reduce spatial heterogeneity as on wind-swept barrens (e.g. desert playas; Vasek and Lund 1980) with few emergent features, but more often wind erosion interacts with rock textures and vegetation structure to create ridges, pockets, and other features that increase heterogeneity.

Animal activities are also surprisingly capable of creating spatial heterogeneity and are perhaps the natural disturbance for which severity is hardest to predict. Animals disturb their environments through movement, feeding, building, digging, burrowing, elimination, and death (Willig and McGinley 1999). Because each disturbance is generally of minimal spatial extent, each adds to small-scale spatial heterogeneity. Exceptions might occur when large animals such as cows homogenize the environment where they concentrate

Figure 7.6   A 10-m deep gully cutting through 2-year-old ash deposits on Kasatochi Island, Alaska (U.S.).

Box 7.1 The very early stages of soil erosion

Erosion is often too slow a process to appreciate, but on the newly deposited volcanic ash surfaces of Kasatochi Volcano in Alaska, erosion is rapid and dramatic. Immediately after the 2008 eruption that buried the pre-existing tundra vegetation and vibrant bird colonies in tens of meters of ash, the slopes were smooth; most salient features had been obliterated under the ash. Almost immediately, however, small erosion rills several centimeters deep began to appear on steep slopes and exposed ridges. Within a year, many of these had anastomosed into gullies several meters deep; within 2 years some cuts resembled small canyons. Cliffs formed at the edge of the beach erosion (Fig. 7.6) and the shoreline that expanded during the eruption gradually retreated. As persistent erosion in this rainy climate removes much of the newly added and still unstable ash—in fact, every rainstorm creates muddy streams—the original, more erosion-resistant volcanic rock will be exposed and erosion will slow. On a geological time scale, Kasatochi Volcano will eventually erode and subside until it is below sea level, a process that takes about 5 million years, as seen in the Hawaiian island chain.

(e.g. around water holes; Rusch 1992). Trampling is a disturbance created by the movement of large vertebrates and can alter water, gas, and root penetration of soils, litter decomposition, and nutrient cycling. By digging burrows, nests, or food caches, vertebrates aerate soils, destroy plant roots, and alter microbial populations (Kilham 1994). Earthworms, termites (see Fig. 5.6), and ants have similar effects (Bardgett 2005). The individual behaviors of animals normally result in increased spatial heterogeneity.

Common natural disturbances in aquatic habitats with significant impacts on spatial heterogeneity include floods, cyclones, and tectonic activity (see Chapter 3). Floods alter freshwater rivers (see Fig. 3.8) and lakes by mixing the water column, introducing sediments, flushing organisms downstream, scouring benthic surfaces, and altering dissolved gases and nutrient cycles (Dent and Grimm 1999). Each of these effects changes the spatial patterns of abiotic and biotic variables. Cyclones can have similar effects on oceans where they churn up the water column to a depth of several meters, break coral reefs, kill fish, and alter sediment loads, dissolved gases, and nutrients. Tectonic activities, including earthquakes and the secondary disturbances that they trigger, are perhaps less frequent than floods or high winds but can result in altered drainage channels for streams, submarine landslides and volcanoes, and other benthic movements including subsidence or uplift. Movements of sediments or rocks to or from an aquatic habitat clearly alter the spatial pattern. Unless a patch is destroyed, spatial heterogeneity is increased.

7.3.2 Anthropogenic disturbances

Humans have altered the spatial effects of disturbances by creating novel disturbances; by increasing the frequency, severity, and extent of disturbances; and by altering the matrix surrounding many patches and the connections among patches. Anthropogenic disturbances often have sharper, straighter **ecotones** than natural disturbances and present barriers to the movement of animals because of their drastic alteration of the environment. Frequent wood cutting in developed countries does not allow forest recovery. The severity of disturbances such as roads, channelized rivers, and dams, for example, provides often insurmountable barriers to animal movements (Forman and Alexander 1998), unless bridges (Underhill and Angold 2000) or tunnels (Cain *et al.* 2003) are constructed. Toxic wastes (see Chapter 4) also present a severe disturbance and can resist invasion, often persisting as uncolonized patches that do not easily blend with the matrix. Anthropogenic disturbances such as agriculture or urbanization (Fig. 7.7) have created extensive, new matrices over many terrestrial surfaces and now determine which colonists are available to disperse into subsequent patches (e.g. into gaps in remnant forests or urban parks). Similarly, over-fishing has led to a rise in some invertebrate populations (e.g. sea urchins) in many disturbed reef habitats. Another aspect of the altered matrix is the increasing proportion of invasive species that are often competitive colonizers of new

Figure 7.7   An urban park in Tokyo, Japan.

disturbances. Patch connectivity (see Section 7.5.2) is altered when disturbances remove habitat corridors among patches (e.g. roads isolate populations of bighorn sheep in the western U.S.; Epps *et al.* 2005) but disturbances can also re-create connections (e.g. among floodplain populations; Ward *et al.* 1999) that have been altered by invasion or succession.

Hotspots are patches that have distinctively high levels of nutrients, biodiversity, or other features that clearly distinguish them from the matrix. For example, heavily grazed areas in eastern Europe are biodiversity hotspots because the grazing reduces dominance by superior competitors (Cremene *et al.* 2005). Point sources of aquatic pollution (e.g. sewage effluent from a pipe) also provide nutrient hotspots (Grimm *et al.* 2008), but sometimes reduce biodiversity. Biodiversity hotspots created by humans are sometimes maintained by high-frequency, low-severity disturbances (e.g. mowing) but are particularly vulnerable because of their limited geographical extent and contrast with the matrix. Pollution sources constitute a disturbance, and because of their limited spatial extent can often be identified and ameliorated.

The small agricultural plots of traditional slash and burn agriculture or small-scale family farms increase spatial heterogeneity and provide many edges between natural and anthropogenic communities. Over time, the forest matrix became so decimated, however, that forests became patches in an agricultural matrix. In some places in the U.S. where farmers abandoned their land to seek more fertile soils (e.g. Vermont) or were drawn to urban lifestyles and jobs (e.g. Puerto Rico), the forest matrix is again dominant. Spatial heterogeneity (e.g. of plant cover) is arguably greatest in a landscape with a mixture of agricultural and forested patches, especially when monocultures and use of pesticides and herbivores reduce heterogeneity in agricultural patches.

7.4 Dispersal creates spatial heterogeneity

The dispersal of organisms into a disturbed habitat can be an important contributor to spatial heterogeneity in emerging communities. This heterogeneity arises from the many potential pathways that can unfold. For a particular propagule to arrive at a given disturbance, many variables must align (Fig. 7.8). First, a potential dispersing organism must be in a reproductive rather than dormant, juvenile, or senescent state. Second, the timing of its reproduction must match the timing of the disturbance. Third, abiotic variables, including climate at various spatial scales, must be favorable for dispersal. Fourth, additional disturbances cannot interfere with the dispersal. Finally, biotic variables such as seed predators cannot disrupt the dispersal. These contingencies plus the large number of potential dispersers and microhabitats provide multiple possible outcomes for each disturbance.

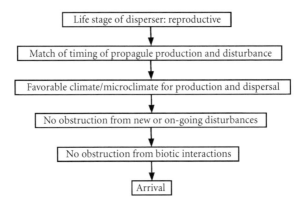

Figure 7.8   Biotic and abiotic variables that must coincide for a propagule to arrive at a new habitat created by a disturbance.

Dispersal can also reduce heterogeneity (e.g. a clonal dominant invades a previously heterogeneous habitat).

Marine organisms that have planktonic transport of larvae can disperse further than terrestrial organisms by several orders of magnitude, up to several hundred kilometers in some cases (Kinlan and Gaines 2003). In contrast, most seed-bearing terrestrial plants (with the exception of light-seeded species such as orchids), spread primarily by diffusion (short-distance dispersal), while long-distance (jump) dispersal is uncommon (see Section 1.6.1). This predominantly short-distance dispersal pattern results in a **leptokurtic distribution** of propagules for most organisms (Lewis 1997). Yet survival of propagules is often poor under a parent plant because of intraspecific competition for resources, increased predation, and disease (Wang and Smith 2002; Howe and Miriti 2004; but see Hyatt *et al.* 2003), so this dichotomy might contribute to a zone of maximal regeneration at an intermediate distance from the parent (Fig. 7.9). Because this pattern is modified by parent size, species-specific dispersal distances, and numerous abiotic factors (e.g. wind patterns), spatial heterogeneity of species distributions, and thus local to regional species diversity, are likely increased (Lewis 2010).

One approach that avoids the dichotomy of intraspecific competition where seeds are most dense is vegetative reproduction. Vegetative expansion can also contribute to spatial heterogeneity. In many environments, including highly disturbed ones, clonal rings expand away from the original center of establishment, often decaying in the center to form a donut-shaped pattern (Adachi *et al.* 1996a; Wikberg and Svensson 2003). Fairy rings of mushrooms present a similar spatial pattern. Sometimes other species can colonize this center, as was found for lupine (*Lupinus*) patches on Mount St Helens (del Moral 1993). Another form of vegetative spread occurs on unstable slopes or in windswept habitats where growth occurs on the downslope

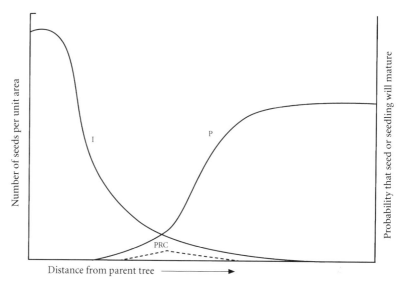

Figure 7.9   Theoretical dispersal distance, germination success, and regeneration zone for many plant species: I = number of seeds per unit area; P = probability of escaping predation and maturing; PRC = population recruitment curve where establishment is most likely. (From Janzen (1970), with permission from the University of Chicago Press.)

or downwind side. Dune communities develop in protected swales (Boorman *et al.* 1997); **solifluction lobes** on tundra slopes support vegetation adapted to higher nutrient levels than undisturbed tundra (Fig. 7.10; Price 1971); desert shrubs expand by splitting along their axes (Schenk 1999); and **krummholz** growth forms develop under windy conditions (Weisberg and Baker 1995). When species do regenerate under the parent plant (e.g. either sexually or vegetatively), spatial heterogeneity can be reduced (e.g. deer herbivory led to declines of forest herbs and increases in *Fagus* tree root sprouts; Carson *et al.* 2005). The variety of growth forms and dispersal modes, the nature of their repetition across the landscape, and their response to the ongoing disturbance regime create a diverse array of spatial patterns in a natural landscape.

A second dispersal dichotomy occurs because long-distance dispersal conveys an advantage to species dispersing into large, disturbed habitats (e.g. with light, wind-blown spores or seeds; Fig. 7.11), yet these species are often not good colonizers or good competitors (Wood and del Moral 1987). However, when a variety of dispersal strategies and competitive abilities are present, spatial heterogeneity in species is maximized (Amarasekare and Nisbet 2001). In primary succession, there may be a lag before self-sufficient species arrive and the impact of the early colonists may be minimal (Walker and del Moral 2003) or at least highly localized (Sikes and Slowik 2010).

Figure 7.10   Solifluction lobes and associated vegetation create stripes across the slope of a volcanic crater in northwestern Argentina. Solifluction, or downward flow, often occurs when surface soils move over more stable, frozen substrates.

Plants that have N-fixing or mycorrhizal symbionts may have heavier seeds (Chapin 1993) or not arrive with their symbiont (Titus and del Moral 1998; Parker 2001). Plant species that do colonize quickly include light-seeded, annual forbs and perennial forbs or grasses, with the perennial plants having more influence on spatial patterns over time (see Chapter 8). Some non-native, invasive species (see Chapter 6) are successful because they not only disperse widely and rapidly but also fill vacant niches such as the N fixation niche in a low-N, volcanic environment (Vitousek and Walker 1989) or a fern that climbs trees and smothers them (Robinson *et al.* 2010). These new species thereby create novel spatial patterns.

A third dispersal dichotomy occurs when both plant and animal species that survive a disturbance (biological legacies) are not adapted to colonize the newly disturbed habitat that may surround them, particularly if resources are severely depleted (e.g. severely burned patches, landslides, lava flows). Legacy communities may serve as important nuclei for expansion into the disturbed parts of a habitat once the abiotic environment is ameliorated (Yarranton and Morrison 1974; Talbot *et al.* 2010), but that process can take decades when the newly disturbed environment is not favorable. Where disturbances are frequent, legacy communities might represent early

Figure 7.11   Fireweed (*Epilobium*) seed dispersal in Alaska.

successional communities adapted to the post-disturbance environment and spread quite readily into the newly available habitat (Clarkson and Clarkson 1983). Alternatively, legacy communities may facilitate establishment of pioneer species suited to the new environment (Fuller and del Moral 2003). Each of these three dichotomies involving dispersal and colonization contributes to the complexity of spatial (and species compositional) patterns in community responses to disturbance.

Dispersal occurs through a variety of mechanisms. In both aquatic and terrestrial ecosystems, passive, short-distance dispersal of spores or seeds by gravity or by currents of water or wind is common. Although passive, long-distance dispersal is less common; when it occurs it reduces the predictability of spatial outcomes. Even species capable of long-distance dispersal rarely travel more than several meters, but only a few individuals have to disperse and establish at a long distance for an invasion to proceed (Clark *et al.* 1999). For example, seeds with plumes dispersed an average of 20 m in North American prairies (Platt and Weis 1977) and seeds on desert flats (**playas**) traveled over 700 m (Fort and Richards 1998). Other mechanisms of plant dispersal include active dispersal (e.g. exploding seed capsules) and passive and active dispersal by animals. Animals inadvertently transport terrestrial plants that cling to them and purposefully move seeds to caches or defecate seeds of fruits that they have eaten. Patterns of animal dispersal can be predicted only in the broadest terms (e.g. along bird migration routes or frequented animal trails) but occasional deviations can result in even more variation, even among normally wind-dispersed plants (Wilkinson 1997). Animals are also dispersed both passively and actively. Plankton (in water or air) are transported by currents and can be widespread. Heterogeneity of abundance within a species declines with time as slower dispersers fill in the new space (Reed *et al.* 2000). Eggs, larvae, and adults can also be transported by water or wind currents. Crickets and spiders, for example, are common colonists of new volcanic surfaces and often feed on each other until plants arrive and more complex food webs develop (Cheng and Birch 1987; New and Thornton 1988; Greenslade 1999). Vertebrates disperse by walking, swimming, rafting, or flying. Vertebrate dispersal often is linked to the availability of favorable habitats to support them during dispersal (Walker and del Moral 2003). Lizards or rodents arrive at remote islands because they are transported on mats of vegetation that provide food and shelter. Bat and bird dispersal can also be limited by food and shelter requirements, much as bird migrations require adequate stopovers. Therefore, animal dispersal is linked spatially to plant community distributions.

The nature of the disturbance clearly affects dispersal. An extensive and severe disturbance may reduce the regional propagule pool and limit colonization to long-distance dispersal. The timing of the disturbance relative to the phenology of the organisms is also critical and will determine survivors and subsequent colonists (Pickett *et al.* 1987b; Walker and del Moral 2003). Gophers survived the Mount St Helens eruption because they were in their burrows when the eruption occurred (Allen and MacMahon 1988). Tens of thousands of bird chicks probably died in the 2008 eruption on Kasatochi Island in Alaska because they were too young to be able to fly away, unlike most of the adult sea birds that survived (Williams *et al.* 2010). For plant succession following the abandonment of agricultural fields, the season of abandonment (Keever 1979) and the last crop sown prior to abandonment (Myster and Pickett 1994) will influence succession.

In addition to dispersal abilities and the timing and severity of disturbance there are other barriers to dispersal (Fig. 7.8). These barriers include abiotic filters (e.g. water or wind for terrestrial organisms; toxic or sediment plumes for aquatic organisms) and biotic filters (e.g. predation, lack of required animal dispersers). The species that pass the various barriers to become the first colonists of a disturbed habitat typically combine diffusion and jump dispersal and have multiple dispersal mechanisms (Clarkson and Clarkson 1983, Clark 1998). Early colonists are therefore not typical representatives of the source populations, a condition called disharmony (Carlquist 1966).

Dispersal is one of the least predictable contributors to spatial heterogeneity, and there is still much to learn about dispersal of specific organisms, particularly for those in remote locations such as limpets on deep-sea hydrothermal vents (Craddock *et al.* 1997). Chance events such as long-distance dispersal can have long-term effects on the patterns of community development. Wide environmental fluctuations can help determine if, when, and where early arrivals survive. As succession proceeds (see Chapter 8) and a disturbed habitat begins to fill with vegetation, the environment becomes modified, there are fewer available sites for colonization, and plant–environment interactions become more predictable, including those governing spatial patterns (Walker and del Moral 2003; Walker *et al.* 2006b). However, more potential colonists compete for fewer sites, and this larger species pool can decrease predictability. Refugia where legacy organisms remain can influence the disturbed matrix around them under favorable conditions where they can colonize it. Under less favorable conditions, refugia will have less effect (Fuller 1999). Dispersal and colonization of animal populations follow similar patterns as they are closely dependent on plant structure for their habitat requirements. In general, predictability of species composition (and therefore spatial heterogeneity) decreases as disturbance frequency, intensity, and extent increase (Turner *et al.* 1989). Dispersal, and its influence on species composition and diversity, is just one of several **stochastic** factors that determine spatial heterogeneity and interact through successional time.

7.5 Patch dynamics

Patch dynamics is the study of relatively discrete spatial patterns or patches (Fig. 7.10) and how they interact with other patches and the matrix that surrounds them (White and Pickett 1985). A matrix is the most spatially continuous element of the landscape mosaic. Patch dynamics does not imply equilibrium of patch distribution in time and space as does the concept "shifting mosaic" (Bormann and Likens 1979). Patches are defined operationally to match the spatial and temporal scales of interest and interact with disturbances at all ecologically relevant scales. Patches present an alteration of structure and resources compared with the matrix and therefore increase spatial heterogeneity. Patches are not always sharply defined, but instead can be indistinct, as when created by diffuse alterations to a habitat. Examples include the dilution of sediments entering the ocean from a river delta or the loss of foliage from a wind storm or a drought. The sharpness of a patch edge (ecotone) is a gradient from sharp to fuzzy, and this range of conditions can itself contribute to spatial heterogeneity.

7.5.1 Creation of patches

Disturbances create patches by the removal of biomass or other alterations of ecosystem structure unless they are diffuse disturbances. Yet even diffuse disturbances can be important at finer scales, as when a branch of a tree or a piece of a coral reef creates a patch for smaller organisms. Patch characteristics, then, depend on the initial abiotic and biotic status of the ecosystems, how they are damaged, and how they respond to a particular disturbance. Patches that are created during one disturbance will likely differ in characteristics from those created later, because the disturbance and the patch response to the disturbance will be altered. The existing network of previously disturbed patches across a landscape can promote or resist disturbances and increase their rate of spread or retard it. For example, a forest damaged by a hurricane may be more susceptible to fire in areas of high disturbance severity but more resistant to fire where damage is minimal (Whigham *et al.* 1991). One model suggests that disturbance frequency is the most critical factor affecting the spread of disturbance across a heterogeneous landscape when patches susceptible to disturbance cover less than half the landscape, and that disturbance intensity is more important when susceptible patches are more widespread (Turner *et al.* 1989).

7.5.2 Connectivity of patches

Patches exchange energy and matter with their surrounding matrix and nearby patches. The nature and rhythms of these exchanges are as variable as the patches themselves. Connections between patches can be beneficial (e.g. providing dispersers, pollinators, mates, and genetic exchange) or not (e.g. transporting diseases and invasive species). In Florida (U.S.), the non-native *Lygodium* fern, for example, connects forest patches across open water, allowing forest fires to cross from one patch to the next (Robinson *et al.* 2010). Sharp patch boundaries generally deter exchanges (e.g. a paved roadway for a desert lizard population) while fuzzy edges typically promote exchanges with the matrix and adjoining patches (e.g. a broad transition between two vegetation types on a gradual elevation gradient). However, too heterogeneous a transition between patches can also decrease the permeability of the patch mosaic.

Corridors connect patches through a background matrix and often serve as important channels for organisms. Corridors, in turn, can have distinct or fuzzy edges with the matrix and be relatively fixed (e.g. a wildlife bridge over a highway) or ephemeral (e.g. a seasonal river in a desert). It is also difficult to generalize about the influence of patch connectivity on the spread of disturbance (Pickett *et al.* 1999), as homogeneous habitats might increase the spread of fire or diseases by providing a steady supply of fuel or prey, but a relatively inflammable or pest-resistant community would resist the same disturbance.

Metapopulations encompass the interactions of populations of a species in various patches across a landscape and generally include relatively ephemeral satellite populations and a more stable source population. Metastability (Hanski 1995) can occur when at a given scale there is stability in the status of an ecosystem, even when the component parts are in flux. For example, total regional (gamma) diversity can remain constant while local (alpha) diversity fluctuates in response to local disturbance regimes.

Recurring disturbances can disrupt or promote patch interactions. Disruptions could occur when corridors needed for crossing between patches are blocked and impede migrating animals, dispersing plants, or the flow of nutrients. For example, river dams stop fish from migrating upstream and sediments from flowing downstream. Some disturbances improve the environment required for connections among patches. For example, natural flooding or the construction of wetlands and reservoirs improves connectivity for migrating aquatic birds (see Box 4.2).

Connectivity occurs in three-dimensional space, and the type and degree of connectivity as well as what is transported can easily vary from aboveground to the soil (Bardgett and Wardle 2010) or the water column to the hyporheic zone of a stream (Valett *et al.* 1994). Therefore, connectivity derives ultimately from the sum of all abiotic and biotic transfers among patches; connectivity is strongly influenced (slowed, promoted, or neither) by the structure of the edge environment (Pickett and Cadenasso 1995).

7.5.3 Duration and loss of patches

Patch duration is influenced by internal ecosystem resistance and resilience and by the frequency and severity of the regional disturbance regime. The temporal dynamics of patches are considered in detail in Chapter 8. Resistance is a function of both the biotic composition of the patch and its physical characteristics. Plant communities that resist successional change tend to extend the duration of a patch (e.g. long-lived thickets; Walker and del Moral 2003) while successional shift in species composition (and resulting structure) removes a patch and replaces it with a different patch. Patches can become less distinct by disturbances that blur patch edges or disappear entirely when colonists from the matrix invade the patch. The disturbance regime also influences the duration of a patch. Extensive, severe disturbances can remove all patch structure, although there is often a part of the original patch structure that survives. For example, cutting down a forest does not immediately alter any soil patterns that existed, perhaps as a result of differential effects of tree species on the soil (Binkley and Giardina 1998). Such soil patterns (fertile soil patches under shrubs) persisted for decades after bulldozers created dirt roads in the Mojave Desert (U.S.; Bolling and Walker 2002). Patches are less resilient following severe than mild disturbances; anthropogenic disturbances are often more severe than natural ones.

Patch duration is relative to the life history of the focal species. Ephemeral patches might be considered as those that last only a small fraction of the life span of the occupying organisms (e.g. hourly periods of nectar availability on a flower relative to the weeks-long life span of a visiting insect). Alternatively, a patch of long duration outlasts the organism of concern (e.g. fertile soil patches in a forest will likely outlast even long-lived *Sequoia* trees) and can therefore be considered a stable feature of that organism's environment. Ultimately, the disturbance regime is controlled by climate change, geomorphic change, invasive species (e.g. a species that is highly flammable increases fire frequency), and humans (e.g. global warming is altering the frequency and severity of many disturbances). The dynamic creation and loss of patches maximizes biotic interactions and species diversity.

7.6 Examples of heterogeneous ecosystems

7.6.1 Soils

Spatial heterogeneity in soil gases, nutrients, acidity, and organisms (see Chapter 5), is generated by both physical and biotic factors, including the prevailing disturbance regime. Geological composition influences patterns of soil development by differential textures, chemical composition, and weathering rates. Uplift and subsequent erosion of the geological strata provide the template upon which shorter-term processes of nutrient cycles, plant growth, and litter deposition lead to soil development (see Fig. 5.2). The spatial patterns of plant colonization and subsequent replacement (succession) are influenced by the variable substrate conditions, as well as climatic conditions, organismal interactions, and the disturbance regime. Plants and animals also modify spatial patterns of the abiotic and biotic features of the soil. The influence of all these factors on the spatial heterogeneity of soil organisms, for example, can be interpreted at various hierarchical levels (Fig. 7.12; Ettema and Wardle 2002). At the scales of the smallest soil organisms (microflora and microfauna; see Chapter 5), bacteria and fungi partition the soil by their relative tolerance of soil acidity, their facility to digest lignin, and other physiological and life-history differences (Bardgett and Walker 2004; Bardgett 2005). Steep gradients of abiotic variables such as concentrations of oxygen, carbon dioxide, or N can influence the spatial heterogeneity of the microflora and microfauna across distances of only a few microns. Such gradients might occur in water films on the surface of soil particles and be affected by very small-scale disturbances. Nematodes and mites are highly specialized in feeding and microsite use, and occupy several trophic levels (Schneider *et al.* 2004; see Fig. 5.7). The spatial distribution of mites, collembola, and other mesofauna or macrofauna (e.g. earthworms, beetles, snails) is determined by nutrient-rich pockets from decomposing

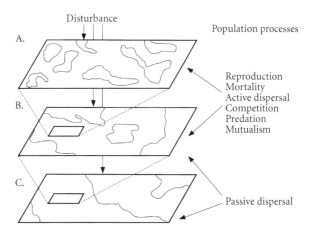

Figure 7.12   Determinants of spatial heterogeneity of soil organisms include a spatial hierarchy of environmental and biological factors as well as disturbance. (From Ettema and Wardle (2002), with permission from Elsevier.)

litter or animal carcasses as well as by various chemical gradients that occur in such places as the root zone at scales of millimeters to centimeters. Disturbances at this scale are easily visualized and include litterfall, burrowing, trampling, herbivory, and animal death. Soil organisms respond in a variety of ways to both biotic and abiotic features of the soil that define the niche for each species and create spatial heterogeneity (Siepel 1996b; Bardgett 2005).

Both the physical and biotic heterogeneity found in soils influence plant and animal populations and also communities at larger spatial scales such as landscapes and biomes. Landscape-level disturbances such as agriculture generally reduce the diversity (Bardgett 2005) and presumably the spatial heterogeneity of soil organisms through vertical soil mixing, the addition of fertilizer, reductions in the diversity of organic inputs, and growth of monoculture crops (Beare *et al.* 1995, 1997). Disturbances can also increase spatial diversity and spatial heterogeneity of soil organisms adapted to the disturbance. A good example of this is in riparian zones where nematode populations aggregated in clusters (detected at about 10–30 m scales) where moisture conditions were favorable (Ettema *et al.* 2000). Populations were reduced by irregular flooding, drought, and erosion but repeated recolonization of favorable habitats created a complex spatial mosaic.

7.6.2 Shorelines

Shorelines present a wide array of spatial patterns that are directly attributable to disturbance (see Fig. 7.2). Meandering river floodplains have dynamic shorelines (Fig. 7.13). Marine shorelines (see Chapter 3) (both aquatic and terrestrial components) can be grouped by substrate. Hard substrates (Fig. 7.14) are typically dominated by coral reefs in the tropics (often with

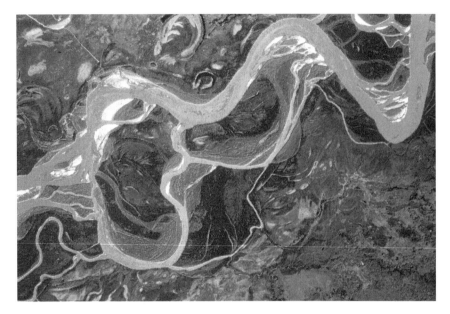

Figure 7.13   A 10-km stretch of the Tanana River floodplain 30 km southwest of Fairbanks, Alaska (U.S.). Note multiple previous river channels and dynamic shorelines. (From a NASA image, NASA-JSC386, Roll 23, Frame 117.)

Figure 7.14   A rocky shoreline in Bali, Indonesia.

sandy shorelines) and kelp communities (with rocky beaches) in temperate regions. Soft substrates can have seagrass beds in the water and salt marsh plants or mangroves on shore. Interactions of marine organisms on rocky, temperate shorelines have been well studied, and there is a growing body of literature on the role of disturbances in these ecosystems (Sousa 1985; Menge and Sutherland 1987; Farrell 1991; Bertness *et al.* 2002), although the explicit role of disturbance in structuring shoreline communities is not frequently considered (McKee and Baldwin 1999).

Abiotic disturbances that impact the heterogeneity of shoreline communities include the excess (flood) or absence (drought) of water, burial by sediments (see Chapter 4) or dead plants, substrate erosion from waves or river discharge, breakage by wind or waves, and fire. Biotic disturbances include grazing, predation, and bioturbation (e.g. burrowing, nesting, animal trails; McKee and Baldwin 1999). Herbivores such as sea urchins maintain a zone cleared of algae around their shelters (Andrew 1993). Floods can physically disrupt the terrestrial biota along the banks of erosion channels and immediate offshore environs, but leave other parts of the shoreline undisturbed, thereby increasing spatial heterogeneity. River floodwaters can further increase heterogeneity by the variable susceptibility of offshore substrates (e.g. due to slope, particle size, or interstitial spaces; Townsend *et al.* 2003). Drought can cause clustering of terrestrial and intertidal organisms around wet depressions in the topography (Duever *et al.* 1994). Sediments or mats of dead seaweed or marsh plants can bury and sometimes kill shoreline organisms, thereby opening up habitat for colonization in the new sediments or when the mats decompose or are displaced (Valiela and Rietsma 1995, van Hulzen *et al.* 2006).

Breakage from wind and waves can occur during storms and tsunamis and break coral reefs, disrupt benthic organisms and defoliate, break, or uproot trees (Hughes 1994; Baldwin *et al.* 1995). Breakage patterns depend on the intensity and direction of the disturbance and the resistance of the organisms and usually leave a more spatially heterogeneous habitat than the pre-disturbance one. Fires can displace animals, kill plants, and increase nutrient-rich run-off and sedimentation into the marine environment. Each of these disturbances generally adds to the complexity of shoreline patches, directly through its initial impact and over time through the variable stages of response and recovery of the shoreline biota (McKee and Baldwin 1999).

7.7 Conclusions

Disturbances generally increase spatial heterogeneity because the disturbance itself is variable and it differentially affects the environment. There are many ecological consequences to spatial heterogeneity, including changes in productivity, biodiversity, and succession. Hierarchical analysis of spatial scales helps focus on the scale of interest. Clarifying the species of interest is also critical to know the scale at which to examine disturbance effects. Dispersal to a disturbed habitat is another factor that adds to spatial heterogeneity. The extent of the disturbance determines the dispersal distance, and the complex life histories of dispersing organisms and variety of dispersal modes and mechanisms will determine when and how propagules are dispersed. Several dichotomies add to spatial complexity, including that the zone of maximal dispersal (e.g. under the parent tree) is often the worst place to germinate, good dispersers are not always good colonizers, and nearby survivors are also not always good colonizers of the newly opened substrate. Chance is also an important factor in dispersal, from rare but important long-distance dispersal, to the distribution of many species among a limited number of safe sites.

Patches differ from the surrounding matrix and their distribution is a measure of spatial heterogeneity. They are created by differential effects of multiple and repeated disturbances and alter a landscape's susceptibility to further disruption. How they are connected with each other or external resources is vital for the organisms that rely on the patch structure. Patches are often temporally unstable and their duration affects which species can survive in them and how disturbances alter them. Patches created by natural disturbances tend to have high edge to volume ratios with an uneven disruption of biota within the patch. Fire, animals, and floods create particularly heterogeneous landscapes. Anthropogenic disturbances, compared with natural disturbances, generally create sharper, straighter ecotones; more uniformly damage the matrix; more severely disrupt natural corridors; and have critical secondary effects (e.g. roads affect microclimates well beyond their edges). Spatial heterogeneity can increase with either natural or anthropogenic disturbances, however, and is often greatest where there is a mix of both types of disturbance.

Some environments, including soils and shorelines, have particularly high levels of spatial heterogeneity. Soils are a matrix of solids, liquids, gases, nutrients, and organisms that are frequently altered by disturbances and display steep gradients of abiotic variables. Numerous soil organisms occupy this complex set of niches and respond variously to the multiple disturbances that soils experience. Shorelines are also spatially heterogeneous. Multiple aquatic communities inhabit the various substrates and zones of water depth and distance from the shore.

8 Temporal dynamics

8.1 Introduction

Disturbance disrupts ecological patterns and processes and triggers responses in affected habitats that continue for some time, often until the next disturbance adds its influence. Habitats can therefore be seen as being in a perpetual state of response to the most recent disturbance. Disturbances are viewed as largely of external origin (allogenic), disruptive, and stochastic. This non-equilibrium approach to temporal dynamics has been in creative conflict with a more equilibrium-oriented approach since the formal beginnings of ecology as a scientific discipline in the late 19th and early 20th centuries (McIntosh 1985). The equilibrium approach emphasizes the more stable periods of ecosystem recovery between disturbances and views drivers of ecosystem change as largely internal (autogenic), integrative, and deterministic. Ecological responses to disturbance are clearly driven by a blend of disruptive and integrative factors, and careful observations and descriptions of the relative importance of each (at a given site or more generally across many sites) provide a pragmatic and balanced approach (Walker and Chapin 1987). Temporal dynamics are as omnipresent and integral as spatial dynamics to the study of disturbance ecology.

Temporal change, like disturbance and spatial dynamics, occurs at many scales. Ecological change typically encompasses scales involving organisms to biomes. Organismal responses to daily and seasonal disturbances (short-term fluctuations) reflect biochemical and physiological traits and do not usually involve a change in species composition (Pickett *et al.* 1987a; Glenn-Lewin and van der Maarel 1992). Succession is the study of the replacement of one group of species by another in the context of population, community, and ecosystem responses to disturbance over time scales of medium-length, from years to centuries. Note that if succession is more strictly defined as directional species replacement, many disturbances do not initiate succession because either biological legacies persist (Platt and Connell 2003) or species accumulate very slowly but do not get replaced (Walker and del Moral 2003). Recent work has extended some aspects of successional studies to millennial time scales (Peltzer *et al.* 2010), although such long time scales are also the context for soil formation, palynology, and other studies of vegetation history (see Fig. 1.3). These divisions into short-, medium-, and long-term time scales are arbitrary and overlap. For example, what might appear to a long-lived organism (e.g. a *Sequoia* tree or a whale) as a short-term fluctuation in climate or species composition can appear as successional change for shorter-lived species (terrestrial herbs or seagrasses). A useful temporal scale for succession is therefore one to ten times the life span of the species. This approach recognizes that succession can occur in bacterial communities in minutes, insects in weeks, herbs in years, mammals in years to decades, forests in decades to centuries, and soil properties in centuries to millennia (Walker and del Moral 2003; Walker *et al.* 2010b).

This chapter focuses on successional change in response to disturbances (years to centuries), with less emphasis on fluctuations or very long-term change. First, I examine the rich body of theory that has developed over the last century. Then, I consider various methods used to study succession, followed by details about establishment and persistence on newly disturbed substrates, species interactions among established populations, successional trajectories, and applications of successional knowledge. Examples will be drawn throughout the chapter from primary and secondary succession (species change following more severe and less severe disturbances, respectively) that occurs in both terrestrial and aquatic habitats.

8.2 Theory

Successional responses to disturbances are readily visible in the growth of weeds in agricultural fields and the recovery of forests after fires and coral reefs after cyclones. Successional change is therefore something immediate and tangible. Humans evolved in fertile habitats that owed their fertility in part to recurring disturbances (e.g. volcanic ash). Even today, farmers face periodic evacuations of their land when faced with volcanic eruptions. Since its beginnings, agriculture (and to some degree fishing) has relied on knowledge about recovery following harvests, diseases, infertility, and climatic shifts. From these early experiences, the formal study of succession developed (Table 8.1; Walker and del Moral 2003). Beginning around 1850, naturalists began to write not only about the practical aspects of land management but about other changes in environments such as floodplains (Reissek 1856), abandoned farms (Thoreau 1860), and dunes (Warming 1895). By the beginning of the 20th century, ecology emerged as a discipline in its own right from its origins in natural history, agriculture, and physiology (Clements 1928). Succession was a central theme during the establishment of the field of ecology and remains so today (Prach and Walker 2011). Margalef (1968) maintained that succession is as critical to ecology as evolution is to biology.

Table 8.1 A century of terms related to disturbance ecology and succession, arranged by decades when they first received substantial attention. (Modified from McIntosh (1999) and Walker and del Moral (2003); see Glossary and text for definitions of some of the terms.)

1900	Association, climax, chronosequence, convergence, disturbance, dynamic, equilibrium, eutrophication, pioneer, progressive, retrogressive, stability, zonation
1910	Biome, competition, ecesis, **hydrarch**, individualistic, migration, mosaic patch, nudation, primary, reaction, secondary, **xerarch**
1920	Allogenic, autogenic, biogeochemistry, gap dynamics, holism
1930	Ecosystem
1940	Energy, trophic dynamics
1950	Continuum, cybernetics, gradient analysis, holist, initial floristics, thermodynamics
1960	Computer models, keystone species, r- and K-**selection**, stable state, strategy
1970	Assembly rule, facilitation, inhibition, intermediate disturbance, nutrient retention, resilience, resistance, tolerance
1980	Resource ratio, river continuum
1990	Complex adaptive systems, complex ecology, ecological law of thermodynamics
2000	Ecosystem services, ecological stoichiometry, novel ecosystems, regime shifts, tipping points

Part of the centrality of succession to ecology is that temporal change is ubiquitous, but another aspect is that succession, despite over 100 yr of close examination, remains poorly understood and is therefore an on-going topic of speculation and debate. Studies of succession emphasize either that disturbed ecosystems approach equilibrium in a deterministic way (holism) or that ecosystems are constantly in flux, as species adjust to on-going disturbances (reductionism). Holism (Patten and Jorgenson 1995) emphasizes autogenic processes, integration, mutualism, and predictability, while reductionism (Glenn-Lewin 1980) emphasizes allogenic processes, disruption, migration, and stochasticity. This dichotomy (and the many approaches that fall somewhere along the continuum between the extremes) has provided an exciting and challenging framework for examining the complexities of temporal change (McIntosh 1985). Even today, the debate continues about whether integrative or disruptive drivers prevail, particularly in light of the relative importance of **facilitation** and competition in community assembly (Callaway and Walker 1997; Brooker *et al.* 2008; Bulleri 2009).

8.2.1 Holism

Clements (1916, 1928, 1936) is largely responsible for the strong temporal emphasis in ecology. His writings (1904–36) provided a comprehensive framework for examining the successional responses of communities after disturbance. He also considered the applied value of ecology, linking his insights to issues of range management and fire ecology; ecology has maintained this practical side (Cook 1996; see Section 8.7).

Table 8.2 Clementsian processes that drive succession and their modern analogs. (From Walker and del Moral (2003) with permission from Cambridge University Press.)

Clemenstian processes	Modern analogs
Nudation	Allogenic disturbances, stochastic events
Migration	Life-history characteristics: dispersal
Ecesis	Life-history characteristics: establishment, growth, longevity
Competition	Competition, allelopathy, herbivory, disease
Reaction	Site modification by organisms, facilitation
Stabilization	Development of climax

Clements carefully documented many types of disturbance and then proceeded to organize what he considered the largely predictable responses to it. He considered succession to be a six-part process (his terms follow in parentheses). First, disturbances denude a surface, creating bare land (primary succession) or at least dramatically altered resources (secondary succession) such as light or fertility (**nudation**). Second, organisms disperse to the disturbed surface (migration). Third, organisms establish successful, reproductive communities (ecesis). Fourth, competition ensues among the newly established organisms and begins to shape structural hierarchies (competition). Fifth, organisms modify the environment, often in ways that do not improve the probability of their own regeneration but that do facilitate colonization by the next suite of colonists (reaction). Finally, the whole system reaches a stable, self-reproducing climax stage (stabilization). Remarkably, these basic processes still underlie most successional studies, probably because they provide a logical framework to delineate the disturbance and a well-ordered temporal framework for measuring biotic responses (Walker and del Moral 2003). Modern modifications (Table 8.2) include the recognition of the often stochastic nature of disturbance (Halford *et al.* 2004); the role that life-history parameters have in influencing species interactions (Walker *et al.* 2003); the importance of other variables that influence species interactions (e.g. herbivory, disease, and **allelopathy**; Pickett *et al.* 1987b); the complex, two-way interaction between site factors and organisms (Bardgett and Wardle 2010); the dynamic balance between competition and facilitation (Callaway and Walker 1997); and a reduced emphasis on the climax stage (DeAngelis and Waterhouse 1987).

Clements believed that within a given climatic zone, largely autogenic processes would eventually drive succession to a predictable, stable, self-replacing climax stage (Clements 1936). When this climatically determined climax was not evident, a plethora of terms was added to the lexicon to explain each case (e.g. a disclimax was a system that was prevented from reaching climax because of a recurring disturbance such as fire; a post-climax described situations such as relict communities on uncommon soil types in an area of shifting climate such as trees on the edge of a prairie). The expanding terminology convinced some observers that the climax concept was not particularly useful (Egler 1951; Colinvaux 1973). Modern examples of this phenomenon include the overuse of the prefix "eco-" (Wali 1999) and multiple modifiers of the terms competition and facilitation (Walker and del Moral 2003). Clements further estranged his followers by comparing community development during succession to organismal development, where the climax was a super-organism, analogous to the adult stage of an organism (Clements 1936). Nonetheless, Clements's legacy is a solid yet flexible framework that recognized disturbances and the temporal response of biotic communities to them as central to an ecological understanding of our environment.

Modern echoes of Clementsian holism are found in the sometimes uncritical acceptance of a climax (Oosting 1948); predictable ecosystem-level consequences of succession (Odum 1969); the emergence of systems ecology (McIntosh 1985); the application of cybernetics to suggest that a climax stage has minimized outside influences (Margalef 1968); complex adaptive systems (Brown 1995); and thermodynamics (Jørgensen 1997). Clementsian holism remains a vital part of disturbance and succession theory, while experimental work has largely adopted a reductionist approach (Walker and del Moral 2003). Modern ecosystem ecology blends the holistic view of an ecosystem as something larger than the sum of its parts with a reductionist acceptance of the role of stochastic disturbances, shifting patches of resources and species, and the pivotal role that some species such as N fixers can have on succession (Vitousek and Reiners 1975; Bormann and Likens 1979; Fisher *et al.* 1982; Hagen 1992).

8.2.2 Reductionism

Cowles (1901), Warming (1909), and Gleason (1917) were early critics of the holistic approach of Clements. They emphasized instead the variation in responses of individual species to disturbance and differences among ecosystem responses across a landscape. Rather than emergent properties (that arise from the sum of all interacting organisms) and deterministic processes, they saw independent responses of organisms and unpredictable patterns. Gleason, in particular, dismissed Clements's organismal analogy and argued that predictable, permanent communities did not exist, as each species varied individually across disturbance or environmental gradients (Gleason 1926, 1939). These attitudes countered the trend (still popular today) to classify and organize communities, but did not imply a lack of interaction or totally random assembly of species (Nicolson and McIntosh 2002) as some claimed. The phytosociological school of ecology that developed in Europe shared elements of this individualistic approach, because it emphasized detailed floristic composition across small scales, not community comparisons across large spatial and temporal scales (Mueller-Dombois and Ellenberg 1974). Phytosociology provided the conceptual basis for many modern analytical tools such as detrended correspondence analysis (Hill 1979) and other ordination techniques that compare similarity of community composition over time. Evidence of transient species composition of plant communities has also accumulated across much longer (historical) time scales. For example, tree species distribution in Europe can be partly attributed to climatic legacies of the Ice Age (Svenning and Skov 2007).

A reductionist approach to succession has prevailed since about the 1950s. Forgoing attempts at global or even regional generalizations, scientists have examined a variety of plant and community characteristics at mostly local spatial scales (Table 8.3). Generalization is still attempted, particularly in conceptual models, but efforts to compare across ecosystems have not been emphasized (but see Messer 1988; Matthews 1992; Schipper *et al.* 2001; Wardle *et al.* 2004; Peltzer *et al.* 2010). What is emerging is a broad consensus that ecosystems are in a constant state of change (non-equilibrium), and that their component parts (populations, species, communities) and the landscapes that ecosystems reside in each respond individualistically rather than in a coherent fashion (Botkin 1990). Succession is therefore not an orderly **progression** of sequentially more competitive species toward a stable climax but instead consists of often unlinked species replacements that vary in their timing and spatially across a variable mosaic of patches (Rohde 2005). There is clearly a continuum of attitudes about how ecosystems respond to disturbance from strongly individualistic to strongly holistic, or indeed how stable any ecosystem is. Intellectual debates are useful exercises when they provide guidance for practical concerns about how to manage ecosystems (Spieles 2010).

8.2.3 Models

The complexities of successional processes make both conceptual and mathematical models appealing because they can highlight which critical pieces of information are missing in our understanding of disturbance responses. Models are limited, however, by our lack of knowledge about succession, the role of disturbance, and the completeness of data sets. Although the task of identifying how species composition changes over time does not sound inherently difficult, it is actually a big challenge because of the many potentially influential variables and their permutations (Walker and del Moral 2003). This complexity has led to failure for those models that attempt to predict entire successional sequences (Franklin *et al.* 1985). More successful models have looked at how disturbance affects resource pools, species pools, and species performance (arrival, establishment, growth, longevity) at a recently disturbed site (Fig. 8.1; Walker and del Moral 2003). Disturbance and succession are inextricably linked, not only by the initial disruptive process, but by on-going disturbances that alter successional trajectories

Table 8.3 Topics considered in reductionist studies of succession. For discussion see text and Walker and del Moral (2003).

Topic	Explanation	Key reference
Patch dynamics[1]	Shifting mosaic of communities across landscapes	Watt (1947)
Initial floristic composition	Sequential conspicuousness of species that all arrive together	Egler (1954)
Metapopulations[1]	Extinction and creation of populations	Levins (1969)
Reproductive traits	r-selection and K-selection	Pianka (1970)
Gradient analysis	Examination of distribution of each species	Whittaker (1973)
Evolutionary sorting	Population-level processes sort along environmental gradients to drive succession	Pickett (1976)
Plants as populations	Modules within a plant govern responses	Harper (1977)
Stochasticity of establishment	Dispersal and survival at a given site unpredictable	Glenn-Lewin (1980)
Life-history characteristics	Relative importance of dispersal, establishment, growth, reproduction, and longevity	Noble and Slatyer (1980)
Disturbance recovery	Succession as a process of recovery, not progress toward a climax	Raup (1981)
Landscape ecology[1]	Spatial patterns across ecosystems	Naveh and Lieberman (1984)
Hierarchy of scales[1]	Constraints from above, mechanisms from below (see Section 7.2)	O'Neill et al. (1986)
Competition	Competition for resources among individuals drives succession	Huston and Smith (1987)
Causes and mechanisms	Detailed, hierarchical list of drivers	Pickett et al. (1987b)
Relative importance of multiple factors	Competition, facilitation, life-history characteristics all interact	Walker and Chapin (1987)
Trajectory analysis	Identification and interpretation of multiple pathways	Walker and del Moral (2003)

[1]Poorly integrated with temporal dynamics.

(Fig. 8.2; Willig and Walker 1999). I will consider both conceptual models that do not assign values to each variable and mathematical models that do (Usher 1992). Conceptual models include those that address evolutionary strategies, incorporate life-history characteristics, or focus on mechanisms that drive successional processes. Mathematical models address invasions (dispersal and establishment), life-history traits, competitive interactions, transition probabilities, and resource ratios.

Conceptual models of succession began with Clements's six processes (Table 8.2) and many descriptions of **seres** have been developed since. Updated, modular flow models of Clements's processes have also been suggested (MacMahon 1981; Pickett *et al.* 2009). A number of studies have been seminal

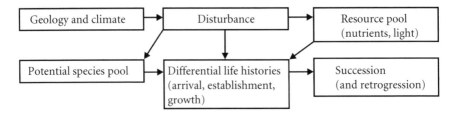

Figure 8.1   General model of factors affecting succession. (From Walker and del Moral (2003), with permission from Cambridge University Press.)

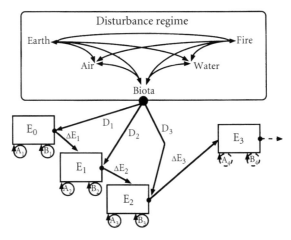

Figure 8.2   Model of the effects of disturbance on succession. Disturbances act as a composite force (D_1) that affect an ecosystem (E_0), causing it to change (move to a new ecological space; E_{1-3}). Successional changes in ecosystem characters are then influenced by subsequent disturbances (D_2, D_3), whose influences are affected by past disturbances and abiotic (A) and biotic (B) forces. (From Willig and Walker (1999) with permission from Elsevier.)

in the "succession" of ideas about succession (Table 8.3). Several models dealt with the evolutionary strategies of organisms that adapted them to particular stages of succession (Pickett 1976). Pianka (1970) developed a dichotomous model to explain how some animals adapt to disturbance by reproducing quickly and rapidly colonizing newly disturbed areas (r-selection), while others evolved a more stationary strategy, becoming good competitors with more investment in each offspring. Grime (1977, 1979) followed with a proposal that plants have evolved different strategies to cope with environments dominated by disturbance (early succession), competition (mid succession), and stress (late succession). Recent application of this model to primary succession on an Italian glacial moraine suggests that adaptations to disturbance and stress are more important that competitive abilities (Caccianiga *et al.* 2006). One complication not addressed by Grime's model is species longevity. Long-lived species that both colonize and persist in succession ("stayers") may experience a disturbance as a stress while short-lived species experience a true disturbance (Menges and Waller 1983).

Several early conceptual models explicitly addressed the role of species longevity in succession. Egler (1954) developed the concept of initial floristic composition, whereby all species colonized as pioneers but succession was then determined by the sequential conspicuousness of species as the fast-growing ones dominated first, followed by slow-growing, long-lived species. Drury and Nisbet (1973) and Noble and Slatyer (1980) considered the importance of a number of life-history characteristics (e.g. arrival times, growth rates, longevity) to succession. Noble and Slatyer (1980) condensed the complexities of life histories into three vital attributes of each species and demonstrated how these attributes could explain post-fire forest succession in Australia. The first attribute considered how a species arrives (or persists) at a site after a disturbance. The second attribute was the ability of a species to establish and grow to maturity. The final attribute was the time each species took to reach reproductive maturity, combined with the duration of the species population and its propagule pool.

A third set of conceptual models integrated species interactions (facilitation, competition) with life-history characteristics to focus on the mechanisms behind successional processes. The first of these (Connell and Slayter 1977) focused entirely on autogenic processes and proposed three trajectories or models: facilitation, inhibition, and tolerance. The models were intended to describe primarily the interaction of the first and second set of colonists following a disturbance (Connell *et al.* 1987), although they were widely seen as attempts to explain whole seres. The facilitation model follows the original concept of Clements, that the first set of colonists improves the environment for the second set of colonists. They purportedly do this by altering the environment in ways that make their own regeneration unlikely but favor establishment of the other species. This process could conceivably drive all transitions until a self-replacing climax stage is reached (Clements 1916, Egler 1954). However, Connell and Slatyer (1977) suggested that later transitions occur because species better adapted to late successional environments outcompete earlier species for resources to germinate or grow. This type of transition can be explained as competitive displacement (Walker and Chapin 1987), or active tolerance (Pickett *et al.* 1987a)—where successful species are those most tolerant of declining resources such as light or nutrients (Tilman 1988). Support for facilitative interactions between species in a successional context has often been found in primary succession where amelioration of a stressful physical habitat is beneficial (Callaway and Walker 1997). However, evidence for obligatory facilitation (where facilitation is a requirement for succession) is rare (Walker 1999b; but see Walker *et al.* 2003).

The inhibition model (Connell and Slatyer 1977) states that the first set of colonists monopolizes resources and thereby alters the environment so completely that later colonists are prevented from establishing. This process is often termed competitive inhibition and can involve plants that form dense mats or thickets (Walker and del Moral 2003) or allelopathy (Tolliver *et al.* 1995). Competition can be for light or nutrients (Tilman 1988). However, as with the facilitation model, later stages rely on other explanations. The first set of colonists eventually dies, perhaps after a time of vegetative reproduction, and other species invade. Although no competition is explicitly invoked, either species are displaced because superior competitors arrive (competitive displacement) or later successional species outlive early species (life-history characteristics). Support for the inhibition model is widespread in both primary and secondary terrestrial succession and inhibition is also a common feature in intertidal succession (Turner 1983; Sousa 1985).

The tolerance model (Connell and Slatyer 1977) states that there is no positive or negative influence of the first set of colonists on later arrivals (passive tolerance, *sensu* Pickett *et al.* 1987b). Instead, the life-history characteristics of the colonists are what determine dominance in early succession. These characteristics typically include arrival times, relative growth rates, and longevities. Transitions in later stages are driven, as in the facilitation model, by competitive displacement of species more tolerant of declining resource levels (active tolerance, *sensu* Pickett *et al.* 1987a). Support for the tolerance model comes from both primary and secondary succession.

The Connell and Slatyer (1977) models provided three theoretically clear choices for explaining transitions between successional stages. The first set of colonists facilitates, inhibits, or has no effect on the next set of colonists. In reality, such a trichotomy is too simplistic, as further studies have shown (Huston and Smith 1987, Pickett *et al.* 1987a,b, Walker and Chapin 1987). Succession results from the net effects of all types of interactions between species (positive, negative, and neutral). These interactions can affect each life-history stage in each successional stage and are in turn affected by the broader context of allogenic processes (e.g. disturbance regimes) and resource availability (Walker 1999b; Walker and del Moral 2003). For example, in Glacier Bay, Alaska, glacial moraines are colonized by *Dryas*, *Alnus*, and *Picea*, usually in that order (Chapin *et al.* 1994). *Picea* can establish at any stage, however, and at each stage there are both facilitative and inhibitory effects (Fig. 8.3). *Picea* primarily establishes in *Alnus* thickets, presumably because at that stage facilitative effects are maximal. River succession in Glacier Bay also exhibited several of the Connell and Slatyer mechanisms (Milner *et al.* 2008). Once dispersal constraints by non-insect taxa were overcome, the tolerance model prevailed. However, nest digging by salmon (family Salmonidae) facilitated the persistence of early colonists and riparian vegetation facilitated colonization by caddisflies and chironomids. Even relatively simple interactions between just two species can vary from mutual (or one-sided) benefit to mutual (or one-sided) detriment, depending on environmental conditions. Interactions can also "turn on" or "turn off" (Table 8.4). It is no surprise that the mechanics of successional transitions are not fully understood.

The Connell and Slatyer models sparked an interest in a more detailed examination of succession that led to several more process models. Several primarily autogenic models include the gradient-in-time and the competitive-sorting models that emphasize the role of species characteristics and competition, respectively (Peet 1992). The gradient-in-time approach suggests that during succession resources inevitably change and the differential responses of species to those changes lead to succession. Competitive sorting suggests that convergence of species composition will increase with successional time after high variability early in succession (Margalef 1968) because the best competitors survive. There is mixed evidence for this in old field succession (Christensen and Peet 1984), because while variability decreased with time, convergence did not necessarily occur. Wilson and Agnew (1992) proposed that positive feedbacks (e.g. between soil conditions and early plant colonists that grew well on that soil type) drive successional changes.

Other conceptual successional models address both autogenic and allogenic influences, specifically incorporating interactions of species and the environment (Gitay and Wilson 1995). Pickett *et al.* (1987b) ably summarized such processes in the context of site and species availability and differential

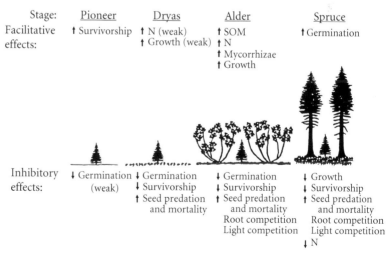

Figure 8.3   Impacts of post-glacial successional stages on the establishment and growth of spruce (*Picea*) seedlings at Glacier Bay, Alaska (SOM = soil organic matter). (From Chapin *et al.* (1994) with permission from the Ecological Society of America.)

Table 8.4 Positive and negative interactions between two species (A and B) are altered by whether or not the interactions are turned on ("Yes") or turned off ("No"). (Modified from Odum (1959).)

Type of interaction	A: Yes	B: Yes	A: No	B: No	Expected results of interaction
1. Neutralism (A, B independent)	0	0	0	0	Neither species affects the other
2. Competition (A, B compete)	−	−	0	0	One species eliminated from niche
3. Mutualism (A, B partners or symbionts)	+	+	−	−	Interaction obligatory for both
4. Proto-cooperation (A, B cooperate)	+	+	−	−	Interaction favorable for both but not obligatory
5. Commensalism (A commensal, B host)	+	0	−	0	Obligatory for A; B not affected
6. Amensalism (A amensal; B inhibitor or antibiotic)	−	0	0	0	A inhibited; B not affected
7. Predation and parasitism (A predator or parasite; B prey or host)	+	−	−	0	Obligatory for A; B inhibited

0, no effect; −, negative effect; +, positive effect.

species performances (Table 8.5) and detailed factors that might modify each process. They considered disturbance, life-history characteristics and competition and added allelopathy, herbivory, and disease. Walker and Chapin (1987) presented a conceptual model of how the relative importance of various mechanisms might change during succession (Fig. 8.4) and across gradients of severe (primary succession) to more favorable (secondary succession) environments. They presented many testable hypotheses (e.g. seed dispersal, facilitation, and stochastic events are more important drivers of change in primary that secondary succession, particularly in early stages; competition, growth rates, and mammalian herbivory dominate in the more favorable environments of secondary succession). Some evidence has accumulated in support of this model but many of these ideas remain untested (Walker and del Moral 2003). Matthews (1992) elaborated on the importance of the favorability of the environment, suggesting that increasing favorability (a temporal trend seen during most seres) leads to a predominance of autogenic over allogenic processes. He also noted that in many disturbed habitats (e.g. glacial moraines), on-going disturbances keep the habitat in a state of disequilibrium. Support for a general amelioration of conditions (indirect facilitation) has come from three decades of work on Mount St Helens (Washington, U.S.), following a catastrophic eruption in 1980 (del Moral 2000a). Proximity of a site to surviving plants or survival of some organisms *in situ* accelerated succession (Fig. 8.5; Walker and del Moral 2003). Supporting Walker and Chapin (1987), the relative importance

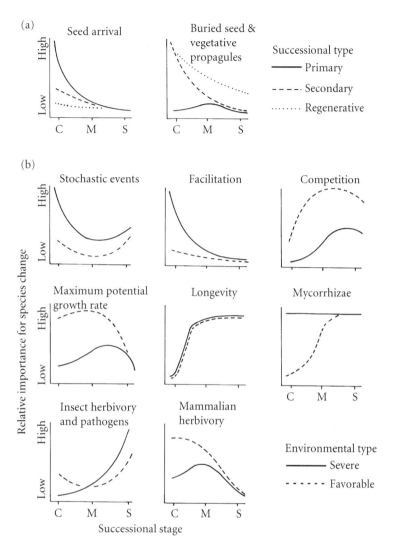

Figure 8.4   Influence of type of succession (top) and environmental severity (bottom) upon major successional processes that determine change in species composition during colonization (C), maturation (M), or senescence (S) stages of succession. (From Walker and Chapin (1987) with permission from Wiley-Blackwell.)

of facilitation declined and that of competition increased with succession on Mount St Helens and other volcanic seres (del Moral and Grishin 1999). Burrows (1990) examined successional processes that others had not emphasized, including fluctuations, cyclical trajectories, direct replacement, and responses to gradual climate change. Direct replacement occurs when a dominant species in the pre-disturbance environment is also a dominant colonist after a disturbance. In this case, there are no species replacements,

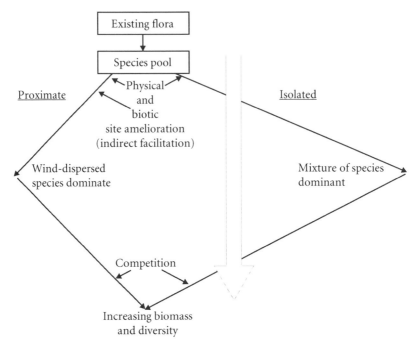

Figure 8.5   A summary of two successional trajectories on Mount St Helens. Isolated sites took longer to develop than sites closer to potential propagules. (From Walker and del Moral (2003) with permission of Cambridge University Press.) The large arrow indicates time.

just a regeneration of the same vegetation. Burrows's first model addressed the occupation and in-filling of newly created surfaces. His second model explored replacements of stands after gradual changes in the environment due to either autogenic or allogenic factors. The third model involved species replacements following senescence of individuals. Each of these process models helped clarify the importance of a changing environment and its influence on autogenic processes such as species interactions. None has become the best explanation for a majority of seres, suggesting that cross-site and cross-gradient generalizations are only possible in broad terms (e.g. in terrestrial primary succession: nitrogen and organic matter increase, pH and P availability go down; Peltzer *et al.* 2010).

One common mathematical approach to modeling succession is to address how species invade a site. This approach ideally requires input about on- and off-site vegetation, past vegetation, current resource and disturbance levels, and stochastic factors (van Hulst 1992). Models frequently build from information about individual plants to predict invasions of disturbed habitats such as forest gaps created from wind-throw, disease, senescence, fire, or logging. The JABOWA models (Urban and Shugart 1992; Bugman 2001)

Table 8.5 A hierarchy of successional causes. (From Pickett *et al.* (1987b) with permission from Springer.)

General causes of succession	Contributing processes or conditions	Factors that modify the processes
Site availability	Coarse-scale disturbance	Size, severity, time, dispersion
Differential species availability	Dispersal	Landscape configuration
	Propagule pool	Dispersal agents, time since disturbance, land use
Differential species performance	Resource availability	Soil conditions, topography, microclimate, site history
	Ecophysiology	Germination requirements, assimilation rates, growth rates, population differentiation
	Life history strategy	Allocation pattern, reproductive timing, reproductive mode
	Stochastic environmental stress	Climate cycles, site history, prior occupants
	Competition	Presence and identity of competitors, within-community disturbance, predators, herbivores, resource base
	Allelopathy	Soil characteristics, microbes, neighbors
	Herbivory, disease, predation	Climate and consumer cycles, plant vigor and defense, community composition, patchiness

and FORFLO (Pearlstine *et al.* 1985) predict species composition based on life-history characteristics (e.g. germination, growth, longevity) and interactions (e.g. competitiveness, shade tolerance, mortality) of forest species, and the results are often used to examine the effects of altered disturbance regimes (e.g. increased flood frequency). SORTIE models add a spatially explicit component (Pacala *et al.* 1993) and have been applied to a variety of habitats from forests to grasslands. Multiple successional scenarios can also be run with various species mixes and disturbance regimes. In one case, competition for light emerged as more important than life-history characteristics (Huston and Smith 1987). Because there are so many combinations of life histories and species interactions, in addition to many potential colonists, predicting specific outcomes is still difficult. The carousel model incorporates the stochasticity of dispersal and establishment, calculating the odds of a given species successfully establishing based on its abundance in the seed rain and its seedling mortality (van Hulst 1992; van der Maarel and Sykes 1993). This model assumes no prior influence of other species so it applies better to primary than secondary succession where there is often a substantial biological legacy (Bellingham and Sparrow 2000), including surviving organisms such as soil microorganisms, seeds, buds, and larvae.

Markov models rely on autogenic processes to predict transitions from one clearly defined successional stage to another (Usher 1981, 1992). They assume that once a starting point is defined, transition probabilities are constant, initial conditions do not influence succession, and eventually equilibrium is reached in species composition. Applications of these models often fail to predict actual conditions because the Markov models do not account for on-going, allogenic effects such as outside disturbances, composition of adjacent plots, fluctuating environments, or the rarity of equilibrium states (Gibson *et al.* 1997; Childress *et al.* 1998).

The resource ratio model (Tilman 1985, 1988, 1994) proposes that species change is driven by declining light and increasing nutrient (typically N) levels. Pioneers are those species competitively superior in high light and low nutrients that characterize many early successional habitats. Part of their competitive strategy is to maximize belowground growth to absorb limiting nutrients. Similarly, late successional species compete successfully in conditions of lower light and higher nutrients. To do this they maximize aboveground growth. This model provides a useful focus on plant physiology (see also Bazzaz 1996) but is perhaps unrealistic in assuming that competing plant species are equally affected by the same resources, that resources necessarily determine competitive success, and that an equilibrium is reached between an organism and resources (van Hulst 1992). In addition, competitive abilities may not lead to success in severely disturbed habitats (Caccianiga *et al.* 2006). Disequilibrium appears to be much more common in a successional environment (Walker and del Moral 2003).

Simultaneous considerations of nutrient dynamics and species interactions have led to recent efforts to understand the ecological role of **stoichiometry** that considers the ratios of nutrients in biology (Moe *et al.* 2005). A recent application of a stoichiometric model to primary succession Marleau *et al.* 2011) used the N to P ratios in plants, soil solution, and dead organic matter to determine competitive and facilitative outcomes of pioneer species on Mount St Helens (Washington, U.S.). The results mirrored field results of succession, indicating co-limitation of N and P and the important facilitative role of a dominant N fixer (*Lupinus*). Despite simplifications in the model that could not address spatial heterogeneity of nutrients, interactions between N and P uptake, and herbivory (Bishop *et al.* 2010), such studies can further our understanding of disturbance responses and succession.

Sudden changes in ecosystem properties provide insights into temporal dynamics. When sudden transitions occur, an ecosystem abruptly crosses a threshold and undergoes a regime shift to an alternative stable state (Suding and Hobbs 2009). Properties that can potentially change (e.g. species composition, nutrient cycles, resistance, resilience) may exhibit increased variance well before the actual transition, as found for P levels in lake waters (Carpenter and Brock 2006) before they became eutrophic. Transitions can lead to increasing deterioration (e.g. simplified food webs, decreased biodiversity) and not be easily reversible, leading to implications for both succession (Walker and del Moral 2009b) and restoration (see Chapter 9; Bestelmeyer *et al.* 2009).

8.3 Methods

8.3.1 Observations

Temporal dynamics following a disturbance can be measured with direct observations, **chronosequences**, experimental manipulations, and models. Direct observations of community responses to fluctuations are possible provided there is adequate (and frequent) access to the study site. Successional studies that involve years to decades can also rely on direct observations, provided there is appropriate standardization of protocols, minimal damage to the study site, and transfer of knowledge across funding cycles and personnel transfers. Permanent plots or observation points facilitate comparisons among repeated visits. Direct observations are a good way to estimate rates of change in species and community composition, although they may be highly variable, as on dunes (Martínez *et al.* 2001). Studies of primary succession on a volcano (del Moral 2000b) and secondary succession in grasslands (Olff and Bakker 1991) suggest that rates of species turnover generally decline with time since a disturbance, in part because the longevity of dominant species increases (Walker and del Moral 2003). Permanent plots are also useful for studying species interactions. One challenge, however, is to anticipate the appropriate placement and size of plots to cover such future possibilities as repeat disturbances or growing vegetation (the 1-m^2 plots in Glacier Bay, Alaska were adequate for early glacial moraine plant communities (Cooper 1923) but inadequate once forests developed; Chapin *et al.* 1994). Observational tools have improved greatly with data loggers that store and integrate information about microclimates, gas fluxes, and plant physiology. Long-distance retrieval of these data is now also possible. Remote sensing from airplanes and satellites greatly extends the spatial scale for monitoring temporal change, allowing such parameters as net primary production (Field *et al.* 1995) and even soil properties (Shepherd and Walsh 2002) to be monitored. Observational studies are receiving renewed attention as a valid method for exploiting natural experiments and hypothesis testing by correlation of variables or comparisons of observed patterns and model outputs (Sagarin and Pauchard 2010), particularly with novel ways of exploring temporal data (Collins *et al.* 2008).

8.3.2 Chronosequences

When time scales extend beyond the life of the investigator, assumptions are made about successional trajectories (see Section 8.6). These assumptions

Figure 8.6   A chronosequence showing succession on the Cooloola Dunes near Brisbane, Australia: (a) surface age of less than 100 yr; (b) 2600 yr; (c) 6000 yr; (d) 50 000 yr; (e) 120 000 yr; (f) more than 500 000 yr. Note increase to maximum biomass at 50 000 yr and then a decline due to severe leaching of soil nutrients.

(e) (f)

Figure 8.6   (*continued*)

generally involve a space-for-time substitution, where it is assumed that
older sites passed through conditions (e.g. species composition; nutrient
status) similar to those represented by increasingly younger stages. This
chronosequence method provides a convenient natural experiment (Pickett
1989; Foster and Tilman 2000) and a way to sample a series of sites of different ages and gain insights into past and future successional changes (Fig.
8.6). Chronosequences have been widely used because few direct observations of succession extend beyond a duration of several decades (Whittaker
et al. 1999; Meiners *et al.* 2007). Verification of temporal linkages between
stages is often possible with such clues as overlapping tree ring patterns, old
stumps, or other physical evidence from the previous stage. However, there
are substantial problems with the use of chronosequences (Walker *et al.*
2010b). The chronosequence method is problematic because of the difficul-
ties inherent in correctly predicting future (or interpreting past) trajectories (Fastie 1995; Johnson and Miyanishi 2008). In addition, some temporal

change can be determined by landscape context or chance rather than by biotic interactions (del Moral 2007). The use of chronosequences is probably most suitable where trajectories are most predictable, including linear, circular, or convergent trajectories, and least suitable for divergent trajectories or complex networks (Walker *et al.* 2010b; see Section 8.6).

8.3.3 Manipulations

Experimental manipulations of successional communities can be useful to determine the mechanisms driving changes in species composition and associated community and ecosystem parameters (Fig. 8.7, Box 8.1). Manipulations also elucidate the relative importance of species interactions such as competition and facilitation in structuring successional communities, and can help disentangle the influences of herbivory, mycorrhizae, land use, and other variables on temporal change. Removal experiments are a typical tool, where the influence of the removed species is determined by comparison with a control. Potential problems with removal experiments include not

Figure 8.7   Experimental removal of scrambling ferns (*Dicranopteris*) from a Puerto Rican landslide.

Box 8.1 Experiments and chronosequences help explain succession

Research on landslides in northeastern Puerto Rico uses a variety of approaches, including direct observations on permanent plots (Myster and Walker 1997), chronosequence studies (Guariguata 1990, Zarin and Johnson 1995a,b), and experimental additions (Shiels and Walker 2003) and removals (Fig. 8.7; Walker et al. 2010a). Although the three approaches share the same goal of understanding plant succession and soil development following a landslide, the assumptions and resulting conclusions differ. Permanent plot research is ideal when continued for many years and on replicated landslides, although there is no guarantee that the landslides chosen are representative of all landslides in the region. For example, some landslides remain unstable for several decades while others develop 8-m high vegetation in that same period. Chronosequences assume that younger landslide processes resemble what older landslides experienced, which is apparently true only in the broadest sense that a suite of 10 common species act as colonists of Puerto Rican landslides but in highly variable proportion. Yet chronosequences provide the best platform from which to extrapolate. Experiments are best for elucidating mechanisms such as the role of birds as seed dispersers or the role of ferns as inhibitors of forest development. However, experiments are also limited to what is measured—perhaps soil organisms are more important than birds or ferns? No single approach to succession is infallible and a combination of approaches is clearly best to help unravel the mysteries of temporal dynamics.

---

conducting them at consistent demographic, phenological, or genetic standards across treatments, the local disruption that a removal (especially of mature organisms) has on a community, and the legacy of the removed species. Each of these problems confounds interpretation of the role of the removed species. Addition experiments avoid legacies and can be better controlled (i.e. addition of genetically similar individuals), but suffer from the many problems that plague transplant efforts (e.g. high mortality of introduced individuals, unnatural densities, inappropriate genomes, novel responses). Exclusion experiments prevent the target species from colonizing the plot. Where controls are invaded by the target species, these experiments are ideal because they eliminate the complications of removals and additions.

8.4 Establishment and persistence

Disturbances create opportunities for invasion by new plant and animal communities and initiate succession. The sequential replacement of these communities is as temporally heterogeneous as is their spatial distribution across landscapes (see Chapter 7). This temporal variability is due to a combination of abiotic and biotic factors. Each disturbance varies in severity so initial physical resources also vary. Dispersal to a disturbed habitat is not only dependent on the spatial distribution of potential propagules (and their dispersers) from the new habitat but also on the timing of the disturbance relative to the production of propagules in the vicinity. Plant colonists must find safe sites in which to germinate in both secondary (Flinn 2007) and primary (Fuller and del Moral 2003) succession, and both plants and animals need adequate environments in which to grow and reproduce. The combination of abiotic and biotic variability thus dictates that the ensuing successional trajectories are asynchronous across a landscape.

8.4.1 Germination

Seeds and spores need microenvironments suitable for germination (safe sites; Harper 1977). These can be pre-existing or created by post-disturbance conditions that ameliorate the environment. Biotic amelioration can come from inputs of organic material (e.g. dead insects, litter, seeds, or spores), inputs of nutrients (e.g. bird feces), or the creation of microhabitats by animal activities (e.g. burrowing; Willig and McGinley 1999). Abiotic amelioration of harsh, barren surfaces such as lava flows can include surface erosion or protective outcrops (Fig. 8.8) that create protected, fertile, moist, shady, or warm microsites. Surface stability and protection from abrasion by wind-

Figure 8.8   A lava boulder provides a safe site for pioneer plant colonization on a cinder slope on Anak Krakatau Volcano, Indonesia.

blown particles are other pre-requisites for germination and vary widely by disturbance type. Where light is limiting, as in a forest gap, well-lit microsites favor germination (e.g. the upraised root surfaces of toppled trees or the upper surfaces of fallen logs that do not get immediately covered by leaf litter). Microsites that provide the requirements for germination (usually a combination of limiting resources) then can become "hotspots" of establishment (Fig. 8.9) as further colonists accelerate amelioration of that site (regeneration niche; Grubb 1977).

Successful dispersal to a microsite that is suitable for germination is a complex lottery of abiotic and biotic filters combined with many chance events. Once germination occurs, there is another lottery—which individuals will survive and grow? Survival of newly germinated seedlings can be random, such that species turnover is common but not directional (Walker and del Moral 2003). Pioneer species typically have a broader range of safe site conditions that they can tolerate (carousel model; see Section 8.2.3) so they dominate succession until microsites that favor later colonists have been created. Species with strict dormancy requirements improve their chances of germinating in a favorable environment (Baskin and Baskin 1988).

Figure 8.9   Plants such as this sorrel (*Rumex*) on the unstable slopes of Mount Fuji, Japan, alter the environment for subsequent colonists.

8.4.2 Growth

Once a propagule (seed or spore) germinates and uses up its reserves, it is dependent on a favorable environment to continue growing. Often it needs a different set of abiotic conditions to prosper than was optimal for germination. The success of a seedling requires adequate moisture and, until roots are well established, drought is a frequent problem, even in normally wet habitats such as floodplains. Site stability continues to be important, and microsites good for germination can become too small to protect a growing seedling. In addition, exposed surfaces can become too hot or cold for survival and excessive salts can damage seedlings in deserts. Pollutants such as heavy metals can also deter growth in anthropogenic habitats (e.g. mine wastes). Biotic limits to growth include physiological conditions resulting from the physical environment and competitive interactions with neighbors. Resource accumulation must be rapid for annuals but sustained for long-lived perennials, so again, optimal microhabitat conditions will vary depending on the life-history stage and size of the organism. The type of growth forms that are successful in early succession vary depending on substrate. Typical matches include lichens and mosses on rock surfaces, rhizomatous grasses on dunes, perennial shrubs on rocky slopes, and fast-growing trees in forest gaps, on abandoned agricultural lands (Fig. 8.10), and on floodplains (Grubb 1987), although evidence for

Figure 8.10   Secondary succession on abandoned sugar cane fields on Oahu, Hawaii featuring fast-growing pioneer trees.

strong correlations between substrate and life form is lacking (Walker and del Moral 2003; Walker *et al.* 2006b). Successful growth and accumulation of biomass generally translates into reproduction and sometimes into dominance of local resources until a species is outcompeted or dies. Biomass increases in succession until all niches are filled or resources exploited, after which it may decline (Kitayama *et al.* 1997).

8.4.3 Persistence

The duration of an organism's presence in a sere depends on its ability to persist, even as the environment is changing. Alternatively, persistence can be related to dominant life forms (e.g. thickets) that arrest successional change. Regional persistence can come from frequent reproduction and survival (e.g. in frequently re-occurring gaps) or from life stages found outside the disturbed area (Townsend *et al.* 2003). Local persistence is mostly attributed to vegetative growth or longevity. Vegetative growth, expansion, and reproduction are ways to monopolize resources and reduce the number of potential regeneration niches for other species (Bellingham *et al.* 1994), thereby winning the competitive lottery—at least as long as the environment does not dramatically change. Longevity allows an organism to endure environmental change but those changes lessen the likelihood that the organism will successfully reproduce sexually. Resource depletion and increases in competitors, herbivores, parasites, or diseases may reduce its regeneration options (Walker and del Moral 2003). Longevity in a forest understory allows herbs, ferns, and mosses to benefit from sunflecks (see Fig. 5.1) and stunted saplings to respond to forest gaps. Extreme longevity of trees reduces the frequency of reproduction needed for replacement and increases the odds of one seedling in hundreds of years growing in to replace a dying tree – again, only in fairly stable environments.

8.4.4 Temporal patterns

The abiotic and biotic filters that govern which species disperse and establish contribute to various temporal patterns following a disturbance. Early successional communities are typically under-saturated, meaning they do not have all the possible species that later successional communities have. Late successional communities can also be under-saturated, as fewer species can tolerate deep shade or heavily leached soil nutrients (Walker *et al.* 2000). The size of the disturbed habitat, as well as its remoteness, influences the rate of colonization and the total number of species that can be supported (island biogeography theory; MacArthur and Wilson (1967). Large habitats near propagule sources will typically support more species than small, remote habitats. Observational and experimental tests of this theory have found mixed support (Whittaker *et al.* 2008), in part because it does not address evolutionary scales, dispersal and establishment are so unpredictable, equilibrium has not been reached between immigration and extinction, competition does not occur in a closed system, and priority effects can be important. Priority effects occur when the order of arrival influences later successional trajectories. Those species that colonize first often grow quickly, monopolize resources, and have a major influence on subsequent replacements (Facelli and D'Anela 1990; Samuels and Drake 1997). Subsequent successional stages can have a substantial lack of congruence between the composition of communities in the disturbed habitat and in surrounding, undisturbed areas. Isolation also contributes to this disharmony.

Our lack of ability to predict temporal patterns of species replacements indicates the complexity of successional drivers. Efforts to explain temporal dynamics (for both intellectual and practical reasons) have included looking for patterns in colonization of higher taxonomic units (e.g. genera or families) and functional groups of species (i.e. species that respond in a similar way to a disturbance; Gitay and Noble 1997; Pausas and Lavorel 2003). Species with distinct functions (e.g. N fixation, high flammability) or structures (e.g. canopy shade providers, keystone predators) might improve the predictability of succession, but few studies have tested the role of functional species in a successional context (Walker and del Moral 2003). Functional groups may be less important than stochastic or other factors to explain sequences of species dominance and replacement (Walker *et al.* 2006b) and are best evaluated in the context of environmental variables (Lebrija-Trejos *et al.* 2010). Another integrative effort is the proposal that there are a set of more or less predictable **assembly rules** that determine how species combine to form communities (Weiher and Keddy 1999). The most predictable sere would have strong assembly rules based on species composition, followed by less predictability if assembly rules existed but were based on growth forms, followed by abiotic (e.g. climate) or biotic (e.g. dispersal limitations) factors that constrain successional directions. Strong stochastic elements appear to be common, however, so assembly rules, with some exceptions, appear to be a neo-Clementsian idea with little general application (Belyea and Lancaster 1999). Exceptions appear to occur either where species replacements are influenced almost entirely by autogenic species interactions (e.g. invertebrate succession on glacial moraines; Kaufmann 2001) or where species exclude members of their own functional groups, maintaining guild proportionality (Holdaway and Sparrow 2006). Another promising approach is to examine causes for species changes at increasing spatial scales over successional time. On several Icelandic lava flows, initially discrete patches of mosses (*Racomitrium*) colonized, followed by expansion to a continuous cover, then colonization by higher plants (*Empetrum, Betula*) in discrete patches, and finally a gradual differentiation of plant communities by topographic variation in the underlying substrate (Cutler *et al.* 2008).

8.5 Species interactions

Species interactions are the principal autogenic driver of both the rates and trajectories of succession; they begin after dispersal and continue throughout the life history of each organism (for plants: germination, growth, reproduction, **senescence**, and death). In the absence of evidence that interactions drive succession, the life-history traits themselves can be the drivers (tolerance model; see Section 8.2.3) or allogenic drivers (e.g. on-going disturbances) can be important. In this section, I examine the successional consequences of interactions of plants and animals with soil organisms and soil structure, plants with plants, plants with animals, and animals with each other.

8.5.1 Soils

Plants and animals interact with soil organisms and impact the physical characteristics of soils (in terrestrial environments) in ways that alter temporal dynamics (Bardgett and Wardle 2010). Plants contribute litter that is an important part of soil formation. Litter adds carbon and nutrients that supply the basis for a detritivore-based food web of soil organisms (see Chapter 5). This food web develops and becomes more complex with time after a disturbance. Soil development, in turn, provides nutrients, water retention, and physical structure for plants to grow in. The variable tolerance of different plant species to these soil properties influences species replacements in succession. For example, in low-nutrient, early stages of succession, N-fixing plants can have a competitive edge that leads to increased soil fertility, nutrient cycling, and decomposition. Other species might then be competitive (facilitation through habitat amelioration), or the N fixer can dominate for a period of time (arrested succession). Some species lower nutrient availability and decomposition through litter low in nutrients or easily decomposable tissue (Chapin *et al.* 1986). The N-fixing *Alnus* thickets on an Alaskan floodplain were replaced by *Populus* trees when *Populus* leaf litter that was rich in tannins decreased N availability and decomposition (Schimel *et al.* 1998). Mycorrhizal fungi are often, but not always, influential in succession. For example, species that require a mycorrhizal symbiont are not typical among pioneer species (e.g. Puerto Rican landslides; Calderón 1993), unless their symbiont is already present in the soil (e.g. dunes in The Netherlands; van der Heijden and Vosatka 1999), yet plants that are facultatively mycorrhizal often do better when infected (Nara and Hogetsu 2004). Other soil organisms such as the fungal endophyte *Neotyphodium* can also influence succession. Along grassland–tree ecotones, tree recruitment decreased when the endophyte was present in association with the dominant, introduced grass (*Lolium*), probably because the endophyte decreased the palatability of the grass and the dominant herbivore (*Microtus*) then preferred to eat tree seedlings (Rudgers *et al.* 2007).

Plants have positive and direct effects on their own fitness when they increase limiting soil resources, which they benefit from, or decrease resource levels below competitors' thresholds for survival (Binkley and Giardina 1998). Other direct effects can include deep litter layers or chemical leachates from the litter that inhibit the germination or survival of competitors. Many indirect effects on their fitness come from other impacts on the soil that may benefit them or their competitors (e.g. increased bacterial populations, soil acidity, soil water, soil stability). Stabilization of soil surfaces, particularly in early primary succession, can facilitate colonization of the target species, or other species that may compete with it, facilitate it, or have no effect on it. Plants create fertile patches by trapping wind-blown sediments, increasing water entrapment and infiltration, and concentrating litter deposition (fertile islands; Garner and Steinberger 1989). These patches can serve as safe sites for later colonists to become established.

Animals that are part of the primary production-based food web alter soil texture by aerating it with burrows, building nests, and digging (e.g. rodents, termites, and elephants, respectively). Large animals also compact soil through surface travel, especially when in large herds. Aeration generally increases nutrient availability and water retention, while compaction does the opposite (except when soils were well aerated to begin with and moderate compaction increases water retention). Animals also alter soil nutrients through the addition of feces, urine, and corpses. Soil texture and nutrients influence germination, establishment, and growth of plants in species-specific ways, thereby influencing succession.

8.5.2 Plants

Plant–plant and plant–animal interactions are pivotal to temporal dynamics following a disturbance. Facilitation (the positive influence of one species on another) increases the rate of species turnover because it promotes the next stage. Facilitation can also alter trajectories by its influence on habitat quality. Facilitation can be direct or indirect, obligatory or facultative. Direct facilitation occurs when an essential service is provided, such as protective cover from the sun, heat, or herbivory. Double facilitation is a mutualism where both species benefit (a N-fixing shrub contributes to soil fertility for a herb that provides leaf litter that benefits N mineralization; Moro *et al.* 1997). Indirect facilitation usually involves any site amelioration such as improvement of soil texture, fertility, or moisture. When facilitation must be present for the transition to the next successional stage to occur it is considered obligatory; when it is not obligatory it is facultative. Most examples of facilitation are facultative. However, red algae were necessary to provide structure for the establishment of surf grass (*Phyllospadix*) on rocky coasts of Oregon (Turner 1983) and *Griselinia* trees did not establish until the N-fixing *Coriaria* shrub was present on a New Zealand volcano (Walker *et al.* 2003). The facilitator generally is not self-replacing (Clements 1928;

Walker and del Moral 2003) because it is short-lived (tolerance model), the environment changes in ways that are no longer suitable for it (facilitation model), or the next colonists outcompete the facilitator (inhibition model).

Facilitation can affect different life-history stages of plants (Table 8.6). Dispersal, germination, establishment, growth, reproduction, and even survival can each be positively affected. Mechanisms of facilitation usually differ for each stage (e.g. trees facilitate dispersal, mycorrhizae facilitate growth, and pollinators facilitate reproduction), but can come from one species. For example, a N-fixing *Myrica* shrub facilitated the establishment, growth, and reproduction of herbs on dunes in the eastern U.S. (Shumway 2000). The environmental and biotic conditions that characterize a particular successional stage can also have multiple facilitative effects on a given species. On an Alaskan glacial moraine, the N-fixing shrub alder (*Alnus*) promoted growth of spruce trees (*Picea*) directly and through increased soil N, soil organic matter, and mycorrhizae (Chapin *et al.* 1994). Net facilitative effects of stages or species on a target species should therefore be interpreted within a defined time frame, as well as in the context of any inhibitory interactions.

Nurse plants facilitate the establishment and growth of other species under their canopy in a variety of ways. Direct facilitation includes protection from herbivory (Niering *et al.* 1963). Joshua tree (*Yucca*) establishment was facilitated by dense, low-lying shrubs in the Mojave Desert in the western U.S. (Brittingham and Walker 2000), presumably because the shrubs protected the *Yucca* from herbivory. Indirect facilitation occurs when the nurse plant increases nutrients and soil organic matter (Charley and West 1975; Schlesinger *et al.* 1996), soil water (Valiente-Banuet and Ezcurra 1991), soil temperature (Jacquez and Patten 1996), and shelter (Carlsson and Callaghan 1991). Nurse plants can also intercept fog drip (Wright and Mueller-Dombois 1988). Succession is promoted when the nurse plant gets outcompeted by the plant that it facilitated, as with the columnar cactus (*Neobuxbaumia*) that outcompetes the *Mimosa* shrub under which it usually colonizes (Flores-Martinez *et al.* 1994). Similarly, the N-fixing shrub *Medicago* facilitated the grass *Lophochloa* but had a negative effect on its own reproduction on an Italian

Table 8.6 Facilitation and competition affect all life-history stages of organisms. (Modified from Walker and del Moral (2003).)

Life-history stage	Mechanism of facilitation	Mechanism of competition
Dispersal	Bird perches, seed traps	Dense canopy
Germination	**Scarification**, site amelioration	Dense leaf litter, soil crusts
Growth	Site stabilization, mycorrhizae, site amelioration	Resource pre-emption, thickets
Reproduction	Alternate food for pollinators	Competition for pollinators, frugivory
Longevity	Reduced disturbance and herbivory	Increased lethal disturbance

shoreline (Bonanomi *et al.* 2008). Sometimes the facilitative effect occurs only when the nurse plant dies, leaving more fertile or moister soil behind (Adachi *et al.* 1996b). Hierarchies of facilitative effects can occur, as when a pine tree (*Pinus*) facilitated a fir tree (*Pseudotsuga*) and a shrub (*Ribes*) by shading them in Montana, U.S. Only shaded seedlings also benefited from wind protection and snow accumulation (Baumeister and Callaway 2006). Despite a renewed interest in positive interactions among plants (Callaway 1995, 2007; Pugnaire 2010), much of the research is not placed in a successional context.

Competition is the negative effect of one plant on another, and in a successional context this can result in arrested succession (competitive inhibition of later colonists) or promotion of succession (competitive displacement of an existing dominant). The former slows succession, the latter speeds it up. The focus here will be on inhibition. As with facilitation, the successional implications of inhibition are not a feature of most competition studies. However, inhibitory interactions are common features of many seres, including many intertidal seres (Sousa 1985). Typically, an inhibitor is an established, mature plant species that resists invasion by other species until it is outcompeted, senesces, or is disrupted by a new disturbance (e.g. shade-tolerant kelp inhibit establishment of algae; Hirata 1992). Its dominance is often dependent on it being the first colonist of its life form that resists further invasion (priority effect). If the inhibitor maintains its stand by vegetative growth or by successful regeneration under its own canopy, the delay can continue for decades. Consequences of slowing or arresting succession depend on the morphology of the inhibitor, its influence on the environment, and the manner of its demise. Terrestrial thickets can accumulate litter that either leads to increased soil development (if easily decomposable), or to decreased nutrient availability (if recalcitrant). Thickets are formed by nearly all types of plants, from algae, mosses, and lichens, to grasses, vines, ferns, shrubs, and trees (Walker and del Moral 2003). Widespread dominance by a single species of tropical tree can be due to lack of disturbance; regeneration under its own canopy, even with conditions of low nutrients and light; good defenses against herbivores; and abundant seed production (Peh *et al.* 2011). The demise of a thicket can lead to altered successional trajectories when nutrient levels or soil stability have been altered. Sometimes successional rates are increased (compared with non-thicket controls) due to environmental amelioration by the thicket. The slow removal of a thicket (e.g. through uneven senescence or selective cutting) can have different implications for succession than a more abrupt removal, because some species may invade under the gradually opening canopy that could not survive under full canopy exposure (Loh and Daehler 2008). Inhibition can affect all life stages of a species (see Table 8.6), from dispersal (a canopy may reduce seed dispersal) to reproduction (frugivores reduce fitness).

Plant–animal interactions also have a role in temporal dynamics, particularly through pollination and herbivory. Pioneer plants are typically either wind-pollinated or self-compatible (Rydin and Borgegård 1991; del Moral and Wood 1993). The successional status of the surrounding vegetation can limit pollinator availability. Late successional stages may not support the pollinators that early successional species require (Walker and del Moral 2003). Decline in populations of some pollinators, particularly tropical vertebrates such as bats, can potentially affect successional trajectories (Whittaker 1992; Compton *et al.* 1994), and is certainly a factor in the survival of already rare species (Walker and Powell 1999). The role of pollinators in succession needs closer examination, particularly regarding their role in improving the regeneration of out-plantings in restoration efforts.

Herbivory is another critical plant–animal interaction that can influence the rate or trajectory of succession (Bond and Keeley 2005). Vertebrate herbivores have their largest impact on early successional stages when biomass is minimal (Fig. 8.11). They can facilitate succession by reducing the competitive dominance of a colonizer that was arresting succession or by simply preferring early to late successional species. For example, the preference by moose (*Alces*) for early successional willows (*Salix*) on an Alaskan floodplain accelerated the transition to spruce (*Picea*) trees (Box 8.2; Bryant and Chapin 1986). Alternatively, vertebrate herbivores can arrest succession by decreasing the fitness of a facilitator or the colonization success of later successional species (Fig. 8.12; Farrell 1991). Herbivorous fish delayed algal succession on coral reefs (McClanahan 1997).

Figure 8.11 An experimental herbivore exclosure on the edge of the Kokatahi River, South Island, New Zealand.

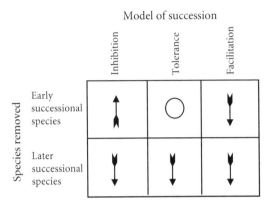

Model of succession

Figure 8.12   Consequences of herbivory in succession. Arrows indicate that herbivores increase (upward arrow), decrease (downward arrows), or have no effect (circle) on the rate of succession. The models are those of Connell and Slatyer (1977). (Modified from Farrell (1991) and used with permission from the Ecological Society of America.)

Box 8.2 Herbivores affect succession

The role of herbivory in succession is still poorly understood and the importance of herbivory varies by ecosystem. In the boreal forests of North America, arctic hares (*Lepus*) are common browsers of early successional willows (*Salix*) and their periodic population increases were once thought to be controlled by their common predator, the lynx (*Lynx*). However, research has more recently suggested that when willows are browsed they produce anti-herbivore compounds in the remaining twigs and leaves, reducing their digestibility to hares (Krebs *et al.* 2001). Less palatable willows could be, in addition to increased lynx predation, a cause of population declines. When studying willows and other early successional plants on an Alaskan floodplain (see Fig. 7.13) during a "hare high" I had to fence my study plots with chicken wire to avoid hare predation. However, the chicken wire fences did nothing to stop the other willow browser, the 2-m tall moose (*Alces*), which walked through the fences as if they were not even there!

The influence of invertebrate herbivores generally increases over successional time because their biggest effect on succession occurs when insect outbreaks damage a widespread species, potentially altering successional trajectories (e.g. by returning it to an earlier stage). Invertebrates alter succession in the same way vertebrates do, by affecting the competitive balances between pioneers and later successional species. Succession was facilitated when mountain pine beetles (*Dendroctonus*) attacked early successional pine (*Pinus*) trees, favoring more beetle-resistant fir (*Abies*) and spruce (*Picea*) (Amman 1977). Succession was arrested when a flea beetle (*Altica*) attacked a willow (*Salix*) on dunes in the central U.S., reduced its competitive dominance, and favored various herbs. Limpet grazing in a rocky intertidal habitat arrested succession by keeping later successional algae in check (Benedetti-Cecchi 2000) and sea urchin grazing arrested algal succession at a turf stage rather than allowing succession to proceed to brown macroalgae (McClanahan 1997). Both vertebrate and invertebrate herbivory can alter succession, but most studies of herbivory do not address the long-term temporal effects.

8.5.3 Animals

Animal–animal interactions occur during heterotrophic succession. Invertebrates are often the first colonists in primary succession, forming a wind-dispersed fauna such as crickets and spiders that then feed on each other (Sugg and Edwards 1998; Thornton *et al.* 1998); alternatively, they may survive a disturbance and feed on residual organic matter before plant communities develop (Hodkinson *et al.* 2002; Sikes and Slowik 2010). Animal succession usually requires plants for food or habitat structure, although the actual plant species composition may be less important (Letnic and Fox 1997). Turnover in animal communities that are closely tied to vegetation structure occurs when that structure is lost or altered (Ferreira and van Aarde 1999). At other times, animal succession is driven by animals that facilitate or inhibit each other (Majer and de Kock 1992). Animal species, like plants, can all arrive in early succession or arrive sequentially, as dispersal, prey availability, and habitat modification allow (Parmenter *et al.* 1985; Fox 1990). Many animal species contribute to habitat structure, thereby facilitating other animals. Termite (*Macrotermes*) mounds (see Fig. 5.6; Dangerfield *et al.* 1998) and ant nests aerate and fertilize the soil and provide food for predators. Tube builders on soft-bottomed, intertidal substrates facilitate the arrival of other organisms (Gallagher *et al.* 1983). Experimental transplant evidence from deep-sea hydrothermal vent communities suggests that vestimentiferan tubeworms that survive physical disruption alter subsequent colonization (Mullineaux *et al.* 2009). These results highlight the importance of legacy effects (Franklin *et al.* 1988) that are not limited to plants or terrestrial environments.

Decomposers perform critical ecosystem services by breaking down dead organic matter. The animals that do this often undergo succession driven by colonization rates, oscillations in predator–prey cycles, and the changing quality of the substrate. The first to arrive at a carcass, for example, in terrestrial environments are mobile flies (Richards and Goff 1997) in aquatic environments, mobile amphipods (Jones *et al.* 1998). Predator–prey cycles led to large population fluctuations among decomposers of sawdust in one study (e.g. bacteria, fungi, and nematodes), but displayed a dynamic equilibrium over time (Wardle *et al.* 1995b). Although decomposition involves many different organisms, their temporal patterns may not easily separate into distinct stages (Schoenly and Reid 1987).

8.6 Trajectories

8.6.1 History

Classifying successional pathways provides a useful way to describe and contrast succession at different sites. Some early studies of terrestrial (Clements 1916) and aquatic (Margalef 1968) succession assumed that a disturbed community eventually returned to its original state, after going through various stages of succession. Another assumption was that in the long-term absence of disturbance, a self-replacing climax community would establish. This climax was considered to be predictable based on the regional climate. Further examination of succession, and the increasing emphasis on non-equilibrium dynamics, led to the description of many types of trajectories, including ones that converge from several starting points, diverge from a single starting point, are circular, or form complex networks (Fig. 8.13; Lepš and Rejmánek 1991; Walker and del Moral 2003). Platt and Connell (2003) noted that most disturbances leave some biological legacy behind that substantially increases the complexity of replacement scenarios.

8.6.2 Convergent trajectories

Clements's climatic climax relied on the convergence of all trajectories within a given climate. Many different trajectories are now recognized, but **convergence** (the reduction in heterogeneity of species composition among sites over time), is still commonly observed in both secondary (Christensen

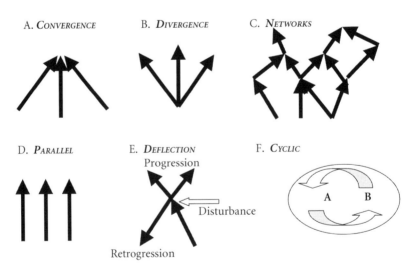

Figure 8.13   Types of successional trajectories. (From Walker and del Moral (2003) with permission from Cambridge University Press.)

and Peet 1984) and primary (del Moral 2007) succession. Convergence is common where there is a substantial biological legacy that initiates a return to pre-disturbance composition or where environmental conditions and biological variables are predictable (Nilsson and Wilson 1991). Environmental predictability occurs when there is a directional pattern of change in a resource, such as steadily increasing shade (Tilman 1988) or decreasing P availability (Peltzer *et al.* 2010). Biological predictability often involves development of a dominant growth form such as grasses, shrubs, or trees. This structural convergence can then contribute to environmental predictability. Convergence of biological traits such as root:shoot ratios, specific leaf area, or bacterial:fungal ratios is sometimes observed in response to environmental changes, but high spatial heterogeneity often obscures temporal patterns (Walker *et al.* 2010b). Convergence of different groups may not coincide in time or space, as when dung beetle convergence did not mimic convergence of trees or birds on mine tailings in South Africa (Wassenaar *et al.* 2005). Disturbance regimes that lessen in severity and frequency and become increasingly predictable over time promote convergence, but many disturbances (e.g. cyclones, earthquakes) remain stochastic.

8.6.3 Divergent trajectories

Divergence refers to the increase in heterogeneity of species composition among sites over time. This type of trajectory can be caused by differential orders of arrival of species that then establish unique successional pathways (priority effects; Lanta and Lepš 2009). High spatial heterogeneity (Matthews and Whittaker 1987; Andrew 1993) in the disturbed habitat (including gradients of severity of disturbance) or highly variable timing and duration (temporal heterogeneity) of one or several overlapping disturbances are other possible causes. Increasing diversity during succession can also lead to divergence compared with low-diversity, early successional stages (Walker *et al.* 2010b). Local convergence can occur where there is structural conformity but high regional diversity can still promote overall divergence (Lepš and Rejmánek 1991). When only two or three species dominate, divergent trajectories can sometimes be predicted.

8.6.4 Complex trajectories

Many trajectories begin with multiple starting points and feature subsequent mixtures of convergent or divergent pathways, collectively called a network (Fig. 8.13). These patterns develop when there are initially heterogeneous habitats and variable drivers that cause alterations between convergence and divergence. Different pioneer communities can lead to different sequences of species replacements, each determining the next stage independently of other parts of the network. On Mount St Helens (Washington, U.S.), post-volcanic succession on a pumice plain developed slowly over 20 yr and there was a complex network of interactions among the 10 communities

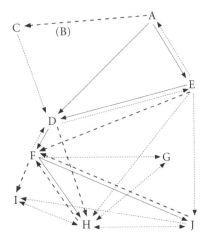

Figure 8.14   A network of successional transitions over a 20-yr period among plant community types (letters) in post-volcanic succession on Mount St Helens: solid lines, more than 100 transitions; dark dashes, 60–99 transitions; light dashes, 30–59 transitions. Fewer than 30 transitions are not shown, but included connections to community type B. Note some reciprocal transitions. (Modified from del Moral *et al.* (2010) with permission from Wiley.)

that developed (del Moral *et al.* 2010; Fig. 8.14). Environmental and biotic drivers were less important than chance, although persistent communities were more likely to develop in protected rills. Stochasticity and spatial heterogeneity in subsequent disturbances can also shape trajectories differently throughout the network in later stages of succession.

8.6.5 Other trajectories

Trajectories can also be parallel to each other, cyclic, deflected by disturbance, progressive in their characteristics, or retrogressive. Parallel trajectories feature analogous but independent development of structure or composition and are likely to occur where priority effects are strong and there is little or no interaction among seres. When sustained, they are the easiest to predict (Walker *et al.* 2010b). Cyclic succession typically occurs when there are few species and frequent disturbances that are not severe (Bokdam 2001). Disturbances can deflect a trajectory from its original direction. Multiple deflections reshape plant and animal communities by keeping them at early stages of succession. Frequent cyclones, for example, keep a large proportion of pioneer species in the flora (Brokaw 1998). Anthropogenic disturbances can also alter rates or trajectories, as when Polynesians burned wetland vegetation in New Zealand (Deng *et al.* 2006) or wood cutting altered landslide succession in Nicaragua (Velázquez and Gómez-Sal 2007).

Progressive and retrogressive trajectories represent an accumulation or loss, respectively, of biomass, structure, nutrient availability, or other ecosystem functions (see Section 5.8 and Fig. 8.6; Wardle *et al.* 2004; Peltzer *et al.* 2010). Progressive seres have been the focus of most succession research and represent the first response to most disturbances. Characterized by an in-filling of available space, development of soils, and active cycling of available nutrients and decomposition (see Fig. 5.2), progressive succession usually incorporates periods of decades to centuries, depending on the longevities of the dominant species. Retrogressive succession occurs as nutrients are depleted through leaching and immobilization into forms unavailable to plants. Soils become infertile and plant communities often change from productive forests to unproductive shrublands or bogs. The time scale for these changes is thousands to millions of years. Old soils can be rejuvenated by disturbances that reset soil development, introduce nutrients such as P, and offer plants and animals a more favorable environment (see Fig. 5.15).

8.7 Applications

The formal study of temporal dynamics began because of concerns about how to best manipulate agriculture and conserve soil (Clements 1916; McIntosh 1985). Experimental studies of succession (Keever 1950; McCormick 1968) were also linked to agriculture—particularly what happened when cultivated fields were abandoned (Cramer and Hobbs 2007). Recent links have been made between succession and management of restoration (L. Walker *et al.* 2007; Hobbs and Suding 2009b), ecosystems (Spieles 2010), landscapes (Lindenmayer and Hobbs 2007), and urban environments (Carreiro *et al.* 2008). Knowledge about temporal dynamics is vital not only to conservation and restoration (see Chapter 9) but also to effects of invasive species on biodiversity (see Chapter 6).

8.8 Conclusions

Disturbance initiates a process of community recovery called succession that has been a focus of ecological studies for over 100 yr. A rich body of theory developed and the initial impetus was to understand practical applications of temporal dynamics to improve farming in Europe and the U.S. Many field and laboratory studies have directly observed change. Some field studies use space for time substitutions (chronosequences) to avoid the problems of studying slowly changing communities. Other studies have examined the mechanistic basis of succession by manipulating species composition, soil fertility, dispersal, establishment, and sometimes even the disturbance regime. Despite numerous studies and some recent efforts to contrast temporal dynamics across gradients involving many sites, there is still not a good, predictive understanding of how communities respond to disturbance. One can argue that the ecological complexity is too great. Not only do species interact with nutrient pools, each other, and disturbances in time and space, but there are large stochastic aspects to every life stage of an organism (e.g. dispersal, germination, growth, reproduction, survival). These complex interactions sometimes result in convergence and sometimes divergence of successional trajectories. Convergence indicates a substantial biological legacy or predictable environmental conditions, while divergence occurs when there is high spatial heterogeneity or highly variable timing and duration of the disturbance. Continued efforts to unravel temporal complexity will rely on improved understanding of disturbance characteristics, species responses, and the growing influence of humans on both.

9 Management

9.1 Introduction

Humans have a wide interest in disturbances. We try to predict their timing and severity, we use them for manipulating our environment, and we attempt to reduce their negative impacts on biodiversity and **ecosystem services** (see Chapters 1 and 4). Disturbances are an integral part of our lives. To survive, humans have adapted to disturbances that we cannot manage and manipulated those that we can. The recent and rapid growth of human populations has exacerbated interactions between disturbances and humans and made modern disaster management a necessary survival strategy (see Fig. 1.1). Part of that strategy involves being flexible to the onrush of environmental and demographic changes that influence our relationships with disturbances. For example, a dependence on traditional techniques in the face of regional and global alterations of the environment may lead to local societal collapse (Diamond 2005). Expanding our knowledge about disturbance regimes and their influences on ecosystem services can help reduce potential future impacts. Wise stewardship of ecosystems includes conservation of natural areas and efficient restoration of those services following a disturbance. It also includes managing the disturbance regime, whether of pasturelands or nature reserves, to maximize not just biodiversity or ecosystem services, but the particular use that has been chosen for that parcel of land (Hobbs and Huenneke 1992). This chapter first addresses historical management of disturbance through agriculture and aquaculture. Then it examines the goals behind conservation and restoration efforts, approaches that have been used for each, and lessons that have been learned from efforts to manage disturbances. The final chapter in the book continues the topic of disturbance management at a global scale and examines how we might extricate ourselves from a purely reactive approach to disturbances.

9.2 Management history

Although humans have manipulated game animals and agricultural products for millennia, proactive responses to disturbances for conservation and reactive ones for restoration are relatively new phenomena in human history. With the removal of forests and the intensification of agriculture, early pastoralists experienced increased erosion and flooding. Responses to these disturbances included the building of terraces, weirs, dikes, canals, and irrigation systems to control flooding and promote agriculture on the fertile floodplains. Four ancient cultures developed from concentrations of agriculture along floodplains (the Tigris-Euphrates, Nile, Indus, and Yellow rivers) but had to struggle with the buildup of silt and salt in the irrigation canals (Chengrui and Dregne 2001). Some indigenous cultures established a more or less harmonious balance with disturbances and focused on the long-term health of the natural world, recognizing their dependence on it and making over-exploitation taboo. Such cultures tended to be stationary, while modern industrial societies have evolved from more exploitative and mobile cultures (Gadgil and Guha, 1992; Primack 2010). Eventual urbanization introduced problems with fires burning closely packed wooden structures and the invasion of disease-bearing fleas and rodents (e.g. the plague). The use of stone as a construction material and the more frequent replacement of straw bedding were some of the responses. Industrialization led to many more manipulations of the natural environment (e.g. levees, roads, prescribed fires, in-filling of wetlands) and novel disturbances (e.g. plowing, paving, building, mining, and trawling). As we create more disturbances daily, we also spend increasing amounts of energy alleviating their side-effects. Roads must be cleared of landslides that road construction triggered; river levees and all the land and buildings they protect must be repaired when floods inevitably overtop them; and fire damage must be repaired when over-protected forests eventually burn. Modern anthropogenic disturbances can favor certain weedy species, can have value for natural processes (e.g. fires and floods promote biodiversity; Reice 2003), and can contain lessons about the overreach of human activities, but they are rapidly destroying biodiversity and transforming ecosystems.

Formal disturbance management began to develop in Europe and its colonies several hundred years ago. The goals of disturbance management and, more specifically, conservation and restoration, depend on one's view of nature. Typical views consider nature as a provider of ecosystem services, as existing independently of humans, or as a combination of the two (Swart *et al.* 2001). Traditionally, humans have viewed the natural world as one to exploit, through fishing, hunting, agriculture, and forestry. Such exploitation reached its extreme in the exploitation of natural resources from European colonies at a time when such resources appeared to be unlimited. More modest versions of this view developed into the resource conservation ethic espoused by some colonial governments. Forest reserves were established in the 18th century in the French colony of Mauritius, throughout India, and in many other colonies. These reserves were largely established to preserve soils and maintain water supplies. Government forestry programs and the Bureau of Land Management in the U.S. were also established with a utilitarian perspective that now includes people who focus on ecosystem services such as clean air and water, flood prevention, soil erosion control, and other practical benefits for humans. In this view, sustainable development purportedly meets human needs without damaging the environment or biodiversity (Czech 2008). Disturbances are considered beneficial when they renew ecosystem services (e.g. floods that bring nutrients to a floodplain), negative when they reduce their delivery (e.g. floods that destroy crops or remove soil), or neutral when they have no effect.

The view that nature is independent of humans emphasizes the conservation of wilderness and (often large-scale) restoration of natural processes when they have been disrupted by humans (e.g. flooding in Europe or large herbivore regimes in North America). This view has most traction where there are large expanses with minimal human impact (e.g. boreal forests and tundra biomes in North America and Asia). Thoreau and Muir were advocates of wilderness preservation in the U.S. and argued that nature had an intrinsic value (Butler and Acott 2007). In this view, natural disturbances are seen as a useful part of regeneration cycles while anthropogenic disturbances are seen as interference to be minimized. This view has suffered from the criticism that it is elitist, espoused only by those who do not have an immediate need for natural resources and can therefore afford to value nature over survival. However, where natural areas have been preserved, everyone benefits in both practical and less tangible ways (Primack 2010). This view is represented in modern conservation biology (see Section 9.3.1).

An intermediate view of nature is that humans co-exist with a patchy landscape including heavily managed fields and woodlots and some less managed areas such as swamps or floodplains. Species diversity is often high in such habitats and human impact can be substantial. Many modern landscapes in eastern Asia, Europe, eastern North America, and New Zealand fit this semi-natural mosaic, at least where urban sprawl has been contained. Modern European conservation efforts often focus on maintaining such landscapes through mowing, draining, pruning, and other labor-intensive efforts (Bakker 1989; Harris and van Diggelen 2006). This attitude views natural disturbances as largely disruptive to the orderly patterns that have been established (Fig. 9.1) and anthropogenic disturbances as either useful for maintenance of the status quo (e.g. mowing) or detrimental to it (e.g. expanding urbanization).

Figure 9.1   Underpass constructed to protect vehicles from constant landslides in the unstable Arthur's Pass region of South Island, New Zealand.

Modern disturbance management has evolved from these various viewpoints and is sometimes an amalgam of all three. The Society for Conservation Biology was established in 1985 and the Society for Ecological Restoration International in 1987. These societies have helped dictate international treaties to conserve biodiversity; spearheaded efforts to protect endangered species; broadened the conservation mission of other organizations such as the World Wildlife Fund; promoted links to industry, consulting, and engineering fields; and led growing education efforts and promoted research in conservation biology and restoration. Unfortunately, critical intellectual and practical links among the fields of conservation, restoration and disturbance ecology (a relatively unorganized subdivision of ecology), and ecological succession (a century-long, central theme of ecology) have not been fully implemented (L. Walker *et al.* 2007; Hobbs and Suding 2009a). Clearly, new approaches to disturbance management that incorporate social and political realities are essential as human influences expand and anthromes replace biomes (del Moral and Walker 2007; Ellis *et al.* 2010).

9.3 Conservation

9.3.1 Goals

Conservation biology attempts to preserve biodiversity and ecosystem processes and reduce untimely extinctions of species. Most conservationists believe that biodiversity has intrinsic value that arises from the evolutionary or ecological role of a species or simply because it exists (Martín-López *et al.* 2007). Conservation biology is also motivated by economic, ecological, medicinal, recreational, or aesthetic values that species might have for humans. Conservation biology is ideally proactive relative to disturbances. It is most successful in areas where disturbances are either minimal or a well-established part of a disturbance regime. Unusually frequent or severe disturbances that endanger species or their habitats trigger reactive conservation activities to save the remaining individuals or habitats. Conservation of biodiversity is often linked to conservation of ecosystem services (e.g. clean air or water) but also to ecological processes (e.g. nutrient cycles, decomposition). Conservation contrasts with restoration, which is generally reactive and attempts to repair the damage disturbances cause to ecosystems and the ecosystem services that they provide (see Section 9.4.1).

9.3.2 Approaches

Conservation biology addresses disturbances that threaten biodiversity. It tries to reduce these threats before species are lost and preserve what remains following those disturbances that do occur. The main causes of biodiversity loss are disturbances that cause the loss or degradation of habitat, followed by over-exploitation of species, invasive species (see Section 6.4), and disease (Primack 2010). Small populations of species are particularly vulnerable to disturbances because of their size and reduced genetic diversity. The approaches that conservation biologists employ focus on the protection, management, and creation of habitats and the expansion and protection of species populations when necessary. I will first discuss the disturbances that threaten biodiversity and then the responses taken by conservation biologists.

Habitat loss, degradation, and fragmentation are the main threats to biodiversity. Organisms can be directly killed when their forest habitat burns, their lake dries out, or their spawning river is dammed. Even when habitats remain, they may be damaged in ways that make them unsuitable to support species. For example, trees in a burned forest may re-sprout, but loss of topsoil (to erosion) may hinder the recovery of understory species; a lake may remain but become too polluted to support fish; or a river may not be dammed but lose upstream nutrient inputs critical to its food webs. Degradation from pollution is insidious because it often spreads well beyond the habitat of origin, weakens the resistance of organisms to other disturbances such as disease, and may not be recognized as deleterious until the degradation is widespread. Examples of anthropogenic pollution include air-borne pollutants such as acid rain from industrial plants and radioactive fallout from bomb tests and power plant meltdowns. Levels of greenhouse gases (e.g. carbon dioxide, carbon monoxide, methane, and nitrous oxide) are also exacerbated by human activities. These pollutants cause well-documented damage to lakes, forests, many other natural areas, and human health. Water-borne pollutants include nitrates, phosphates, pesticides, and plastics that impact whole food chains of fresh and salt water bodies in various ways. Soils get polluted as well, often from liquid wastes such as oil spills and runoff from waste disposal sites. Fragmentation occurs when the spatial integrity of habitats is disrupted due to disturbance (Wilcox and Murphy 1985). Species can then be negatively affected by edge effects that result in altered microclimates; disruptions of movement and gene flow; and invasions of competitors, predators, or diseases. Roads are typical disturbances that produce fragmentation (Forman *et al.* 2003; see Section 4.5), but suburbs, industrial construction, dams, canals, power line right-of-ways, and many other anthropogenic corridors also intensify habitat fragmentation that leads to loss of biodiversity. The effect of fragmentation depends on the scale of the fragments and the matrix they are embedded in relative to the scale at which organisms perceive and use the landscape (see Chapter 7; Lord and Norton 1990).

Although natural disturbances are widespread, human activities are now the major cause of habitat loss, degradation, and fragmentation. Eighty per cent of all mammals, birds, amphibians, and gymnosperms are threatened by habitat loss and degradation (IUCN 2004). The proportion of the area within terrestrial biomes already converted to human use (e.g. agriculture and silviculture) ranges from 50 to 70% (Mediterranean-type shrublands, temperate forests, and temperate grasslands), to 25 to 50% (tropical dry forests, savannas, and deserts), to less than 25% (tropical rainforests, temperate and boreal coniferous forests, and tundra) (Millennium Ecosystem Assessment 2005). Within biomes, human population density is positively correlated with habitat damage. In the U.S., for example, although an average of 42% of natural vegetation remains, nearly all is gone along the heavily urbanized east coast (Noss *et al.* 1995). The main threats to endangered species in the U.S. are (in descending order) agriculture, commercial development, water projects, outdoor recreation, livestock grazing, pollution, roads, escape of prescribed fires, and logging (Stein *et al.* 2000; Wilcove and Master 2005). This list reflects the relative wealth of the U.S., its land-use history (logging was certainly a more extensive disturbance in the 19th century), and the location of endangered species (often growing in fertile or moist conditions).

Tropical forests and coral reefs host the highest numbers of species, and both biomes are rapidly being damaged or destroyed. Tropical dry forests have already been extensively logged or cleared, often because of expansion of human settlements and agriculture. Tropical rainforests have long been affected by small-scale, slash-and-burn agriculture where crops were grown in the ashes of small patches but then abandoned as farmers shifted to a new area of forest. Deforestation by small farmers is still expanding, particularly in Africa. However, in recent decades, huge swaths of tropical rainforests have been cleared for growing livestock (to feed developed countries: the so-called 'hamburger connection'), soybeans (often for feeding cattle), and palms for oil production. Deforestation of tropical rainforest is reducing the original area of 16 million km^2 by about 1% per year (Laurance 2007) and, at current rates, most of these centers of biodiversity will be gone in a few decades.

Coral reefs are as threatened as tropical rainforests, with 20% lost and 20% degraded; future damage from humans and from climate change is expected to continue current trends (Millennium Ecosystem Assessment 2005). The destruction of coral reefs comes largely from pollution, sedimentation from runoff following coastal deforestation, and over-exploitation of fish and other reef animals. Cyclone damage (Adjeroud *et al.* 2009), bioerosion (see Figure 3.12; Glynn 1997), and disease (Eakin *et al.* 2010) also contribute to coral damage.

Many other habitats are vulnerable to degradation and loss of species by anthropogenic disturbances. These include wetlands (Fig. 9.2), mangroves, and grasslands. Their vulnerability may arise from one predominant disturbance, but more often it is the positive feedback among multiple disturbances. For example, the loss of two amphibians (golden toad, *Bufo*, and harlequin frog, *Atelopus*) from a cloud forest in Costa Rica appears to be a combination of high temperature, unusually dry conditions, micro-parasites, and perhaps atmospheric contaminants (Pounds and Crump 2002). Extreme conditions tip an ecosystem past a threshold of self-recovery (Fig. 9.3), at which point restoration is needed to re-establish initial levels of biodiversity and ecosystem function.

Over-exploitation is a pervasive anthropogenic disturbance that often directly impacts populations of vulnerable species. Hunting and fishing are the usual methods of over-exploitation, but sales of wildlife to zoos and for the pet trade as well as sales of plants such as orchids and cacti for plant nurseries also endanger the targeted species. Both plants and animal parts (e.g. elephant tusks, tiger bones) are also collected and sold for traditional medicines. Furthermore, coral is collected to make jewelry and decorate aquaria. Hunting has negatively affected animal populations wherever human populations have expanded, but is particularly egregious today because of the efficiency of guns, mechanized transport to and from hunting areas, and the **bushmeat** trade. Fishing has a similar history, and populations of many of the world's food fish have been

(a)

(b)

Figure 9.2   Springs are vulnerable habitats due to their limited extent, water table fluctuations, and invasion by non-native species. (a) Hot spring in Ash Meadows National Wildlife Refuge, California, U.S. (b) Devils Hole; this spring is a part of Death Valley National Park, California, U.S. and is the only home of the Devils Hole pupfish (*Cyprinodon*).

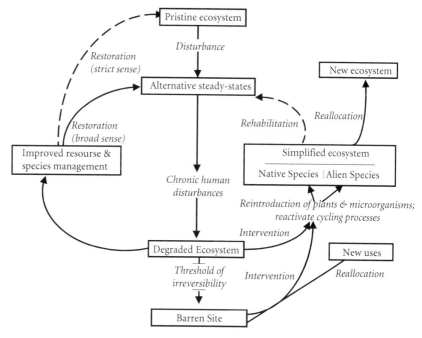

Figure 9.3   Model of ecosystem degradation and potential management responses. (Modified from Aronson *et al.* (1993) with permission from Wiley.)

over-exploited. Whales, dolphins, and turtles are also exploited. Hunting and fishing pressures can be positively correlated, as when coastal African nations export fish to Europe but then increase the harvest of bushmeat to supplement their protein needs (Brashares *et al.* 2004).

Disease is a biotic disturbance that can threaten biodiversity. The spread of diseases is facilitated by other disturbances that degrade a habitat and weaken the health of endangered populations (Gilbert and Hubbell 1996). The extent and mobility of humanity has created a ready vector for diseases that might previously have remained more geographically isolated. Humans and domesticated and wild animals can exchange diseases, particularly where there is extended contact such as through hunting or wildlife captive rearing programs. Many contagious diseases have spread from humans to captive animals, including tuberculosis, measles, and influenza (Szentiks *et al.* 2009). Disruption of food webs can promote diseases (e.g. mouse hosts for Lyme disease thrive in suburban habitat fragments) and diseases can also disrupt food webs through their effects on critical species (e.g. causing the death of a keystone species). For example, the North American chestnut tree (*Castanea*) was nearly obliterated from the eastern U.S. by a fungus introduced from Chinese chestnut trees.

Responses of conservation biologists to the loss, degradation, or fragmentation of habitats focus on the protection and management of existing habitats and the creation of new habitats when necessary. In practice, many of these approaches overlap. Habitat protection begins with the identification of the disturbance, analysis of its effects on the habitat, and clarification of the conservation goals. The loss of particular species (indicator species) can suggest when a habitat is in need of conservation (Dufrêne and Legendre 1997; Medellín *et al.* 2008). Much conservation work requires involvement in political processes to stop the bulldozers, redesign future developments, and educate the public about the values of the remaining habitat. Ecotourism can increase public interest in a site (Fig. 9.4; see Section 4.8). Conservation goals usually focus on preserving current biodiversity, but historical biodiversity (e.g. fossil beds), geological wonders, and culturally important sites are also targeted by conservationists. Biological reserves are created to preserve both habitat and biodiversity by limiting exploitation and establishing protective measures. The optimal number, placement, size, and shape of biological reserves are widely debated (Shafer 1990; Halpern and Warner

Figure 9.4   Banna Wild Elephant Valley near Jinghong, Yunnan Province, China. Tourists can stay overnight in these aerial huts and hope to see the elephants.

2003). Single, large reserves offer maximal area to edge ratios and are best for organisms that have wide home ranges and for landscapes with high habitat diversity. Several small reserves can best preserve localized populations and hotspots of biodiversity while minimizing cost and reduction in human use of the land.

Habitat management involves proactive protection of relatively pristine areas. Such protection is rare outside of established reserves, except to the extent that there are national laws or cultural taboos about exploitation (e.g. no smoking in the forest, no camping near streams). Biological reserves do not reduce natural disturbances so these are ideally incorporated into management practices. In fact, global climate change is likely to reduce the conservation value of many reserves. In response, management of reserves needs to be flexible to maximize conservation value, ecological processes, and resilience of the reserve ecosystem to expected future disturbances (Table 9.1; Chapin *et al.* 2006; Hobbs *et al.* 2010). Linking reserves with wildlife corridors on both elevational and latitudinal gradients is another approach (Lindenmayer and Nix 1993; Laurance and Laurance 1999). Habitat creation is the province of restoration (see Section 9.4).

Responses to the over-exploitation of species often include efforts to protect *in situ* populations of exploited species, expand their affected populations, and explore *ex situ* alternatives. *In situ* efforts are the first priority and involve the establishment of reserves where hunting and fishing are restricted. These measures are most successful when they are designed and implemented in consultation with local human populations and efforts to

Table 9.1 Broad strategies for promoting resilience and specific ways to promote each strategy. (From Chapin *et al.* (2006) and Hobbs *et al.* (2010).)

Reduce vulnerability	Sustain soil resources and species pools that accumulate slowly and provide buffers
	Mitigate stresses that drive change
Enhance adaptability	Foster diversity (ecological, economic, cultural, management) that protects future options
	Create capacity for learning and innovation at multiple scales
Enhance resilience	Strengthen stabilizing feedback loops but allow disturbances that help adjust to change
	Sustain legacies (ecological, cultural) to retain system memory
	Build linkages across scales (e.g. adaptive governance and landscape connectivity)
Foster transformability (ability to move to desired novel state rather than passively degrade)	Think creatively about novel solutions
	Treat crises as opportunities for constructive change

recognize the economic value that the areas of a reserve have for indigenous people. Local involvement can be done by providing special dispensation to continue low-intensity traditional hunting, fishing, and harvesting, or through participation in law enforcement (e.g. anti-poaching efforts) within the reserves. Development of new products harvested sustainably by locals from the reserves is another approach that provides both an economic and a conservation incentive. These products may be labeled and marketed as certified and sustainable (Butler and Laurance 2008).

Population expansion is an active conservation alternative that overlaps with the field of restoration (see Section 9.4). Bottlenecks to population growth are first identified (Briskie and Mackintosh 2004). Often the first task is to improve reproductive success. This is sometimes accomplished through the collection of eggs in the wild and release of young animals (e.g. sea turtles, condors), artificial insemination, or manual pollination. Other species need food or water supplements to survive harsh environmental conditions. Treatment for debilitating diseases, protection from predators or herbivores, and removal of human influences such as trampling, streetlights, or pets can be helpful to endangered organisms.

Ex situ preservation of over-exploited populations is important when population numbers are severely reduced and *in situ* conservation appears to be insufficient. Game reserves, zoos, arboreta, botanical gardens (Fig. 9.5), **seed banks**, and even cryopreservation (freezing) of sperm or other genetic material are all based on the rather tenuous assumption that rare species can eventually be returned to an expanded and healthier habitat. Nevertheless, species that are extinct in the wild (or are so rare that they no longer have any obvious ecological function) can often be maintained through *ex situ* preservation. Examples of *ex situ* preservation include the Franklin tree (*Franklinia*), Père David's deer (*Elaphurus*), and Przewalski's horse (*Equus*), although re-introductions of the latter have begun in Mongolia (Primack 2010). Additionally, there are large portions of most landscapes where over-exploited populations may continue to survive, albeit often in small or scattered populations. Examples of these intermediate habitats between nature reserves and zoos include agricultural land, pastureland, silvicultural stands, suburbs, and urban parks. Within these areas many species survive, sometimes even thriving when they learn to take advantage of suburban lawns, garbage dumps, reservoirs, and other anthropogenic habitats (Clark *et al.* 2009; Vermonden *et al.* 2009; Primack 2010).

Responses to invasive species include analysis of the nature of the invasion, limiting the invasion through eradication, **biocontrols**, or disruption of dispersal vectors and facilitating mechanisms. For example, invasion of aquatic organisms via ships' hulls and ballast water can be reduced with open-ocean ballast exchange and biocides (Ruiz and Carlton 2003). Other approaches include the promotion of native community resistance or resilience, and

Figure 9.5   Eka Karya Botanic Garden, near Denpasar, Bali that features collections of local herbs and gymnosperms from around the world and also has a tropical forest restoration project.

sometimes establishment of replacement communities that may incorporate invasive species (see Section 9.4.2). Eradication can include direct removal (e.g. hunting, uprooting), or indirect efforts such as the introduction of competitors or diseases. Eradication can lead to undesirable complications when the invasive species has a dominant role in the community. Biocontrol uses predators or parasites to control invasive species. Biocontrol has had mixed success and sometimes backfires when the introduced species negatively impacts species other than the intended target species. Prediction of the full ecological consequences of species additions is still in its infancy, so some conservationists argue for caution in the use of biocontrols (Louda and Stiling 2004).

Disruption of dispersal, establishment, or growth might mean restricting or modifying human movements (e.g. washing boat hulls to remove invasive mussels), the removal of pollinators or symbionts (host plant for an insect), building barriers such as walls, plowing the soil, burning the forest, fertilizing the substrate or making it less fertile. Community resistance can be promoted by many of the above activities while promoting resilience might also include adding competitors to the native community. Replacement communities may be necessary when an invasive species has dominated a community and is then removed. For example, the removal of extensive

---

Box 9.1 Unknown consequences of restoration

Salt cedar (*Tamarix*) was introduced from Eurasia over 100 years ago and now dominates floodplain vegetation along most rivers in the western U.S. The hybrid swarm that forms *T. ramosissima* and that now also includes *T. aphylla* (Gaskin and Shafroth 2005; Walker *et al.* 2006a) has negatively affected floodplain vegetation, wildlife diversity, and water conservation, and has increased fire frequency. Despite some perceived benefits (bank protection, habitat for several bird species), many agencies are attempting to control it. A recent release of an apparently successful biocontrol (the salt cedar leaf beetle, *Diorhabda*) will now likely have many unknown consequences (Hultine *et al.* 2010). Concern about this biocontrol agent having negative effects on native species appears to be minimal, dwarfed instead by other concerns. When removal of a dominant species occurs, how do ecosystems respond? Will other non-native species invade? Will native species return? What about the fate of birds now accustomed to using *Tamarix* habitat? Will banks erode, reducing overall floodplain vegetation and therefore food and habitat for many other species? Restoration is still an exciting gamble because we do not yet understand the cascading consequences of species removals.

---

stands of the invasive tamarisk (*Tamarix*) in floodplains of the southwestern U.S. is accompanied by planting of salt-tolerant native species because soil salinity has increased due to the decades-long dominance of the invader (Box 9.1). Alternatively, invasive species may become a desirable part of the community, such as when they are valued components of pastures or useful in the stabilization of eroded areas.

Reponses to diseases vary with the nature and vector of the disruptive organism. Insect-borne diseases are counteracted with attempts to kill or weaken a particular life stage of the insect. Coral bleaching can be partly addressed by reducing nutrient enrichment (Bruno *et al.* 2003a). People working with endangered animals (captive or wild) can sterilize clothing, wear masks, and otherwise avoid disease transmission and sick individuals can be separated from the population (Sandmeier *et al.* 2009). Finally, attempts can be made to develop disease-resistant strains and re-introduce these back into wild populations.

9.3.3 Lessons

The rise in extinction rates due to human activities highlights our responsibility to learn how to conserve the remaining biodiversity. Although aware of our role in the loss of species, we do not always realize the pervasive influence we have on ecosystem dynamics. Obstacles to conservation success include not only such ignorance, but also insufficient knowledge about the life histories of endangered species and their responses to disturbances. However, more frequently, the principal barrier to success is a lack of political will in society to effect the changes needed for proper conservation.

Rationalizations for not implementing proper conservation actions are wide-ranging. In addition to not admitting our central role in causing the damage to habitats, we tend to profess ignorance about the links between habitats and vital ecosystem services, prioritize jobs and economic growth over nature, and argue that human welfare, including the right to expand our population centers anywhere, trumps conservation. Future efforts in conservation biology need to counter these concerns and proceed with clearly designed conservation strategies that include not only appropriate participation of the community, but also long-term monitoring, innovative experiments, and modeling (Lindenmayer 1999).

Conservationists do not have all the answers, just as no one fully understands the ecological implications of the rapid rise of anthropogenic disturbances on biodiversity and the health of the habitats that support it. However, it is mostly through the century-long persistence of far-sighted individuals and the organizations that they have assembled that approximately 5% of the earth's terrestrial surface (and a growing proportion of its oceans) has been placed in nature reserves. Whether this tiny fraction is all that will remain, at least until rendered moot by climate change, depends on the immediate actions of conservationists and their success at expanding their base of support (Balmford and Cowling 2006).

9.4 Ecological restoration

9.4.1 Goals

The broadest goal of restoration ecology is to achieve sustainable, resilient, and inter-connected ecosystems that provide ecosystem services and habitat for humans and other organisms (van Andel and Aronson 2006). These lofty goals are approached through efforts to manipulate successional dynamics on disturbed habitats to reach a desired stage, typically one that resembles pre-disturbance conditions (L. Walker et al. 2007). Restoration can also be seen as an attempt to provide the ecosystem services normally provided by succession (Prach and Walker 2011). One criterion is the re-establishment of levels of genetic, species, and ecosystem diversity that adequately supply the desired ecosystem services. Problems abound in reaching these goals. An understanding of the genetic diversity of even a fraction of organisms is still in its infancy; it is estimated that less than 10% of species have been identified; and ecosystems have numerous ways in which their diversity can be measured (see Chapter 6). Ecosystem services are also complex and their relative values can be defined by their social and cultural context (Table 9.2). Although an anthropocentric context is critical to get support for restoration activities, the long-term sustainability of ecosystem services may rely on better understanding the intrinsic values of ecosystems as providers of habitats for all organisms. Sustainability and interconnectedness of ecosystems are also challenging goals, given the non-equilibrium nature of most ecosystems and the frequent loss of landscape connectivity from disturbance (Suding *et al.* 2004). Baseline or pre-disturbance reference ecosystems are likely to have changed substantially before any restoration can be accomplished. These many challenges mean that restoration in a strict sense (returning an ecosystem to its original state) is rarely accomplished and other, more modest, goals (restoration *sensu lato*) have to be identified (Table 9.3). Throughout this discussion, I will use restoration *sensu lato* unless otherwise indicated. Realistic restoration goals address specific attributes of an ecosystem that contribute to

Table 9.2 A general restoration goal is to re-establish ecosystem services that are variously categorized as values, functions, or indices of ecosystem health (summarized from Harris and van Diggelen 2006). Examples are in parentheses.

Value[1]	Function[2]	Index of health[3]
Direct use (drinking water)	Regulation (maintain essential ecological processes)	Vigor (metabolism, productivity)
Indirect use (filtration of pollutants)	Habitat (living space for wildlife)	Organization (diversity and interactions)
Optional use (recreation)	Production (resources to make consumables)	Resilience (ability to maintain function during stress)
No use (intrinsic: species survival)	Information (opportunity for cognitive development)	

[1] Dabbert *et al.* (1998).
[2] De Groot *et al.* (2002).
[3] Rapport *et al.* (1998).

Table 9.3 Variable approaches to restoration. (After Aronson *et al.* (1993) and Walker and del Moral (2003).)

Approach	Description
Reclamation	Stabilization of landscape and increase in its utility for cultivation or natural vegetation recovery; may involve abiotic (e.g. filling in mine pit) or biotic (e.g. sowing grass on new surface) modification
Reallocation	Deflection of land use and successional trajectories to increase functionality; a type of reclamation (e.g. flooded mine pit to swimming hole)
Rehabilitation	Repair of ecosystem structure and functions targeting native species (e.g. mine wastes planted with early successional species to initiate succession)
Bioremediation	The use of plants and microbes to reduce site toxicity; a type of rehabilitation (e.g. use of plants tolerant of toxins in mine wastes)
Restoration *sensu stricto*	Return to pre-disturbance ecosystem structure and function and community composition (e.g. mine wastes to forest with original composition)
Restoration *sensu lato*	Reversal of degradation and effort to return to part of the original state (e.g. mine wastes to forest of different composition or structure from the original)

overall sustainability, resilience, biodiversity, and ecosystem services (Palmer *et al.* 2005). These attributes include biotic structure, function, composition, food webs, and successional dynamics in approximate order of increasing difficulty of attainment. A variety of approaches are used to restore these attributes.

9.4.2 Approaches

Restoration of biotic structure is an explicit goal that can be attained at least for some macro-organisms. The most diverse ecosystems on earth (tropical rainforests and coral reefs) are also among the most structurally complex. Fast-growing trees and coral can at least partially replace the basic structure of damaged ecosystems within relatively short time periods after a disturbance. However, fast-growing annual plants can leave a site vulnerable to periodic erosion. Heterogeneity in biotic structure is often a useful goal that can minimize competitive exclusion by a few successful colonists, maximize use of existing abiotic patchiness, and provide optimal vertical stratification to attract pollinators, dispersers, and increase the biodiversity of organisms from all trophic levels.

The most appropriate tools for re-establishing biotic structure depend on the severity of the disturbance. Little needs to be done when disturbances are minimal in extent or severity, because adjacent organisms fill in the gaps or survivors re-establish using existing structures and nutrient sources. For example, gaps in tropical forests are recolonized quickly (Aide *et al.* 1996), often from in-growth of surrounding trees and tree seedlings when gaps are small (Walker 2000). Burned boreal forests regenerate spontaneously from serotinous seeds following fires (Greene *et al.* 1999). Active intervention accelerates restoration of structure following more severe disturbances, and can include seeding, out-plantings (particularly of nurse plants), and small dams and flow deflectors to increase habitat heterogeneity in channelized rivers (Muotka *et al.* 2002). Growing multiple cohorts in silvicultural stands can help imitate the structure of natural forests and improve biodiversity (Palik *et al.* 2002). Indirect efforts include fertilization or control of herbivores and competitors of desired colonists. Manipulations that address several goals at once (e.g. unpalatable nurse plants; Callaway *et al.* 2005) are advantageous. Severely damaged sites might initially need physical reclamation to improve stability, establish minimum fertility levels, reduce toxicity, or otherwise ameliorate harsh conditions (Walker and del Moral 2003). Under such conditions, too much fertilization can reduce opportunities for slow-growing, low-nutrient-adapted native species that often provide a more diverse structure than fast-growing, high-nutrient-adapted competitors.

Restoration of some types of ecological function is also relatively straightforward, in part because of functional redundancy among species. Establishing minimum levels of primary productivity, for example, can be accomplished by promoting the growth of a variety of types of species. Many species of algae, grasses, and herbs respond rapidly to early successional conditions with low levels of competition and abundant light and nutrients. Sustained productivity depends on either self-maintenance of these rapidly growing colonists or promotion of successional change to longer-lived and larger organisms (kelp, shrubs, trees). One concern in this regard is that the fast-growing species often form monospecific thickets that can arrest succession (Walker 1994; Slocum *et al.* 2004). Restoration of other types of functions is increasingly difficult. Establishing nutrient cycles, nutrient availability, and adequate but not excessive fertility levels are complex goals that require site-specific approaches. Abandoned agricultural fields provide perhaps the least challenge for restoration of fertility (Aide *et al.* 1996). In contrast, greater challenges are faced when attempting to restore proper fertility levels to old soils in retrogressive stages of succession (Walker and Reddell 2007), soils in infertile or arid conditions (Prober *et al.* 2009), or soils where nutrient availability is severely restricted by toxic conditions (Bradshaw and Chadwick 1980; Walker and del Moral 2003). Site amelioration is likely to be required under such circumstances, first by abiotic fertilization (or nutrient reduction), contouring, decontamination, and other means, followed by introduction of species able to tolerate extreme conditions and substantial follow-up efforts and potentially repeated introductions. Mine tailings provide extreme, low-nutrient but stable conditions, for example, and are first colonized by species able to reproduce vegetatively (Řehounková and Prach 2010). Most sites will present challenges to the restoration of fertility levels that are intermediate to these extremes, but the introduction of organisms, their temporal replacement, and on-going disturbances continue to have important effects on the nutrient status of disturbed habitats. More complex ecological functions such as the flow of energy or nutrients through several trophic levels are not easily re-created but occur as a byproduct of many more straightforward restoration activities.

Species composition is often slower to recover than structure or function, in part due to the variable dispersal and growth rates of organisms, the complexity of species interactions, and successional dynamics. Essential information about the **phenology**, dispersal, and growth of targeted organisms is often lacking (del Moral *et al.* 2007), so restoration can be by trial and error. Predictions of species composition based on assembly rules derived from species traits are potentially applicable in areas of relatively high fertility and competition (Weiher and Keddy 1999; Fukami *et al.* 2005), but less useful where low fertility and severe disturbances predominate (Walker *et al.* 2006b). Critical (keystone) species may be needed to re-establish certain functions, food webs, temporal and spatial patterns, and overall heterogeneity (Hobbs and Norton 1996; Ehrenfeld 2000). Alternatively, functional redundancy of species can result in a variety of successional trajectories rather than a return to a desired reference community. Such unpredictability suggests that species composition is not an ideal measure of restoration success. Alternatives such as structural complexity, biodiversity, or reproductive success are perhaps more appropriate (del Moral *et al.* 2007). In general, heterogeneous mixtures of species can improve productivity, resistance to invasion, and enhance ecosystem functions over more homogeneous mixtures (HilleRisLambers *et al.* 2004; Fargione and Tilman 2005).

Targets of species composition for a particular restoration project are dependent on such local conditions as the degree of damage (severity of disturbance), the availability and proximity of potential natural colonists, or the viability of introduced species. More extensive disturbances will take longer to recover because appropriate species are more distant (Fig. 9.6). Different mixtures of native or non-native species may be desirable. Local plants are best suited genetically for restoration, but if the disturbed habitat no longer supports them, mixtures of genotypes from various sources may be appropriate (Lesica and Allendorf 2002). If erosion or other on-going disturbances are severe, non-native plants may be appropriate in the short term, even when they inhibit native recolonization in the long term (Cargill and Chapin 1987). In addition, the original, pre-disturbance community may be difficult to re-create if it existed only in the distant past or has shifted in composition since the disturbance. Changes in background environmental conditions (e.g. climate, hydrology, human impacts) may require the use

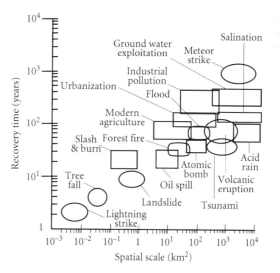

Figure 9.6   The relationship between the spatial scale of natural (ovals) and anthropogenic disturbances (rectangles) and the approximate duration of ecosystem recovery. (From Dobson *et al.* (1997) with permission from the American Association for the Advancement of Science.)

of a new reference community that is better able to cope with current conditions (Harris and van Diggelen 2006).

Restoration of food webs is a critical but difficult step toward self-maintenance of ecosystems. Establishment of productivity, adequate nutrient levels, and biodiversity are intertwined with the reorganization of food webs and nutrient cycles. For example, eutrophication of lacustrine ecosystems by P from sewage and fertilizers results in reduced aeration, increased water temperature and turbidity, increased algal growth, and, when extreme, loss of fish populations. Restoration depends on stopping the immediate inputs of P and also decreasing long-term inputs from soil and lake sediment pools (Carpenter 2003). Shallow lakes are particularly difficult to restore to their clear-water condition with submerged vegetation and invertebrates that support birds and fish because the turbid, algal-dominated condition of a eutrophic lake is self-reinforcing (Scheffer 1998). Without large zooplankton (that use vegetation as protection) to control phytoplankton, large vegetation decreases, exposed sediments increase turbidity, and sediment-dwelling invertebrates predominate, along with fish that feed on that kind of prey. Restoration usually requires more than a reduction in P, including such efforts as alteration of water level or removal of fish (Carpenter and Scheffer 2009).

Other food web disruptions in lakes occur with invasive fish species, such as the Nile perch (*Lates*) that resulted in the loss of many species of native cichlids in Lake Victoria, Africa (Goudswaard *et al.* 2008) or the rainbow smelt (*Osmerus*) that threatens piscivores such as walleye (*Stizostedion*) in the Great Lakes in North America (Mercado-Silva *et al.* 2006). The invader, often supplemented by changes in fishing pressures and water quality, can disrupt native food webs by creating shifts from dominance by forage fish to piscivores (Lake Victoria), or vice versa (Great Lakes). This negative relationship is mediated by the balance between consumption of juvenile piscivores by forage fish and consumption of adult forage fish by piscivores (Carpenter and Scheffer 2009).

Soil–plant–animal feedback involves complex food web interactions that are important to consider for terrestrial restoration. Plant species influence the growth and composition of soil organisms through the quality, quantity, and timing of litter inputs that they supply to decomposers (Eviner and Chapin 2003; Porazinska *et al.* 2003). In turn, soil organisms directly affect plants through associations with plant roots (e.g. mycorrhizal fungi, root herbivores, root pathogens) and indirectly through decomposition and nutrient availability (Van der Putten 2005). Disturbances alter these relationships immediately by damaging plants and soil organisms, thereby disrupting litter production and soil habitats. Longer-term consequences occur during the successional response to the initial disturbance. Shifts in plant species composition, for example, trigger shifts from microbial to fungal domination, from *r*-selected to *K*-selected nematodes, and from arbuscular to ectomycorrhizal fungi (Wardle and Peltzer 2007). Disruptions by herbivores such as deer or moose can either retard or accelerate succession, depending on whether they stimulate decomposition through high-quality litter or reduce decomposition through low-quality litter, respectively (Bardgett and Wardle 2003).

Non-native invasive plants, usually responding to disturbances, often have critical impacts on soil food webs. Invasive plants that add nutrients through N fixation (Vitousek *et al.* 1987) or P accumulation (Chen *et al.* 2003; Bellingham *et al.* 2005b) can shift nutrient availability in their favor. Restoration attempts to reduce nutrient availability (e.g. by adding sawdust that leads to microbial immobilization of nutrients) are not generally effective (Wardle and Peltzer 2007). Sometimes local pathogenic fungi or nematodes control non-native invasions (Reinhart *et al.* 2003), while soil biota can either aid or deter invasive plants (Packer and Clay 2004). Restoration is aided by these various plant–soil organism linkages when they are predictable and susceptible to manipulation, but this is not always the case. More awareness of the presence of these feedbacks and their potential influence on succession and invasions is a first step toward increasing our understanding about how plants, soils, and herbivores interact following disturbances.

Restoration of successional dynamics is a challenging but worthwhile goal for restoration—in part because it envisions an ecosystem that undergoes change with minimal or no human intervention. However, it is probably the most difficult restoration task and assumes some level of success in the preceding and more achievable steps of restoring productivity, biodiversity, nutrient cycles, and food webs. A multi-pronged approach can address the successional dynamics at several points in the process, much like greasing a machine or igniting a fire at several places. For example, a disturbed site can be physically reclaimed, seeded (to bypass potential dispersal barriers), and then have the ecosystem manipulated until a desirable successional trajectory is initiated (Fig. 9.7). However, exact mimicry of natural succession is unwise because it often proceeds too slowly to meet the frequently urgent goals of re-establishing ecosystem services (Dobson *et al.* 1997) or does not result in desired levels of biodiversity or heterogeneity (del Moral *et al.* 2007). Nevertheless, at intermediate levels of productivity and with anthropogenic disturbances, spontaneous succession can be the most successful approach to restoring successional processes (Prach 2003; Prach *et al.* 2007). Successional research has provided numerous lessons about disturbance, species interactions, and trajectories of recovery that are useful to the practice of restoration (Table 9.4). If disturbances are allogenic, then restoration must adjust to the realities of the disturbance regime, but if they are autogenic, species interactions can be manipulated. Intermediate disturbance frequency and severity tend to promote successional change by thinning dominant species, and high or low levels of frequency or severity tend to have more stable and predictable trajectories. Restoration responses

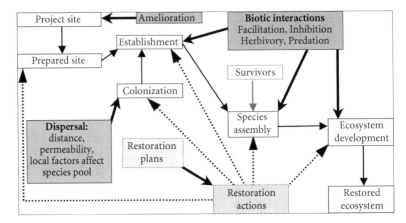

Figure 9.7   Effects of natural and human (restoration) mechanisms that drive ecosystem development through succession (thin arrows). The three darkest boxes represent natural processes and their influences (thick arrows). Dotted arrows represent where restoration activities affect ecosystem development. (From del Moral *et al.* (2007), with permission from Springer.)

range from a focus on competitive interactions where both frequency and severity are low to questionable success of any restoration where both are high (Table 9.4; Walker and del Moral 2009a). When disturbance interactions are prevalent, restoration can focus on providing a range of life spans and disturbance adaptations, and when novel disturbances are present, species in naturally analogous situations can be utilized. Finally, if maximum biodiversity and spatial heterogeneity are desirable, one could promote intermediate levels of disturbance that spatially overlap.

Succession also provides lessons about species interactions that have restoration consequences (Table 9.4). To improve on the largely stochastic process of colonization, the addition of multiple species is usually desirable. To promote successional replacements, one must know the type of replacement sequence that normally occurs and try to emulate it. Restoration must also address dominant species and decide if they are to be allowed or avoided. When positive (facilitative) interactions among species predominate, these can be cultivated to promote desired change, but when negative (competitive) interactions predominate, care must be taken to avoid arrested succession, unless that is the goal. Shrub thickets, for example, require minimal input (e.g. power-line right-of-ways where use of herbicides can be minimized; Niering 1987).

Successional trajectories can also be manipulated with judicious use of restoration tools. Most restoration projects want succession to proceed in a normal fashion, but the rate of change can be accelerated or slowed by the life spans of the species that are introduced. Retrogression may be desirable when practitioners of restoration would rather reduce diversity, productivity,

Table 9.4 Succession-related topics, lessons from succession, and applications to restoration. Modified from Walker and del Moral (2009b) with permission from Island Press.

Topic	Lessons from succession	Applications to restoration
Disturbance		
Allogenesis	Off-site factors can alter trajectories	Adjust goals to realities of the disturbance regime
Autogenesis	Within-site factors drive small changes.	Consider species interactions
Low F, low S[1]	Trajectory predictable	Minimal impact by disturbance; address competition instead
Low F, high S	Impact unpredictable	Difficult to plan for; introduce stress tolerators?
High F, low S	Fast-growing pioneers dominate	Utilize or address problems of weedy species
High F, high S	Little biological legacy	Evaluate if restoration is even possible
Interactions	Complex impacts on trajectories	Aim for a range of life spans and disturbance adaptations
Novel types	There are often natural analogs	Experiment with species found in natural analogs
Diversity	High at intermediate disturbance	Promote or allow some disturbance
Patchiness	Supports multiple stages	Use spatially explicit methods; overlapping stages OK
Species interactions		
Colonization	Survival low, dispersal stochastic	Add excess of desired species from multiple life forms
Replacement	Direct, sequential, cyclic, fluctuating	Use methods best for dominant type of replacement
Dominance	Often but not always arrests change	Avoid unless dominant species belongs in desired community
Facilitation	Importance varies with fertility	Use selectively to promote change
Competition	Can increase diversity or arrest change	Avoid thicket-forming species
Trajectories		
Progressive	Most often desired	Promote short-lived stages
Retrogressive	Sometimes desired	Reduce fertility or diversity
Arrested	Relative to time scale; common	Promote only if matches goal
Hysteresis	No system lacks history	Actions impacted uniquely by each system

[1] F = disturbance frequency; S = disturbance severity.

structural complexity, or nutrient availability. Such conditions may require intervention with disturbances such as targeted fires or manipulations of ground water, carbon additions to reduce fertility, or manipulation of grazing regimes (Schrautzer *et al.* 2007).

Disturbance regimes, both past and future, will aid in understanding the context for any restoration project (Chapman 1999). The history of a site determines its structure, fertility, composition of available propagules, and resilience. Future disturbances will provide both opportunities and challenges for restoration. Restoration of low-diversity ecosystems adapted to frequent disturbances can be relatively straightforward, with the frequent disturbances resetting any errors that are made or at least providing a fresh template to try again. Naturally resilient ecosystems such as rivers can also sometimes be readily restored (Muotka *et al.* 2002). Restoring rivers means promoting processes that can again be self-sustaining, such as channel movement, river-floodplain exchanges, retention of organic matter, and biotic dispersal (Palmer *et al.* 2005) (Box 9.1). These efforts lead to improved spatial and temporal heterogeneity and connectivity among different elements of a floodplain that were initially promoted by natural disturbance regimes and destroyed by anthropogenic ones (Fig. 9.8; Ward 1998). Fire and floods are necessary disturbances for the maintenance of many successional communities (e.g. birds; Brawn *et al.* 2001) and can introduce heterogeneity that promotes biodiversity (del Moral *et al.* 2007). Active use of disturbances such as fire is a part of many restoration efforts (Baker 1992), although cultural and practical concerns limit its use. In addition, fields are sometimes plowed to promote meadows (Kohler *et al.* 2011) or rewetted to favor conversion to carrs (Schrautzer *et al.* 2007). Species are removed to reduce competitive exclusion (Walker and del Moral 2003) or species are introduced to enhance restoration goals with seeds or seedlings (Prach *et al.* 2007; Antonsen and Olsson 2005), animals (Sarrazin and Barbault 1996; Seddon *et al.* 2007), or mycorrhizae (Allen *et al.* 2005). However, re-occurring disturbances, whether natural or a feature of the disrupted site, can be problematic for restoration projects when they kill vulnerable colonists, decrease critical functions, or alter successional trajectories.

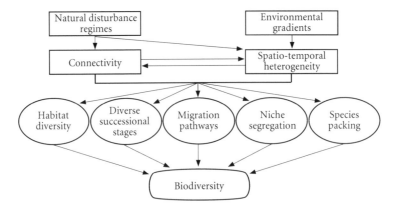

Figure 9.8   Proposed interactions that structure biodiversity patterns in riverine landscapes. (From Ward (1998), with permission from Elsevier.)

Measures of restoration success are largely site-specific and reflect the many goals that direct restoration efforts. Generally, but not always, these goals include increased primary (and secondary) productivity, structure, productivity, and fertility. Sometimes increased biodiversity, complexity of food webs, re-establishment of natural successional transitions, and resilience to on-going disturbances are desirable. Decreasing intervention is certainly a laudable goal, as an ecosystem becomes more independent of manipulations, but continual monitoring and additional manipulations are the norm. Rapidly changing disturbance regimes, including the prevalence of often unpredictable anthropogenic disturbances, make restoration an on-going process of adjustment. Even the resilience of a restored ecosystem is best defined in terms relative to its own performance and the current disturbance regime, rather than in comparison with a reference ecosystem that may no longer exist or that may have an altered resilience due to a new disturbance regime. Trade-offs are likely among different goals, and success can best be measured in terms of each one, even when these conflict. For example, biodiversity may have increased due to restoration activities, but rare pioneer species may have become locally extinct. Productivity may have increased at the cost of spatial heterogeneity. Any measure of restoration success, however, must consider the response of the manipulated ecosystem to the current disturbance regime.

9.4.3 Lessons

Restoration success is dependent on the utilization of all available tools, from theoretical contributions from various fields (Table 9.5) to proper adjustments to local realities. Restoration projects are undertaken to improve or sustain ecosystem services, and these can be ecological (e.g. re-establish productivity or maintain biodiversity), to societal (e.g. improve aesthetics or recreational opportunities) (Hobbs and Norton 1996). Ecological restoration is the manipulation of temporal dynamics (i.e. succession) that may be only partially understood. Therefore, while restoration relies on what is known about succession, restoration efforts also inform us about succession and can contribute to the development of ecological theory (L. Walker et al. 2007). Successional lessons for restoration include considerations of broader temporal and spatial scales than local projects might otherwise consider (Kauffman et al. 1997); guidance about expected trajectories due to initial site fertility; implications of species interactions with the landscape (e.g. for dispersal or herbivory) or with each other (e.g. competitive exclusion or facilitation); and increased awareness of the role of invasive species (Hobbs et al. 2007). Restoration goals often involve accelerating successional processes, but longer-term consequences of manipulations need to be monitored and used to adjust not only further restoration practices but also sometimes the goals. Understanding disturbance regimes is critical to both the initial restoration activity and subsequent adjustments, particularly in light of on-going disturbances.

Table 9.5 Fields of study that influence restoration, the impact each has on restoration, and potential problems conducting restoration without addressing the contributions from each field. (Modified from Hobbs *et al.* (2007) with permission from Springer.)

Field	Impact	Problems if omitted
Succession	Predictions of temporal change and target-setting	Wrong trajectories attempted, wrong species mixtures used, arrested trajectories, goals too narrow
Assembly	Filters	Difficulty starting, mismatch of propagules and environment
Landscape ecology	Regional information exchange	Efforts too spatially limited, lack of information on role of surrounding biota
Disturbance ecology	Initiation and boundaries	Multiple restarts, loss of biomass and organic matter, inadequate site stabilization
Climate change	Shifting reference systems	Old communities not suitable as reference sites or propagule sources, complications of novel communities
Historical ecology	Legacies	Wrong targets used, belowground influences ignored
Environmental ethics	Aesthetics, goals relating to naturalness and place	Societal values not met, visually unappealing, ill-fitting targets, lack of public support

9.5 Conclusions

Human alterations of the environment create new and widespread challenges for the management of disturbances. The rapid destruction of biodiversity means that we are facing the challenge of conserving increasingly smaller patches and populations. Furthermore, many of the remaining patches are spatially and temporally isolated. Spatial isolation limits dispersal and connectivity; temporal isolation removes variability in representation of different successional stages (Boughton and Malvadkar 2002) and makes establishment of successional trajectories less likely. Are some populations so small or isolated that they are ecologically, or at least evolutionarily, extinct? Should the limited resources for conservation be directed at saving those species that have broad public support (e.g. charismatic megafauna), that provide critical ecological functions (e.g. keystone species), or ecosystem services (e.g. clean air and water), or those that contribute maximal benefit to human health (e.g. medicinal plants)? Ultimately, political support for conservation will have to increase to offset the current rapid rate of destruction. Unfortunately, connecting human actions with their ecological consequences has not been humanity's strong suit, despite decades of accumulating evidence of both our immediate and long-term effects.

We are just beginning to understand how to restore severely disturbed habitats, such as abandoned urban areas or waste heaps. Restoration is many times more difficult than conservation because of the complexity of ecosystems. Even with large budget increases, restorationists would not be able to repair all damaged habitats. In addition, climate change is likely to impact on restoration trajectories (Harris *et al.* 2006), expanding our challenges in unknown ways. While some restoration activities targeted at widely recognized ecosystem services (e.g. erosion control on slopes above urban residences or habitat improvement to avoid species extinctions) will continue to be supported, it appears that the best strategy for restorationists is to increase their support for conservation. Ecosystems have remarkable abilities to recuperate, even after radiation releases (e.g. the Chernobyl power plant accident; Davydchuk 1997) or toxic mine tailings (Řehounková and Prach 2010), provided there are natural areas nearby. However, the scale of anthropogenic disturbances suggests that many habitats have reached tipping points where natural regeneration is unlikely and human intervention is essential (Hobbs and Norton 1996). In such habitats, the skills of restorationists are still needed, even if only to reclaim some of the original functions of the damaged sites and reduce the length of time for which habitats remain degraded (Dobson *et al.* 1997).

Ecological carrying capacities are approached either gradually or rapidly. Human population growth, although slowing, is still out-pacing most solutions to the problems it creates. This situation suggests that humanity may face many self-created and large-scale disturbances in the next few decades. Our ability to respond effectively to current large-scale, natural disturbances (e.g. cyclones, tsunamis, blizzards) is marginal because we live on an increasingly crowded planet. Anthropogenic disturbances exacerbate the natural ones and provide additional challenges. The fundamental tools for managing disturbances are known but societal priorities do not seem aligned with either conservation or restoration at present, particularly at regional to global scales.

10 Global concerns and future scenarios

10.1 Introduction

Humankind is careening toward a potentially dangerous future. Having so far avoided the ultimate disturbance of nuclear annihilation, we appear to be headed inescapably toward another severe disturbance, but one with a more subtle onset: the collapse of human civilization due to the abuse of resources. To avoid this collapse, we must learn how to manage our natural resources in a more sustainable manner. On the one hand, the rich have "affluenza," the illness of over-consumption that is rapidly squandering natural resources that are extracted from around the world in pursuit of a comfortable lifestyle. On the other hand, the poor reduce local resources to find food and fuel necessary for daily survival. Human-aggravated changes in global climate patterns exacerbate and intensify natural disturbances, leading us into an uncertain future characterized by increased variability and intensity of tornadoes, cyclones, floods, droughts, fires, and insect outbreaks. Humans have also contributed many novel disturbances that add complexity to disturbance regimes. Pumping aquifers dry for short-term gains in agricultural productivity deprives future generations of water security. Losses of irreplaceable topsoil to erosion from agricultural practices and expanding urbanization reduce our ability to feed a growing population. Extensive agriculture has also led to human alteration of the global N budget (Vitousek *et al.* 1997a) and depletion of the world's readily available P (Gilbert 2009). Deforestation reduces carbon sequestration and increases global warming through the production of the greenhouse gas carbon dioxide. Over-fishing, tourism, and acidification from industrial pollution damage once pristine oceans and lakes. The buffering capacity of ecosystems against disruption is threatened by human pressures on many fronts. The number of humans the earth can support depends on how many resources each of us uses and the trend is not toward reduction but rather toward increased consumption, particularly as once poor nations (e.g. India, China) emulate the consumption patterns of the richer nations.

To some, the signals of increasingly intense and unpredictable disturbances are not the sign of impending global chaos but a challenge to be met by human ingenuity. Technologists propose large-scale solutions such as shading the earth to reduce global warming or fertilizing the oceans to increase carbon dioxide consumption, in addition to pursuing dreams of using fusion for energy or improving the efficiency of photosynthesis to feed a growing population. Humanists argue that technological approaches will always be inadequate and that what is needed is a fundamental change in attitudes toward nature and the consumer lifestyles that threaten our natural resources. Fatalists assert that humankind is not going to solve its problems and that we are inevitably headed for the collapse (through wars and famines) that often occurs when a population outgrows its resource base (exceeds its carrying capacity; Diamond 2005). Any approach that slows rampant development and resource abuse will lessen the intensity and severity of the altered disturbance regimes that will characterize our future.

In this chapter, I explore the links between the spiraling intensification of disturbances and the two most pressing challenges humans face: global climate change and overpopulation. Then, after reviewing lessons garnered from disturbance ecology, I present several pathways toward a more hopeful future where our interactions with disturbance regimes are more balanced and we become less vulnerable to disturbances than we are now. I argue that essential to any responsible future is an expanded understanding of how humans directly disturb their environment and indirectly exacerbate natural disturbances. We need to integrate what we have learned about the predictability of disturbances and the damage they cause with our growing insights about how to conserve ecosystems or help them to resist damage as well as how to restore or improve ecosystem resilience. With that integrated understanding we can begin to reverse negative trends or even evade our dangerous future.

10.2 Global climate change

Perhaps the most dramatic example of an anthropogenic disturbance, and the one with the most far-reaching consequences, is the alteration of global climate. Beginning with the Industrial Revolution in the 18th century, human activities have substantially raised the concentrations of greenhouse gases (particularly carbon dioxide, methane, and nitrous oxide) in the atmosphere, leading to greater heat entrapment and rising air and water temperatures. Ocean acidification is a direct effect of increasing carbon dioxide levels and is discussed in Section 5.6.2. Warmer air and water are having global repercussions on ecosystem processes and constitute a critical anthropogenic disturbance. Temperature changes and related changes such as increases in the frequency and intensity of wildfires are likely to affect between 10 and 50% of terrestrial vegetation (Gonzalez *et al.* 2010). Regional and local effects are still difficult to predict, but temperature increases will probably be most pronounced at high latitudes and in continental regions. Particularly vulnerable are arctic and alpine tundra and boreal forests. Methane releases from melting permafrost increase global warming, leading to further melting in a positive feedback loop (Walter *et al.* 2006). Melting of arctic ice from rising temperatures decreases the reflectivity of northern latitudes, leading to further warming of the Arctic Ocean and accelerated melting in another feedback loop. One currently high-profile consequence of this loss of ice is the plight of the polar bear that depends on the arctic ice pack for hunting seals. Higher temperatures create many other disturbances, including melting of alpine glaciers and ice sheets in Greenland and Antarctica. Agriculture in much of Asia is dependent on snowmelt from the Himalayas and neighboring mountain ranges, and would be devastated if rivers run dry. Given that India and China produce more wheat and rice than other countries, widespread food shortages are a likely outcome (Brown 2009). Greenland's ice sheet is melting much faster than expected as melt water seeps deep into the ice, warming it and lubricating its movement toward the sea (Zwally *et al.* 2002). Additions of large amounts of fresh water to the North Atlantic Ocean may alter ocean circulation in ways that could have profound effects on global climate (Vellinga and Wood 2002). Substantial melting of ice sheets could inundate many islands, coastal wetlands, and urban areas. Rises in sea level are already affecting coastal Inuit villages at northern latitudes.

Higher temperatures will also intensify many weather patterns. Recent evidence suggests that wind speeds over the world's oceans are increasing, leading to higher waves and perhaps increasing atmospheric humidity (Young *et al.* 2011). In some regions, cyclones and floods are likely to become more intense and possibly more frequent, in part due to more moisture in the warmer air. In other regions, droughts and fire may intensify, due to both warmer temperatures and reduced precipitation (Solomon *et al.* 2009; Veblen *et al.* in press). In many locations, the phenology and behavior of plants and animals is already responding to seasonal shifts. Migratory birds are arriving earlier and plants are flowering up to several weeks earlier than even 40 years ago (Fitter and Fitter 2002). Movement of ecotones is expected, as some forests become savannas or grasslands, as in central North America (Frelich and Reich 2010), and deserts turn into dust bowls affected by megadroughts (Seager *et al.* 2007). Climate change may also facilitate the spread of invasive species, due to their competitive advantages at higher temperatures and levels of atmospheric carbon dioxide, increased N levels, and more habitat fragmentation (Weltzin *et al.* 2003).

There are many unknowns about the specific implications of climate change, but the human contribution to climate change and its causes are clear. Management of the problem thus involves altering the anthropogenic disturbances that have led to global warming, the extraction and burning of fossil fuels. Oil production has probably peaked (Bardi 2008), but extraction and burning are likely to continue for some time, augmented by natural gas, oil extraction from Canadian shale, and continued coal production in China and the U.S. Reducing carbon dioxide emissions by 80% by 2020 is a goal set by climate scientists to keep carbon dioxide levels below 400 ppm, setting the stage for reductions to 350 ppm—a level needed to avoid expected runaway effects of climate change (Hansen *et al.* 2008). In early 2011, carbon dioxide levels were 391 ppm and rising faster than many had expected. Tools for managing this global disturbance are well known, but the political will to fully implement them is present in only a few nations. A transition to renewable energy sources presents some novel types of disturbances (e.g. visual pollution and disruption of bird and bat flight by large wind turbines; reliance on various limited resources to manufacture solar panels; disruption of natural and sacred areas by new power-line corridors; uncontrolled releases from geothermal and nuclear power plants); and plenty of economic and technical challenges, in addition to political ones. However, the environmental benefits of using less-polluting renewable sources versus polluting fossil fuels are clear. A massive, international campaign is needed immediately to begin this important transition. Such a change is technically and economically feasible, with renewable energy sources replacing 100% of coal and oil and 70% of gas now used to generate electricity and 70% of fossil fuels used in transportation by 2020 (Brown 2009).

10.3 Overpopulation

Cohen (1995) asked how many people the Earth can support (Fig. 10.1, Box 10.1). The answer is complicated because it depends on the time frame and level of resource consumption. Levels of resource consumption, in turn, reflect the fluctuations of human populations and our effects on disturbance frequency, intensity, and severity. Assuming a time frame that implies a concern for the well-being of seven generations into the future, following some Native American cultures (Graham 2008), the answer then revolves around what lifestyle (consumption level) we are willing to contemplate. Estimates range from one to several hundred billion, but most converge between 7 and 12 billion people living at a modest level of consumption (perhaps analogous to consumption levels of eastern Europeans before the collapse of the Soviet Union; Cohen 1995). My view, and that of many ecologists, is that these estimates are overly optimistic, given the unsustainable levels of resource use we currently maintain.

Carrying capacity can be seen as a fixed number of individuals that the earth can support. Population growth then approaches that value in one of several ways. A sigmoidal growth curve gradually approaches capacity and may or

Figure 10.1   An "umbrella jam" that occurred when crowded pedestrians at the Gion Festival in Kyoto, Japan all raised their umbrellas simultaneously as it started to rain.

may not oscillate around it. A sharp overshooting of capacity will theoretically result in a subsequent population crash or a stable or unstable oscillation. However, a more realistic view is that the human carrying capacity of the earth is altered by human activities (Cohen 1995). Anthropogenic disturbances that increase carrying capacity (at least in the short-term—because more resources lead to more people who use more resources) include mining of ores, fossil fuels, and other resources, as well as improvements in agricultural scope and efficiency. Examples of increased scope include expansion of farming and grazing to former forests; increased efficiency includes reduction of erosion through contour plowing, rain harvesting in arid lands, and the near universal application of fertilizers to improve production. Agriculture can be seen as a complex mixture of managed anthropogenic disturbances including regular soil disruption through plowing, fertilizing, and harvesting. Anthropogenic disturbances that reduce capacity include over-use of natural resources (e.g. over-fishing, deforestation) or damage of resources (e.g. pollution). Carrying capacity can also fluctuate, either in or out of synchrony with human populations. Humans have expanded their carrying capacity through technological advances that allowed increased exploitation of soil, water, and fossil fuels, but reductions in over-exploitation of natural resources can only be achieved by stabilizing population growth in combination with a reduction in per capita resource use.

Box 10.1 The world is getting crowded

People have different tolerances of crowds and definitions of adequate personal space, often based on culture and upbringing. But what we can all agree on is that the world is getting increasingly filled with people (Fig. 10.1). If you live in a city you may measure crowding by how often you get a seat on the subway or how long you wait for a table at your favorite restaurant. Suburbanites may feel crowded when they can't find a parking place in front of their house or their commute to work takes another half hour. Rural dwellers notice when vacant lots fill with new houses and there is increased traffic on the one road through town. When we go on vacation, we also notice that crowds are growing. Tourists jostle each other to snap the perfect photograph of the breaching whale, cascading waterfall, or performing dancer. With my rural upbringing, I feel uncomfortable packed sardine-like in a Tokyo subway or an Egyptian tomb and prize the tour guide who can avoid the crowds. I am always looking for a natural area such as a park or arboretum in which to recoup my sanity apart from the jangling noisiness of the city streets. In my home region, I notice lakes, canyon trails, and mountain glens are becoming increasingly crowded every year with both tourists and locals. Overcrowding can be merely an aesthetic disappointment (e.g. seeing anyone else at all in a remote wilderness area) or a direct reduction in quality of life (e.g. no dinner because the soup kitchen ran out before you got to the head of the line). Either way, finite resources are being shared among more people and we are all beginning to feel crowded. When will our human population stabilize and at what level?

10.3.1 Water

Available fresh water is a resource that limits human population growth in various ways. About half of all humans suffer from diseases related to a lack of adequate levels of uncontaminated fresh water (WHO 1992); these diseases are a leading cause of childhood and adult mortality. Water shortages, often due more to wasteful transport and usage than to lack of supply, reduce food production and shorten human lives. Disturbances that disrupt water supply and quality include natural processes such as melting glaciers, flooding, and erosion that add silt and can alter drainage patterns. Anthropogenic disruptions arise from the direct and indirect use of ground and river water, water contamination, and landscape to global climate alterations. We use water directly for drinking, bathing, cleaning, sewage disposal, and recreation and indirectly for agriculture, aquaculture, power generation, manufacturing, and industrial cooling. In addition to sewage disposal, we contaminate water through additions of fertilizers, herbicides, pesticides, medicines, cleaning solutions, and other chemicals. Diseases spread in contaminated water. We reduce the quantity and quality of water at larger scales when deforestation reduces water retention, agricultural irrigation increases soil salinity (Hillel 1991), and rising oceans flood fresh water aquifers, as in Bangladesh. Food production is water-intensive, using approximately 2000 liters of water to produce the food each person consumes in a day (Kijne 2003). Irrigation often relies on depleting aquifers, thereby favoring food production for this generation at the expense of water and food security for future generations (Brown 2009). Cohen (1995) estimated upper limits of world population depending on several assumptions of supply and demand. When enough water was used to irrigate wheat production to feed everyone an adequate diet (that supplied at least 2000 kilocalories of energy per day), upper limits ranged from 5–8 billion. In 2011, the human population reached 7 billion, suggesting we are at or very near carrying capacity in terms of water resources.

Water management begins with disturbance management such as through reforestation of slopes that reduces flooding and erosion, and reduction of pollution that improves water quality. Additionally, recycling reduces the water needs for manufacturing and power generation; increased efficiency of water delivery to homes, agriculture, and industry reduces water waste; and economic approaches such as increased pricing can reduce usage and reflect the ecological importance of water. Desalinization plants are options, but have drawbacks in their high energy use and heat pollution of coastal waterways. Geothermally heated water reduces the energy needed for heating domestic water for bathing and cooking (Fig. 10.2). Extensive ground water pumping, canals to divert rivers, and shipping icebergs from cold, wet regions to warm, dry ones are all stopgap measures with high financial and sometimes environmental costs. The expanding demands of a growing population for clean, affordable water are offsetting management efforts, and water availability will undoubtedly increase in importance as a factor regulating human population growth.

10.3.2 Soils

Human population growth led first to extensive use of soils for agriculture in fertile valleys and then to farming on increasingly marginal soils often prone to erosion or desiccation. Industrialization then led to more intensive disturbances using cultivation methods that focused on irrigation that led to saline soils, deep plowing that led to a loss of soil structure, and chemical fertilization that led to eutrophication of drainage waters. Conversion of grassland into cropland is often associated with losses in organic matter (which add to atmospheric carbon dioxide) and water-holding capacity (Evrendilek *et al.* 2004). The intensification of agriculture has made soils increasingly susceptible to erosion; topsoil is now eroding faster than it is being formed on about one-third of the world's croplands (Wali *et al.* 1999). Furthermore, overgrazing contributes to desertification as seen across huge swaths of northern China and northern Africa (White *et al.* 2003). The final blow has been the

Figure 10.2   Water heated by a volcano at Changbai Shan Nature Reserve, Jilin Province, China on the North Korean border is used to cook eggs and corn for tourists.

loss of agricultural lands and soils to expanding urbanization, a disturbance that increases pressures on the world's remaining breadbaskets. Mongolia, Lesotho, and Ethiopia are but a few cases of countries once self-sufficient in grain production that now import most of their grain (Brown 2009). When productivity drops in places like Australia or Russia, as it did during droughts in 2010, food prices around the world go up. The degradation and loss of soil is a short-sighted blunder that makes us vulnerable to food shortages and is one step toward creating a future that begins to look like a giant, unstable house of playing cards or Ponzi scheme. We destabilize our future by stealing (exploiting) resources to pay for our present luxuries (Flannery 2002) and call it gross domestic product. Why not invest in an economy that heals and builds a healthy future rather than steals from it (Hawken 2009)?

Soil management is essential, and when soils are lost, civilizations collapse (Diamond 2005). The Middle East, where soils are now largely degraded or gone, was once a flourishing, self-sufficient region, i.e. the Fertile Crescent. Many cultures have sustained agriculture for centuries without substantial soil losses, but changing climates and politics can wreak havoc with even the best management plans. Agricultural expansion into the Great Plains of North America, Kazakhstan, and the Brazilian cerrado has resulted in soil erosion from the plowing of grasslands. The U.S. managed to conserve some prairie soils through extensive tree planting and strip cropping, but in Kazakhstan grain production is now greatly reduced (Christian 1997). Tree planting, conversion to tree crops that use less water, minimal or no tillage practices, and compensating farmers to allow croplands to lie fallow all help reduce soil losses from erosion, allowing soil structure and organic matter to recover. Reductions in the movement and size of grazing herds help conserve grasslands. Massive tree belts are being planted across the Sahel in Africa and around the Gobi Desert in China (Brown 2009). Yet growing population pressures and global climate changes can offset these ambitious schemes. Are incremental, regional approaches sufficient, or are substantial lifestyle changes (e.g. shifts to vegetarianism and less water-intensive agriculture) also required? The success of efforts to counter resource abuses and protect future generations will ultimately rely on education and political persuasion that realigns the priorities of the governments of the world. A more equitable distribution of resources among developing and developed countries may also help stabilize resource extraction and preserve soil resources. Brown (2009) estimated that to restore critical natural resources would cost about $110 billion per year (tree planting to sequester carbon and conserve soil, $23 billion; protecting topsoil $24 billion; restoring rangelands and fisheries $22 billion; protecting biodiversity $31 billion; and stabilizing water tables $10 billion). That is about 4% of what is spent in pursuit of an elusive "national security" each year by world governments. Real security lies in protecting water, soil, and other resources, while simultaneously reining in population growth.

10.4 Lessons from disturbance ecology

Humans can learn how to address climate change, overpopulation, and other disturbances by studying lessons from disturbance ecology. Disturbances are an integral part of ecosystem dynamics that have many beneficial effects. These benefits include energy and nutrient recycling, maintenance of biodiversity and spatial heterogeneity, and initiation of successional responses. Humans began their intimate interaction with disturbances by manipulating natural ones (e.g. fires, floods) to hunt animals, improve forage, and irrigate crops. However, as human populations have expanded, our manipulations have become far more sophisticated. It has been known for several decades that we utilize about half of all global primary productivity (Vitousek *et al.* 1986), but we appear to be reaching the upper limits of resource extraction (Gomiero *et al.* 2010). We also tend to focus on products rather than ecosystem consequences (i.e. the generation and fate of waste streams have generally been ignored), and now detrimentally affect most habitats on earth. To guide our future interactions with the environment, we must start to pay attention to how we alter disturbance regimes. We need to better understand what triggers both natural and anthropogenic disturbances, how they change ecosystems, and how ecosystems respond to them in the short and long term. Furthermore, we must address the entire, interactive disturbance regime, not simply focus on one disturbance at a time. Then, we need to develop scenarios to maximize gains for human needs while minimizing disruption of critical ecosystem processes such as nutrient and hydrological cycles and soil conservation. Presumably, we would develop plans where the disturbances most disruptive to ecosystem services are prevented or, if that is impossible (e.g. volcanic eruptions), then try to minimize their effects (e.g. no development in their vicinity). Of course, volcanoes and other highly destructive (severe) disturbances have a critical role in larger-scale ecosystem processes. Therefore, part of our strategy needs to include determining which disturbances we can and want to affect. To establish such (local to global) disturbance strategies, we need to take into account the general principles that are being discovered about disturbances.

Disturbances are highly variable in cause and consequence, frequency and severity. Spatial heterogeneity generally increases due to disturbance, further complicating the study of disturbances. Patterns of recovery following disturbance are also highly variable due to the stochastic nature of the dispersal of colonists of disturbed habitats, the many possible interactions of colonizing organisms, and the frequency and nature of subsequent disturbances. Nonetheless, general patterns of recovery include early dominance by fast-growing, weedy plants and animals with high rates of reproduction, and the accumulation of nutrients and organic matter in soils, followed by the dominance of slow-growing plants and animals and immobilization of soil nutrients. Some successional trajectories converge on a similar biotic composition from different starting points (when biological legacies prevail), while other trajectories diverge (where there is high spatial heterogeneity).

Human alterations of natural disturbances and the creation of novel anthropogenic ones add additional complexity. Anthropogenic disturbances tend to exacerbate the effects of natural ones by increasing their duration (e.g. year-round grazing of cattle replacing seasonal grazing of migrant ungulates), frequency (e.g. reducing arid land aquifers leading to decreasing intervals between droughts), severity (e.g. warmer ocean temperatures resulting in more severe cyclones), or extent (e.g. fire suppression leading to bigger fires). Anthropogenically disturbed habitats tend to have relatively sharper edges and less patchiness than naturally disturbed habitats. Yet the distinction between the effects of natural and anthropogenic disturbances is sometimes not obvious because of the many spatial and temporal scales at which effects can be measured. For example, plowing a field may release more labile nutrients, but microbial immobilization may also increase, leading to no net change in fertility. Similarly, non-native, invasive species introduced by humans may outcompete native species but promote introduction of other new species, so total biodiversity remains constant. Clearly, a plowed field or an invasive weed patch will have different characteristics from adjacent, undisturbed habitats, but by some measures (fertility, diversity), the two habitats remain comparable.

Management of novel anthropogenic disturbances is difficult because we are still trying to understand how such ecosystems will respond to conservation or restoration efforts. New species mixtures lead to changes throughout the ecosystem, including everything from a new soil microbial composition to new competitive hierarchies. When disturbances are exceptionally severe, thresholds can be crossed that result in the need for active restoration efforts to restore some semblance of the pre-disturbance conditions. Consequences of novel anthropogenic disturbances may be more pervasive than first realized. The effects of large oil spills may linger for decades beyond the initial clean-up (e.g. population declines in killer whales, *Orcinus*, after the 1989 *Exxon Valdez* spill in Alaska; Matkin *et al.* 2008). Budgetary concerns require triage to determine which ecosystems to conserve: those with the highest biodiversity, those most likely to survive global climate change, those most or least healthy, or those most resistant or resilient to future disturbances. Given the generally low level of public support for either conservation or restoration, education about disturbances may have to focus on the more dramatic changes that are expected (e.g. sea level rise) to begin to focus attention on larger questions of disturbance ecology.

10.5 How to reduce our vulnerability to disturbance

Humans have been incredibly adept at ignoring the biophysical limits that our environment places on us. First, we eliminated most potential predators, increasing our longevity over our hunter-gatherer ancestors by reducing mortality from wild animals. Next, manipulation of our environment through the development of agriculture fed us so well that birth rates climbed. Subsequent advances in medicine further increased longevity by vanquishing many diseases and extending life spans. Now, as nations pass through the demographic transition from high to low birth and death rates, their environmental impacts and vulnerability to disturbances change again. No longer limited by predators, famine, or disease, both our population and our resource consumption have increased dramatically. We now rely, in rich nations particularly, on a web of supply lines that criss-crosses the globe. We likely live in, or at least pursue our leisure in, a coastal city prone to some combination of earthquakes, drought, cyclones, and sea level rise. Much of our personal fortune is probably invested in a house and a vehicle that both rely on diminishing supplies of fossil fuels (e.g. to heat, cool, and furnish the house or run the vehicle). As previously noted, we further aggravate our vulnerability to disturbances by depending for our food on agribusinesses that rely on fossil fuels for tilling, harvesting, fertilizing, and transportation, in addition to cheap water and fertile, salt-free soils. Many economies depend on a steady stream of tourists transported by fossil fuel-dependent cars, trains, boats, and airplanes (Walker and Bellingham 2011). Our house of cards is vulnerable to the first temblor of an earthquake, the first kink in the oil or water delivery systems, or the slightest rise in sea level. No wonder we are finally waking up to the fact that we cannot control, but in fact are profoundly reliant on, our environment.

The biotic world, which we have dominated for several hundred years, is a source of vital services such as water retention, carbon sequestration, pollination, food, medicine, and decomposition. The biotic world also provides less tangible benefits, including aesthetic ones, and serves as a repository for still unknown advantages such as crop diversity. The physical world, which we have only partially manipulated, also provides services, in part by circulating nutrients, water, and gases through natural disturbances including volcanoes, earthquakes, floods, and fires. Biotic recovery depends on these patterns of circulation, whether by recycling N and P in soils, dispersing propagules, aerating water columns, or disrupting oil spills. As we gain more understanding of our intimate links with biophysical processes, we recognize the urgency to better understand these relationships. To the extent that we can reduce our vulnerability to disturbances, we will improve the quality of our lives and of those seven generations into the future. I will discuss three approaches to this reduction, a technological one, a cultural one, and an ecological one.

10.5.1 Technology

Technological advances are behind every tangible human development. Much technology improves our material quality of life, but negative consequences include more efficient resource extraction and accompanying increases in resource depletion and environmental pollution. Dependence on technology has now become overwhelming with the ubiquity of cars, personal computers, and cell phones. Certain rare earth minerals used in their manufacture are becoming scarce (Ragnarsdóttir 2008). Disruptions of highways, electricity lines, or cell phone towers abruptly sever many people from their habitual forms of working and communicating. Nonetheless, our commitment to technology appears total, so while we become increasingly vulnerable to technology, we are also busy harnessing ways to ameliorate its effects. I will discuss the feasibility of several technological approaches to problems that arise from natural and anthropogenic disturbances and their interactions (see Chapters 2–4). Some technology addresses how to improve hazard prediction to enable those in danger to evacuate. Better predictability would certainly save lives, as in the 2011 earthquakes in Christchurch, New Zealand and Sendai, Japan. Most technology focuses on how to minimize the frequency, intensity, and severity of disturbances for human populations.

Humans have little control over severe natural disturbances such as volcanoes, earthquakes, tsunamis, and spreading dunes, but that does not stop us from trying to ameliorate their consequences. Spraying water on them is rarely effective in stopping lava flows, but buildings can be constructed to resist damage by earthquakes, tourist centers can install tsunami warning systems, and slow-moving dunes can sometimes be stabilized by plantings. Erosion can be managed on small scales by plantings, altering drainage patterns, channelizing rivers, and building levees. However, as the devastation caused by Hurricane Katrina to New Orleans (U.S.) demonstrated, there is always one flood that overtops or breaks through the levee. Years of reduced sediment deposition in the Mississippi River delta and widespread destruction of protective coastal wetlands to improve barge transport exacerbated the effects of the storm. The U.S. government is now considering spending billions of dollars to rebuild the protective wetlands and has already overhauled the levee system. But with the expected increases in cyclone intensity and sea level, will these efforts be enough to stave off future storms?

Other natural disturbances, such as floods, droughts, and fires, are also exacerbated by human activities. Flood stages will be higher where deforested watersheds retain less ground water. Droughts are intensified by global climate change. Fires burn hotter and spread more readily where decades of fire suppression have led to increased fuel loads, especially in forest understories. Technology can help alleviate our vulnerability to such disturbances at local scales (helicopter evacuations, fire breaks). However, when widespread, these disturbances are much less manageable. Larger-scale measures such as dams, irrigation canals, and cloud seeding to alleviate problems usually only postpone serious damage (until the dams break, the canals overflow, or the drought intensifies). Natural disturbances are largely outside of human influence except at local scales.

Technological amelioration of anthropogenic disturbances is more realistic, at least at local to regional scales. Military bases can reduce the sound levels and flight patterns of their jets and employ the latest techniques in restoring damaged land. Mining companies can minimize surface disturbances and build environmental amelioration into their pre-mining (Cooke 1999) and post-mining (Nordstrom and Alpers 1999) plans. City developers are busy trying to create more sustainable urban environments by reducing vehicular traffic and water usage and by promoting bicycles, rapid transit systems, and community gardens (Brown 2009). Reductions in pollution can be rapid when urban waste disposal (e.g. Tokyo), fireplace soot (e.g. London), or sulfur emissions (e.g. Manchester, U.K.) are controlled. Technological improvements in transportation continue apace with the commercial development of electric and hybrid fuel vehicles, although many countries still rely on relatively fuel-efficient trains and buses. Forestry and agriculture are reducing runoff through selective cutting and minimal tillage. Oil spills are being treated with dispersants and fertilizers to accelerate bioremediation, among other measures (see Chapter 4). Banning two-stroke engines for taxis and motor scooters in Asian cities

(Hopke *et al.* 2008) and recreational boats in marine environments (Warnken and Byrnes 2004) has reduced air pollution. Fishing technologies have focused on the development of faster ways to find, catch, and process fish (see Chapter 4), but techniques have also been developed to minimize by-catch, especially by improving opportunities for seabirds, dolphins, and turtles to avoid nets (Bull 2007). Despite these various palliative techniques, amelioration of anthropogenic disturbances is probably not offsetting their environmental effects.

Global-scale disturbances such as loss of biodiversity and productivity, and climate change are even less amenable than regional disturbances to amelioration with technology. Nevertheless, many tools are being explored. Remote sensing technology has enabled conservationists to better pinpoint where biodiversity and productivity hotspots are found (Carlson *et al.* 2007). Advances in molecular systematics improve our understanding of phylogenetic relationships among species and ideally can guide conservation efforts. Attempts to improve photosynthetic efficiency could, if successful, help alleviate hunger (Parry and Hawkesford 2010). Likewise, crops are constantly being developed with improved tolerance of drought or cold and resistance to common diseases. Efforts to offset global climate warming are under way with proposed technological fixes of such large magnitude that they are called geo-engineering. These proposals include efforts to either reduce incoming radiation by shading the earth from outer space or increasing atmospheric **albedo**, or efforts to reduce levels of carbon dioxide in the atmosphere through ocean fertilization, carbon burial, or other techniques (Fig. 10.3; Wigley 2006). However, there are many uncertainties about the feasibility and potential for success of such large-scale projects (Bengtsson 2006). Ultimately, technological fixes are limited to partial and temporary amelioration of local disturbances (e.g. dams prevent flooding during the finite life of the dam) and are more effective counteracting anthropogenic than natural disturbances. The immense scale of climate change, the stochastic nature of natural disturbances, and the dual pressures of population growth and resource demands indicate that technology by itself will not protect us from disturbances.

10.5.2 Culture

A cultural approach to reducing our vulnerability to disturbances involves changes not in technology but in lifestyle and attitude. In this view, humans are part of the problem and can address their vulnerability by examining their role in the creation of risks that threaten us today. Lifestyles that minimize disturbance threats will vary with the disturbances and scales of concern. At regional and global scales, permanent settlements in areas of high risk would be discouraged. Such areas would include volcanic slopes, earthquake faults, erosion-prone hillsides, river floodplains, coastal flats, deserts, and fire-prone shrublands and woodlands. For example, Egypt is facing multiple challenges

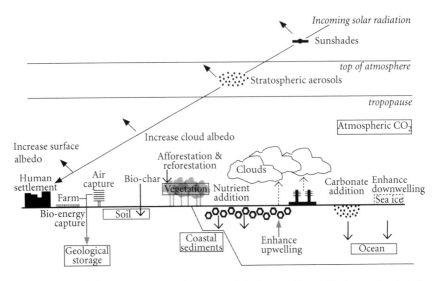

Figure 10.3   Overview of climate geo-engineering proposals being considered. Black arrowheads indicate shortwave radiation; open arrowheads show the enhancement of natural flows of carbon; gray arrowheads illustrate the engineered flow of carbon (to geological storage) or the engineered flow of water; dotted arrows show the sources of cloud condensation nuclei; and solid boxes represent carbon stores. (From Vaughan and Lenton (in press) with permission from Springer.)

on its major floodplain, the Nile River. Its delta is shrinking because annual sediment deposition has been halted by the Aswan Dam (Bohannon 2010). Combined with expected increases in sea level, much of this fertile area may be threatened. Other problems in Egypt include increased salinization from year-round irrigation and the spread of schistosomiasis from parasitic worms that thrive in the irrigation canals. Rather than move people out of the delta, the government is instead creating a new delta by diverting 10% of the Nile River to a previously uninhabited region. This example highlights the interwoven nature of technological, cultural, and ecological issues that have no simple solution. Because no area is completely free of natural disturbances, global relocation of settlements would be based on risk assessments. Of course, places that are popular as areas to live in developed because of readily available resources (e.g. fresh water for drinking and washing; running water for waste disposal; oceans and large rivers for transportation; volcanic soils and floodplains for fertile soils). Therefore, any alteration of large population centers would need to consider resource access among many other factors. If such a plan seems drastic (and it would mean displacement of most of the world's people), then we should evaluate the upcoming cost of sea level rise in terms of urban destruction. Moving everyone several hundred meters away and up from sea level would likely result in huge cost savings, although our ancestors sometimes dealt with ocean rise and increases in terrestrial flooding by building villages on stilts (Brahic 2009). Initially, such migrations to "safe" locations inland will only happen among those individuals most aware of and concerned about current and future large-scale disturbances.

Other lifestyle choices involve an evaluation of individual and community roles in supporting habits that sustain natural and anthropogenic disturbances. Resource consumption indirectly lies behind most types of anthropogenic disturbances. If one is concerned about the destruction of habitat or biodiversity, or other disturbances triggered by industrialization, then one should consider not supporting (e.g. through purchases), military activities, mining operations, fossil fuel extraction, and commercial fisheries, forestry, and agriculture. This logic leads to more reliance on recycling and reusing as well as an emphasis on local sources of food, fuel, and recreation (e.g. community gardens, solar panels, local parks; Carreiro and Zipperer in press). Economic incentives are important agents of change, because local customers support local farmers. Such lifestyles make one immediately less vulnerable to the economic misfortunes that globalization has recently wrought, but would only provide amelioration for large-scale disturbances if they were widely adopted (e.g. when local, sustainable forestry practices reduce flooding in a populated valley). Lifestyle choices are currently so regulated by marketing traditions, advertising, and cheap pricing of food and merchandise that breaking out of the consumerist, disturbance-supporting mold is difficult, especially in urban and large suburban communities. Nonetheless, many communities around the world are beginning to explore how to live more sustainably, become less vulnerable, and thereby reduce their reliance on the increasingly unstable house of cards (Walker 2010).

Changes in attitude provide another cultural approach to reducing human vulnerability to disturbance. Attitudes can be changed through increased knowledge and understanding of the causes and effects of disturbances. Either with or without accompanying lifestyle changes, attitudes can lead to political action, changes in educational institutions and workplaces, and the development of different family values. Where biodiversity is valued, for example, voters are more likely to support the creation of nature reserves. Where regular fires are supported, ecologically based logging practices can help reduce the occurrence of large and damaging conflagrations. Where forest conservation or reforestation is supported (Fig. 10.4), healthy forests decrease the incidence of droughts and reverse the process of desertification. Where traditional taboos on resource extraction such as fishing are observed, fish populations thrive. Where local connections are emphasized, bicycles begin to replace cars. Where fossil fuel use is eschewed, telecommuting and local holidays are likely to prevail. Attitudes lead to action which can, when properly directed, reduce our vulnerability to disturbances at many scales.

The global scope of the problem of resource abuse is not going to be solved by bottom-up, cultural solutions alone. Top-down, governmental policies

Figure 10.4   A sign in China reflecting an admirable attitude toward the natural world.

are also essential in efforts to reign in international corporations that prefer short-term profits to long-term environmental health. Such policies are implemented when individuals take the responsibility to espouse an approach that considers the vital links between natural and cultural resources. Energy policies that include some form of carbon tax and commodity pricing that reflects environmental costs are two steps that governments can take (Friedman 2008). Cultural attitudes can accelerate such changes to ecologically smarter policies that will reduce our vulnerability to global scale disturbances such as climate change.

10.5.3 Ecology

Ecology is the study of the relationship between organisms and their environment, and therefore readily encompasses the relationship between humans and disturbance. The science of ecology is just over 100 yr old, yet ecological knowledge is much older. Hunter-gatherers needed detailed knowledge of the habits of their prey and habitats where food plants grew. Early agriculturalists succeeded by applying selective disturbances to soils, manipulating nutrient cycles and competitive interactions, combating invasive plants and herbivores, and intimately understanding the life cycles of their targeted crop plants. Fishermen also had to understand when and how to fish and when to cease fishing to sustain populations of critical food species. An ecological approach to minimizing disturbance effects on humans relies on a body of knowledge developed and refined in recent years that has built on these early human experiences. Disturbances and organisms interact at local and regional scales (e.g. the type and amount of forest cover influences cyclone wind patterns) and understanding these relationships can help decrease human vulnerability to disturbances.

Ecologists who study disturbance use the scientific method to gather data and test hypotheses about disturbance effects (Turner 2010). They incorporate appropriate technology and are often part of a cultural group concerned about the roles humans have in supporting or damaging healthy ecosystems. In that sense, the ecological approach represents a partial combination of the technological and cultural approaches already discussed. There are many opinions about the relationship that humans have or should have with their environment (Easton 2009) and human ecology, landscape ecology, and restoration ecology are examples of subdisciplines that are particularly interested in disturbance–human interactions. Characteristics of an ecological approach therefore include a reliance on extensive data gathering to answer specific questions, an openness to possible interpretations of the data, and a mixture of both reductionist and holistic approaches to be able to place details about organisms or resources in a broader picture of ecosystem interactions.

Ecologists are concerned about human impacts on resources and their exacerbating effect on disturbance regimes. Humans are considered as vulnerable to growth limits as are other organisms. While humans are not limited by dispersal and have little trouble establishing resident populations following disturbances, we are susceptible to displacement and mortality from severe disturbances. However, disturbances also have a recycling and sometimes rejuvenating value for ecosystems, and humans have harnessed those processes in agriculture for centuries. As our anthromes, or human-dominated biomes (Ellis *et al.* 2010), increasingly occupy former agricultural lands, our apparent dependence on natural cycles declines. After all, residents of suburbs and cities do not care about rainfall or planting cycles or soil fertility. Or do they? Ultimately, of course, urban populations do rely on those ecosystem processes somewhere to provide them with food and other services (Ruffo and Kareiva 2009). In the U.S., the percentage of the population that are farmers has declined from about 35% in 1900 to 2% in 2010 (Lobao and Meyer 2001). Globally, farmers are losing the battle for water to urbanites, further destabilizing our house of cards. Disruption of food sources is much more likely when fewer farmers are involved.

Ecologists recognize the interwoven links between the various activities (e.g. agriculture, forestry, fisheries), resources (e.g. water, soil, biodiversity), and processes (e.g. succession, restoration) that support humans at our current standard of living. Ecologists are particularly aware of how human-dominated ecosystems no longer rely on current solar income, resilience from biodiversity, and complete recycling of wastes. Instead, human ecosystems now exploit fossilized solar energy (fossil fuels) and produce much toxic waste that is not recycled (Orr 2004). Disturbances that disrupt one process can affect many others in a complex series of poorly understood cascading events. A drought not only destroys crops that we want to eat, but also leads to loss of herbivores that relied on those plants (above and below ground); loss of soil retained by the roots of crop plants and other species; and loss of spatial mosaics and temporal patterns. There may even be feedbacks to broader climatic conditions or cycles of degradation, depending on the frequency and intensity of future droughts. Each of these disturbance responses can be evaluated at different spatial and temporal scales, but when considering the effects of disturbance on humans, we tend to think in hectares and years. Such scales limit creative problem solving at larger or longer scales that can be highly relevant when considering the effects of global changes. Modeling is then essential to predict, for example, the effects of climate change on global primary productivity (Geider *et al.* 2001). Some ecologists do consider such scales and interact with paleobiologists and anthropologists to explore what lessons past patterns may have for future scenarios (Swetnam *et al.* 1999; Cohen 2011).

As an ecologist concerned about humanity's future, I suggest that all three approaches are essential to lessening the vulnerability of humans to disturbances. Scale-appropriate technology should certainly be pursued, particularly where results are immediate and helpful, as with small-scale irrigation (Hill and Woodland 2003). Collections of large bodies of synchronous environmental data are now possible by downloading field measurements to satellites (Keller *et al.* 2008). Global geo-engineering may help tackle climate change, although the larger the scale, the more room there is for errors (Bengtsson 2006). For example, preliminary additions of iron to the South Pacific Ocean have not led to as big or as prolonged an increase in phytoplankton productivity as was expected (de Baar *et al.* 2008) and may have deleterious side-effects (Smetacek and Naqvi 2008). Solely relying on technology to fix our anthropogenic disruptions of global conditions does not seem reasonable. I also encourage a cultural shift to emphasize a realignment of values and appropriate lifestyles. In a process analogous to how plants invade severely disturbed habitats, communities that increase their sustainability can become models or nuclei for further expansion to other communities. Such a cultural progression will be more rapid if there is experimental evidence that these attempts at increased sustainability prove successful in creating an improved quality of life, reducing costs, reducing over-reliance on imported goods, and practicing land management that conserves soil resources (Walker 2010). Finally, I encourage a synthesis of the best of technological and cultural adaptations with an unbiased, scientific approach that evaluates approaches based on evidence and impartial measures of success. Another critical aspect of an ecological approach is the flexibility to adapt to new information about both changing conditions and methods used to manipulate disturbances. This research-based approach can serve us well as new and more severe disturbances increasingly challenge the status quo. Gradual change now is much more socially desirable than last-minute panic as we plan for our altered planet. Humans are affected by disturbances, many of our own creation, and we must address both how to ameliorate them and how to adapt to them in our near future.

Glossary

Glossary terms appear in bold type within the text at their first use.

Abiotic: pertaining to non-living factors such as wind, temperature, erosion

Actinorhizal: vascular plants with an N-fixing bacterium that resembles an actinomycete fungus; non-leguminous N fixers (e.g. alder)

Albedo: reflectivity of a surface

Allelopathy: a form of competitive inhibition based on release of chemicals

Anoxia: lack of oxygen

Anthrome: a biome heavily modified or even constructed by humans

Anthropogenic: created or influenced by humans

Archaea: a domain of single-celled microorganisms

Assembly rules: predictions concerning mechanisms of community organization

Autotroph: an organism that produces organic compounds from the energy provided by light (photosynthesis) or chemical reactions (chemosynthesis)

Benthic zone: the ecological region at the bottom of a body of water, including surface sediments

Biocontrol: use of organisms to control other, undesirable organisms

Bioerosion: erosion of hard substrates by organisms (e.g. corals by fish; rocks by lichens)

Biological legacy: organisms that survive a disturbance

Biomass: the mass of organisms (living or dead)

Biome: a geographical region with similar vegetation and climate (e.g. tropical forest)

Biotic: pertaining to biological factors

Bioturbation: disruption of sediments or soil by organisms (e.g. burrowing)

Boreal: a biome dominated by coniferous forests with cold winters (i.e. taiga)

Bushmeat: meat of wild animals killed for subsistence or sale

Chronosequence: a space-for-time substitution that allows study of long-term succession

Competition: a negative effect of one organism on another

Coppicing: traditional woodland management relying on resprouting of cut trees or shrubs

Convergence: a successional trajectory where communities become more similar with time

Denitrification: the conversion of nitrite to N gas under low oxygen conditions

Desertification: the conversion of rangeland to desert with low value

Dispersal: movement of propagules (e.g. seeds, offspring) away from parent (e.g. by steady "phalanx" advance or by expansion of scattered "guerilla" populations via nucleation)

Disturbance: a relatively abrupt loss of biomass or structure from forces within (autogenic) or outside (allogenic) the habitat; characterized by its extent (area disturbed), frequency (return interval), intensity (measure of its force), and severity (damage caused)

Divergence: a successional trajectory where communities become less similar with time

Dystrophic: water rich in organic matter but low in nutrients

Ecesis: establishment of organisms following a disturbance

Ecosystem service: ecosystem benefit for humans (e.g. clean water)

Ecotone: the boundary between two ecosystems; the edge of a patch

Enchytraeids: a group of aquatic and terrestrial worms related to earthworms (e.g. pot worms)

Eutrophication: a process of nutrient enrichment of aquatic ecosystems that often has negative consequences for aquatic organisms

Facilitation: a positive influence of one organism on another

Fluctuation: impermanent changes as part of natural variation (e.g. seasonal temperatures)

Fractal dimension: a high fractal dimension occurs when a fractal (geometric shape) fills a given space more completely; in landscape ecology, the degree of dissection

Fragmentation: the dividing of a landscape or habitat by disturbances such as roads

Holism: an approach that emphasizes connectivity and that the whole is greater than the sum of the parts; *see* Reductionism

Hydrarch: pertaining to aquatic habitats (e.g. succession)

Hyporheic zone: the region beneath a stream where ground water and stream water mix

Hysteresis: delay in adjustment of a process in response to a change in an associated process

Invasional meltdown: an interaction (usually facilitative) among invading species

Iron pan: a hard, water-impermeable layer of soil rich in leached iron oxides

K-selected: a species with traits adapted to competitive, late-successional environments

Krummholz: stunted growth due to physical factors such as wind or low temperatures

Lahar: a slurry of mud and debris created by ice melted during a volcanic eruption

Leaf area index: a measure of leaf area equal to projection of the surface area of all leaves onto the ground; often greater than 1 because leaves overlap

Leptokurtic distribution: a steeply declining curve (e.g. pattern of seed dispersal away from the parent)

Loess: fine, unconsolidated, wind-blown sediments

Microclimate: the climate of a small area relevant to individual organisms:

Midden: a refuse pile created by humans or rodents

Monoterpenes: volatile substances in plants that have a role in defense from herbivory

N: nitrogen

Nitrification: oxidation of ammonium to nitrite and nitrite to nitrate by aerobic microorganisms

Novel ecosystem: a new ecosystem created by human influences

Nucleation: the invasion process in which isolated founder populations serve as foci for later expansion

Nudation: the process of denuding the landscape by a disturbance

Nurse plant: a plant that facilitates another plant, particularly under its larger canopy

Nutrient use efficiency: the amount of a nutrient taken up by a plant compared to the quantity of that nutrient available in the soil

Oligotrophic: a low-nutrient environment

P: phosphorus

Pelagic: pertaining to the open water column, away from the shore or the benthic zone

Phenology: the life stages of an organism

Photoinhibition: reduced photosynthetic capacity (e.g. from a sudden increase in light intensity)

Playa: a desert basin with no outlet that is sometimes flooded and has little vegetation cover

Polychaete: a class of generally marine worms

Progression: a successional state of increasing biomass, diversity, or function

Pyroclastic flow: volcanic material that moves rapidly down slope

***r*-selected:** a species with traits adapted to disturbed and early successional environments

Reductionism: an approach that emphasizes individual processes; *see* Holism

Resilience: the ability of an ecosystem to recover from a disturbance

Resistance: the ability of an ecosystem to withstand a disturbance without changing

Restoration: returning a damaged ecosystem to at least some of its former biological state

Retrogression: a successional state of decreasing biomass, diversity, or function

Rhizosphere: the zone around a root

Scarification: scratching (e.g. of a seed coat)

Seed bank: the pool of all seeds found on a plant, on the soil surface, or buried in the soil

Senescence: aging of organisms with reduction in function

Sere: a successional sequence

Serotiny: seed release due to an environmental trigger such as fire

Solifluction lobes: slow creep of soil down slope over a more rigid frozen substrate

Stoichiometry: in ecology, the relationship between chemicals (e.g. nutrient ratios)

Stochastic: unpredictable

Stability: a community characteristic expressing lack of change or resistance to disturbance

Stress: any factor that limits productivity (e.g. heat, cold, drought)

Succession: species and community replacement over time following a disturbance that leaves little (primary) or substantial (secondary) biological legacy and may reach a relatively equilibrial state (climax)

Tipping point: a state of disrepair of an ecosystem that, when reached, leads to a cascade of further disruption that then requires active intervention to offset

Xerarch: pertaining to dry environments (e.g. succession)

References

Aarssen, L.W. (1997). High productivity in grassland ecosystems: effected by species diversity or productive species? *Oikos*, **80**, 183–184.

Abbott, P.L. (1996). *Natural Disasters*. Brown, Dubuque, IW, U.S.

Abe, T., Bignell, D.E., and Higashi, M., eds (2000). *Termites: Evolution, Sociality, Symbioses, Ecology*. Kluwer, Dordrecht, Netherlands.

Adachi, N., Terashima, I., and Takahashi, M. (1996a). Mechanisms of central die-back of *Reynoutria japonica* in the volcanic desert on Mt Fuji. A stochastic model analysis of rhizome growth. *Annals of Botany*, **78**, 169–179.

Adachi, N., Terashima, I., and Takahashi, M. (1996b). Central die-back of monoclonal stands of *Reynoutria japonica* in an early stage of primary succession on Mount Fuji. *Annals of Botany*, **77**, 477–486.

Adjeroud, M., Michonneau, F., Edmunds, P.J., *et al.* (2009). Recurrent disturbances, recovery trajectories, and resilience of coral assemblages on a South Central Pacific reef. *Coral Reefs*, **28**, 775–780.

Adler, P.B., D'Antonio, C.M., and Tunison, J.T. (1998). Understory succession following a dieback of *Myrica faya* in Hawai'i Volcanoes National Park. *Pacific Science*, **52**, 69–78.

Agee, J.K. and Skinner, C.N. (2005). Basic principles of forest fuel reduction treatments. *Forest Ecology and Management*, **211**, 83–96.

Aide, T.M., Zimmerman, J.K., Rosario, M., and Marcano, H. (1996). Forest recovery in abandoned cattle pastures along an elevational gradient in northeastern Puerto Rico. *Biotropica*, **28**, 537–548.

Ali, S.H. (2003). *Mining, the Environment and Indigenous Development Conflicts*. University of Arizona Press, Tucson, AZ, U.S.

Ali, S.H. (2009). *Treasures of the Earth: Need, Greed and a Sustainable Future*. Yale University Press, New Haven, CT, U.S.

Al-Kaisi, M.M. and Yin, X. (2005). Tillage and crop residue effects on soil carbon and carbon dioxide emission in corn-soybean rotations. *Journal of Environmental Quality*, **34**, 437–445.

Allen, E.B. and Allen, M.F. (1980). Natural re-establishment of vesicular-arbuscular mycorrhizae following stripmine reclamation in Wyoming. *Journal of Applied Ecology*, **17**, 139–147.

Allen, E.B. and Allen, M.F. (1990). The mediation of competition by mycorrhizae in successional and patchy environments. In J.G. Grace and D. Tilman, eds *Perspectives on Plant Competition*, pp. 367–389. Academic Press, New York.

Allen, M.F. and MacMahon, J.A. (1988). Direct VA mycorrhizal inoculation of colonizing plants by pocket gophers (*Thomomys talpoides*) on Mount St. Helens. *Mycologia*, **80**, 754–756.

Allen, M.F., Crisafulli, C., Friese, C.F., and Jenkins, S. (1992). Reformation of mycorrhizal symbioses on Mount St. Helens, 1980–1990: Interactions of rodents and mycorrhizal fungi. *Mycological Research*, **96**, 447–453.

Allen, M.F., Allen, E.B., Zink, T.A., *et al.* (1999). Soil microorganisms. In L.R. Walker, ed. *Ecosystems of Disturbed Ground, Ecosystems of the World 16*, pp. 521–544. Elsevier, Amsterdam.

Allen, M.F., Allen, E.B., and Gomez-Pompa, A. (2005). Effects of mycorrhizae and nontarget organisms on restoration of a seasonal tropical forest in Quintana Roo, Mexico: factors limiting tree establishment. *Restoration Ecology*, **13**, 325–333.

Allen, R.B., Bellingham, P.J., and Wiser, S.K. (1999). Immediate damage by an earthquake to a temperate montane forest. *Ecology*, **80**, 708–714.

Alpert, P. (2000). The discovery, scope, and puzzle of desiccation tolerance in plants. *Plant Ecology*, **151**, 5–17.

Altieri, M. A. (1995). *Agroecology: The Science of Sustainable Agriculture*. Westview Press, Boulder, CO, U.S.

Amarasekare, P. and Nisbet, R.M. (2001). Spatial heterogeneity, source-sink dynamics, and the local coexistence of competing species. *The American Naturalist*, **158**, 572–584.

Amman, G.D. (1977). The role of the mountain pine beetle in lodgepole pine ecosystems: impact on succession. In W.J. Mattson, ed., *The Role of Arthropods in Forest Ecosystems*, pp. 3–18. Springer, New York.

van Andel, J. and Aronson, J., eds (2006). *Restoration Ecology*. Blackwell, Oxford.

Anderson, A. (2009). The rat and the octopus: initial human colonization and the prehistoric introduction of domestic animals in Remote Oceania. *Biological Invasions*, **11**, 1503–1519.

Anderson, D.M. and Garrison, D.J., eds (1997). The ecology and oceanography of harmful algal blooms. *Limnology and Oceanography*, **42**, 1009–1305.

Andrew, N.L. (1993). Spatial heterogeneity, sea urchin grazing and habitat structure on reefs in temperate Australia. *Ecology*, **74**, 292–302.

Anthony, K.R.N., Kline, D.I., Diaz-Pulido, G., Dover, S., and Hoegh-Guldberg, O. (2008). Ocean acidification causes bleaching and productivity loss in coral reef builders. *Proceedings of the National Academy of Sciences USA*, **105**, 17442–17446.

Antonsen, H. and Olsson, P.A. (2005). Relative importance of burning, mowing and species translocation in the restoration of a former boreal hayfield: responses of plant diversity and the microbial community. *Journal of Applied Ecology*, **42**, 337–347.

Aplet, G.H. (1990). Alteration of earthworm community biomass by the alien *Myrica faya* in Hawaii. *Oecologia*, **81**, 414–416.

Aplet, G.H., Anderson, S.J., and Stone, C.P. (1991). Association between feral pig disturbances and the composition of some alien plant assemblages in Hawaii Volcanoes National Park. *Plant Ecology*, **95**, 55–62.

Archer, S., Scifres, C., Bassham, C.R., and Maggio, R. (1988). Autogenic succession in a subtropical savanna: conversion of grassland to thorn woodland. *Ecological Monographs*, **58**, 111–127.

Aronson, J., Floret, C., LeFloc'h, E., Ovalle, C., and Pontanier, R. (1993). Restoration and rehabilitation of degraded ecosystems in arid and semiarid regions. I. A view from the South. *Restoration Ecology*, **1**, 8–17.

Ash, H.J., Gemmel, R.P., and Bradshaw, A.D. (1994). The introduction of native plant species on industrial waste heaps: a test of immigration and other factors affecting primary succession. *Journal of Applied Ecology*, **31**, 74–84.

Auerbach, N.A., Walker, M.D., and Walker, D.A. (1997). Effects of roadside disturbance on substrate and vegetation properties in arctic tundra. *Ecological Applications*, **7**, 218–235.

Ayres, M.P. and Lombardero, M.J. (2000). Assessing the consequences of global change for forest disturbance from herbivores and pathogens. *Science of the Total Environment*, **262**, 263–286.

Baird, A.H., Campbell, S.J., Anggoro, A.W., *et al.* (2005). Acehnese reefs in the wake of the Asian tsunami. *Current Biology*, **15**, 1926–1930.

Baker, R.J. and Chesser, R.K. (2000). The Chernobyl nuclear disaster and subsequent creation of a wildlife preserve. Environmental Toxicology and Chemistry, **19**, 1231–1232.

Baker, W.L. (1992). The landscape ecology of large disturbances in the design and management of nature reserves. *Landscape Ecology*, **7**, 181–194.

Bakke, T.A. and Harris, P.D. (1998). Diseases and parasites in wild Atlantic salmon (*Salmo salar*) populations. *Canadian Journal of Fisheries and Aquatic Sciences*, **55**, 247–266.

de Baar, H.J.W., Gerringa, L.J.A., Laan, P., and Timmermans, K.R. (2008). Efficiency of carbon removal per added iron in ocean iron fertilization. *Marine Ecology Progress Series*, **364**, 269–282.

Bakker, J.P. (1989). *Nature Management by Grazing and Cutting*. Kluwer, Dordrecht.

Baldwin, A.H., Platt, W.J., Gathen, K.L., Lessmann, J.M., and Rauch, T.J. (1995). Hurricane damage and regeneration in fringe mangrove forests of southeast Florida, USA. *Journal of Coastal Research*, **21**, 169–183.

Baldwin, D.S. and Mitchell, A.M. (2000). The effects of drying and re-flooding on the sediment and soil nutrient dynamics of lowland river-floodplain systems: a synthesis. *Regulated Rivers: Research and Management*, **16**, 457–467.

Balmford, A. and Cowling, R.M. (2006). Fusion or failure? The future of conservation biology. *Conservation Biology*, **20**, 692–695.

Bancroft, W.J., Garkaklis, M.J., and Roberts, J.D. (2005). Burrow building in seabird colonies: a soil-forming process in inland ecosystems. *Pedobiologia*, **49**, 149–165.

Barbercheck, M.E., Neher, D.A., Anas, O., El-Allaf, S.M., and Weicht, T.R. (2009). Response of soil invertebrates to disturbance across three resource regions in North Carolina. *Environmental Monitoring and Assessment*, **152**, 283–298.

Bardgett, R.D. (2000). Patterns of below-ground primary succession at Glacier Bay, south-east Alaska. *Bulletin of the British Ecological Society*, **31**, 40–42.

Bardgett, R.D. (2005). *The Biology of Soil: a Community and Ecosystem Approach*. Oxford University Press, Oxford.

Bardgett, R.D. and Griffiths, B.S. (1997). Ecology and biology of soil protozoa, nematodes and microarthropods. In J.D. van Elasas, E.M.H. Wellington, and J.T. Trevors, eds *Modern Soil Microbiology*, pp. 129–163. Marcel Dekker, New York.

Bardgett, R.D. and Walker, L.R. (2004). Impact of coloniser plant species on the development of decomposer microbial communities following deglaciation. *Soil Biology and Biochemistry*, **36**, 555–559.

Bardgett, R.D. and Wardle, D.A. (2003). Herbivore mediated linkages between aboveground and belowground communities. *Ecology*, **84**, 2258–2268.

Bardgett, R.D. and Wardle, D.A. (2010). *Aboveground–Belowground Linkages: Biotic Interactions, Ecosystem Processes, and Global Change*. Oxford University Press, Oxford.

Bardgett, R.D., Bowman, W.D., Kaufmann, R., and Schmidt, S.K. (2005). A temporal approach to linking aboveground and belowground ecology. *Trends in Ecology and Evolution*, **20**, 634–641.

Bardi, U. (2008). Peak oil: the four stages of a new idea. *Energy*, **34**, 323–326.

Barwick, R.D., Kwak, T.J., Noble, R.L., and Barwick, D.H. (2004). Fish populations associated with habitat-modified piers and natural woody debris in Piedmont Carolina Reservoirs. *North American Journal of Fisheries Management*, **24**, 1120–1133.

Baskin, J.M. and Baskin, C.C. (1988). Endemism in rock outcrop plant communities of unglaciated eastern United States: an evaluation of the roles of the edaphic, genetic and light factors. *Journal of Biogeography*, **15**, 829–840.

Basnet, K., Scatena, F.N., Likens, G.E., and Lugo, A.E. (1993). Ecological consequences of root grafting in tabanuco (*Dacryodes excelsa*) trees in the Luquillo Experimental Forest, Puerto Rico. *Biotropica*, **25**, 28–35.

Batterbury, S. and Warren, A. (2001). The African Sahel 25 years after the great drought: assessing progress and moving towards new agendas and approaches. *Global Environmental Change*, **11**, 1–8.

Baumeister, D. and Callaway, R.M. (2006). Facilitation by *Pinus flexilis* during succession: a hierarchy of mechanisms benefits other plant species. *Ecology*, **87**, 1816–1830.

Bazzaz, F.A. (1979). The physiological ecology of plant succession. *Annual Review of Ecology and Systematics*, **10**, 351–371.

Bazzaz, F.A. (1996). *Plants in Changing Environments: Linking Physiological, Population, and Community Ecology*. Cambridge University Press, Cambridge.

Beare, M.H., Coleman, D.C., Crossley, D.A., Hendrix, P.F., and Odum, E.P. (1995). A hierarchical approach to evaluating the significance of soil biodiversity to biogeochemical cycling. *Plant and Soil*, **170**, 5–22.

Beare, M.H., Vikram Reddy, M., Tian, G., and Srivastava, S.C. (1997). Agricultural intensification, soil biodiversity and agroecosystem function in the tropics: the role of decomposer biota. *Applied Soil Ecology*, **6**, 87–108.

Beavers, A.M. and Burgan, R.E. (2002). *Analysis of Fire History and Management Concerns at Pohakuloa Training Area*. The Center for Environmental Management of Military Lands, Colorado State University, Fort Collins, CO.

Becher, H.H. (1985). Compaction of arable soils due to reclamation or off-road military traffic. *Reclamation and Revegetation Research*, **4**, 155–164.

Beisner, B.E., Haydon, D.T., and Cuddington, K. (2003). Alternative stable states in ecology. *Frontiers in Ecology and the Environment*, **1**, 376–382.

Bellingham, P.J. and Sparrow, A.D. (2000). Resprouting as a life history strategy in woody plant communities. *Oikos*, **89**, 409–416.

Bellingham, P.J., Tanner, E.V.J., and Healey, J.R. (1994). Sprouting of trees in Jamaican montane forests, after a hurricane. *Journal of Ecology*, **82**, 747–758.

Bellingham, P.J., Tanner, E.V.J., Rich, P.M., and Goodland, T.C.R. (1996). Changes in light below the canopy of a Jamaican montane rainforest after a hurricane. *Journal of Tropical Ecology*, **12**, 699–722.

Bellingham, P.J., Peltzer, D.A., and Walker, L.R. (2005). Contrasting effects of dominant native and exotic shrubs on floodplain succession. *Journal of Vegetation Science*, **16**, 135–142.

Bellingham, P.J., Tanner, E.V.J., and Healey, J.R. (2005). Hurricane disturbance accelerates invasion by the alien tree *Pittosporum undulatum* in Jamaican montane rain forests. *Journal of Vegetation Science*, **16**, 675–684.

Bellwood, D.R. and Choat, J.H. (1990). A functional analysis of grazing in parrotfishes (family Scaridae): the ecological implications. *Environmental Biology of Fishes*, **28**, 189–214.

Bellwood, D.R., Hughes, T.P., Folke, C., and Nyström, M. (2004). Confronting the coral reef crisis. *Nature*, **429**, 827–833.

Belnap, J. (2003). The world at your feet: desert biological soil crusts. *Frontiers in Ecology and the Environment*, **1**, 181–189.

Belnap, J. and Gillette, D.A. (1998). Vulnerability of desert biological soil crusts to wind erosion: the influences of crust development, soil texture and disturbance. *Journal of Arid Environments*, **39**, 133–142.

Belyea, L.R. and Lancaster, J. (1999). Assembly rules within a contingent ecology. *Oikos*, **86**, 402–416.

Benedetti-Cecchi, L. (2000). Predicting direct and indirect interactions during succession in a mid-littoral rocky shore assemblage. *Ecological Monographs*, **70**, 45–72.

Bengtsson, L. (2006). Geo-engineering to confine climatic change: is it all feasible? *Climatic Change*, **77**, 229–234.

Bénito-Espinal, F.P., and Bénito-Espinal, E., eds (1991). *L'Ouragan Hugo: Genese, Incidences Géographiques et Écologiques sur la Guadeloupe*. Imprimerie Désormeaux, Fort-de-France, France.

Bennett, B.A., Smith, C.R., Glaser, B., and Maybaum, H.L. (1994). Faunal community structure of a chemoautotrphic assemblage on whale bones in the deep northeast Pacific Ocean. *Marine Ecology Progress Series*, **108**, 205–223.

Bennett, E.M., Carpenter, S.R., and Caraco, N.F. (2001). Human impact on erodible phosphorus and eutrophication: a global perspective. *BioScience*, **51**, 227–234.

Bentz, J.B., Régnière, J., Fettig, C.J., *et al.* (2010). Climate change and bark beetles of the western United States and Canada: direct and indirect effects. *BioScience*, **60**, 602–613.

Berman, T. and Bronk, D.A. (2003). Dissolved organic nitrogen: a dynamic participant in aquatic ecosystems. *Aquatic Microbial Ecology*, **31**, 279–305.

Bertness, M.D., Trussell, G.C., Ewanchuk, P.J., and Silliman, B.R. (2002). Do alternate stable community states exist in the Gulf of Maine rocky intertidal zone? *Ecology*, **83**, 3434–3448.

Bertness, M.D., Trussell, G.C., Ewanchuk, P.J., Silliman, B.R., and Crain, C.M. (2004). Consumer-controlled community states on Gulf of Maine rocky shores. *Ecology*, **85**, 1321–1331.

Bestelmeyer, B.T., Havstad, K.M., Damindsuren, B., *et al.* (2009). Resilience theory in models of rangeland ecology and restoration: the evolution and application of a paradigm. In R.J. Hobbs and K.N. Suding, eds *New Models for Ecosystem Dynamics and Restoration*, pp. 78–96. Island Press, Washington, DC.

Binkley, D. and Giardina, C. (1998). Why do tree species affect soils? The warp and woof of tree-soil interactions. *Biogeochemistry*, **42**, 89–106.

Binkley, D., Suarez, F., Stottlemeyer, R., and Caldwell, B. (1997). Ecosystem development on terraces along the Kugurorok River, northwest Alaska. *Ecoscience*, **4**, 311–318.

Birk, E.M. and Simpson, R.W. (1980). Steady state and the continuous input model of litter accumulation and decomposition in Australian eucalypt forests. *Ecology*, **61**, 481–485.

Birkmann, J., ed. (2006). *Measuring Vulnerability to Natural Hazards: Towards Disaster Resilient Societies*. United Nations University, New York.

Birt, G., Rodwell, L.D., and Richards, J.P. (2009). Investigation into the sustainability of organic aquaculture of Atlantic cod (*Gadus morhua*). *Sustainability: Science, Practice and Policy*, **5**, 4–14.

Bishop, J.G., O'Hara, N.B., Titus, J.H., Apple, J.L., Gill, R.A., and Wynn, L. (2010). N–P co-limitation of primary production and response of arthropods to N and P in early primary succession on Mount St. Helens volcano. *PLoS ONE*, **5**, e13598.

Biswas, S.R. and Mallik, A.U. (2010). Disturbance effects on species diversity and functional diversity in riparian and upland plant communities. *Ecology*, **91**, 28–35.

Blood, E.R., Anderson, P., Smith, P.A., Nybro, C., and Ginsberg, K.A. (1991). Effects of Hurricane Hugo on coastal soil solution chemistry in South Carolina. *Biotropica*, **23**, 248–255.

Blundon, D.J. and Dale, M.R.T. (1990). Dinitrogen fixation (acetylene reduction) in primary succession near Mt. Robson, British Columbia. *Arctic and Alpine Research*, **22**, 255–263.

Bohannon, J. (2010). The Nile delta's sinking future. *Science*, **327**, 1444–1447.

Bokdam, J. (2001). Effects of browsing and grazing on cyclic succession in nutrient-limited ecosystems. *Journal of Vegetation Science*, **12**, 875–886.

Bolling, J.D. and Walker, L.R. (2000). Plant and soil recovery along a series of abandoned desert roads. *Journal of Arid Environments*, **46**, 1–24.

Bolling, J.D. and Walker, L.R. (2002). Fertile island development around perennial shrubs across a Mojave Desert chronosequence. *Western North American Naturalist*, **61**, 88–100.

Bonan, G.B. and Shugart, H.H. (1989). Environmental factors and ecological processes in boreal forests. *Annual Review of Ecology and Systematics*, **20**, 1–28.

Bonanomi, G., Rietkerk, M., Dekker, S.C., and Mazzoleni, S. (2008). Islands of fertility induce co-occurring negative and positive plant-soil feedbacks promoting coexistence. *Plant Ecology*, **197**, 207–218.

Bond, W.J. and Keeley, J.E. (2005). Fire as a global "herbivore": the ecology and evolution of flammable ecosystems. *Trends in Ecology and Evolution*, **20**, 387–394.

Bond, W.J. and van Wilgen, B.W. (1996). *Fire and Plants*. Chapman and Hall, London.

Bongers, F., Poorter, L., Hawthorne, W.D., and Sheil, D. (2009). The intermediate disturbance hypothesis applies to tropical forests, but disturbance contributes little to tree diversity. *Ecology Letters*, **12**, 798–805.

Bongers, T. (1990). The maturity index: an ecological measure of environmental disturbance based on nematode species composition. *Oecologia*, **83**, 14–19.

Boorman, L.A., Londo, G., and van der Maarel, E. (1997). Communities of dune slacks. In E. van der Maarel, ed. *Dry Coastal Ecosystems, Ecosystems of the World 2C*, pp. 275–295. Elsevier, Amsterdam.

Boren, L.J., Gemmell, N.J., and Barton, K.J. (2002). Tourist disturbance on New Zealand fur seals (*Arctophalus forsteri*). *Australian Mammalogy*, **24**, 85–96.

Bormann, F.H. and Likens, G.E. (1979). *Patterns and Process in a Forested Ecosystem: Disturbance, Development, and the Steady State Based on the Hubbard Brook Ecosystem Study*. Springer, New York.

Bornkamm, R., Lee, J.A., and Seaward, M.R. D., eds (1982). *Urban Ecology*. Oxford University Press, Oxford.

Botkin, D.B. (1990). *Discordant Harmonies: a New Ecology for the Twenty-first Century*. Oxford University Press, New York.

Boughton, D. and Malvadkar, U. (2002). Extinction risk in successional landscapes subject to catastrophic disturbances. *Ecology and Society*, **6**, 2 [online]. URL: http://www.consecol.org/vol6/iss2/art2 (accessed 18 March 2011).

Bradbury, I.K. (1999). Disturbance and primary production in terrestrial ecosystems. In L.R. Walker, ed. *Ecosystems of Disturbed Ground*, Ecosystems of the World 16, pp. 571–583. Elsevier, Amsterdam.

Bradshaw, A.D. and Chadwick, M.J. (1980). *The Restoration of Land: the Ecology and Reclamation of Derelict and Degraded Land*. Blackwell, Oxford.

Bragg, J.R., Prince, R.C., Harner, E.J., and Atlas, R.M. (1994). Effectiveness of bioremediation for the *Exxon Valdez* oil spill. *Nature*, **368**, 413–418.

Brahic, C. (2009). Ancient sea level rise. *New Scientist*, **202**, 40–41.

Brander, K.M. (2007). Global fish production and climate change. *Proceedings of the National Academy of Sciences USA*, **104**, 19709–19714.

Brashares, J.S., Arcese, P., Sam, M.K., Coppolillo, P.B., Sinclair, A.R.E., and Balmford, A. (2004). Bushmeat hunting, wildlife declines, and fish supply in West Africa. *Science*, **306**, 1180–1183.

Brawn, J.D., Robinson, S.K., and Thompson, F.R., III (2001). The role of disturbance in the ecology and conservation of birds. *Annual Review of Ecology and Systematics*, **32**, 251–276.

Briskie, J.V. and Mackintosh, M. (2004). Hatching failure increases with severity of population bottlenecks in birds. *Proceedings of the National Academy of Sciences USA*, **101**, 558–561.

Brittingham, S. and Walker, L.R. (2000). Facilitation of *Yucca brevifolia* recruitment by Mojave Desert shrubs. *Western North American Naturalist*, **60**, 374–383.

Brokaw, N.V.L. (1987). Gap-phase regeneration of three pioneer tree species in a tropical forest. *Journal of Ecology*, **75**, 9–19.

Brokaw, N.V.L. (1998). *Cecropia schreberiana* in the Luquillo Mountains of Puerto Rico. *The Botanical Review*, **64**, 91–120.

Brooker, R.W., Maestre, F.T., Callaway, R.M. *et al.* (2008). Facilitation in plant communities: the past, the present and the future. *Journal of Ecology*, **96**, 18–34.

Brooks, M.L., D'Antonio, C.M., Richardson, D.M., *et al.* (2004). Effects of invasive alien plants on fire regimes. *BioScience*, **54**, 677–688.

Brown, J.H. (1995). Organisms and species as complex adaptive systems: linking the biology of populations with the physics of ecosystems. In C.G. Jones and J.H. Lawton, eds., *Linking Species and Ecosystems*, pp. 16–24. Chapman and Hall, New York.

Brown, J.H. and Kodric-Brown, A. (1977). Turnover rates in insular biogeography: effect of immigration on extinction. *Ecology*, **58**, 445–449.

Brown, J.R. and Archer, S.R. (1999). Shrub invasion of grassland: recruitment is continuous and not regulated by herbaceous biomass or density. *Ecology*, **80**, 2385–2396.

Brown, L.R. (2009). *Plan B 4.0: Mobilizing to Save Civilization*. Norton, New York.

Brown, V.K. (1985). Insect herbivores and plant succession. *Oikos*, **44**, 17–22.

Bruno, J.F., Petes, L.E., Harvell, C.D., and Hettinger, A. (2003a). Nutrient enrichment can increase the severity of coral diseases. *Ecology Letters,* **6**, 1056–1061.

Bruno, J.F., Stachowicz, J.J., and Bertness, M.D. (2003b). Inclusion of facilitation into ecological theory. *Trends in Ecology and Evolution*, **18**, 119–125.

Brussaard, L., de Ruiter, P.C., and Brown, G.G. (2007). Soil biodiversity for agricultural sustainability. *Agriculture, Ecosystems and Environment*, **121**, 233–244.

Bryant, E. (2001). *Tsunami: the Underrated Hazard*. Cambridge University Press, Cambridge.

Bryant, J.P. and Chapin, F.S., III (1986). Browsing-woody plant interactions during a boreal forest plant succession. In K. Van Cleve, F.S. Chapin III, P.W. Flanagan, L.A. Viereck, and C.T. Dyrness, eds *Forest Ecosystems in the Alaska Taiga, a Synthesis of Structure and Function*, pp. 213–225. Springer, New York.

Buckley, R. (2004). *Environmental Effects of Ecotourism*. CABI, Wallingford, U.K.

Bugman, H. (2001). A review of forest gap models. *Climatic Change*, **51**, 259–305.

Bull, L.S. (2007). Reducing seabird bycatch in longline, trawl and gillnet fisheries. *Fish and Fisheries*, **8**, 31–56.

Bulleri, F. (2009). Facilitation research in marine systems: state of the art, emerging patterns and insights for future developments. *Journal of Ecology*, **97**, 1121–1130.

Bullock, J.M. and Clarke, R.T. (2000). Long distance seed dispersal by wind measuring and modeling the tail of the curve. *Oecologia*, **124**, 506–521.

Burdon, J.J., Thrall, P.H., and Ericson, L. (2006). The current and future dynamics of disease in plant communities. *Annual Review of Phytopathology*, **44**, 19–39.

Burgin, A.J. and Hamilton, S.K. (2007). Have we overemphasized the role of denitrification in aquatic ecosystems? A review of nitrate removal pathways. *Frontiers in Ecology and the Environment*, **5**, 89–96.

Burkepile, D.E. and Hay, M.E. (2006). Herbivore vs. nutrient control of marine primary producers: context-dependent effects. *Ecology*, **87**, 3128–3139.

Burleigh, J.S., Alfaro, R.I., Borden, J.H., and Taylor, S. (2002). Historical and spatial characteristics of spruce budworm *Choristoneura fumiferana* (Clem.) (Lepidoptera: Tortricidae) outbreaks in northern British Columbia. *Forest Ecology and Management*, **168**, 301–309.

Burrows, C.J. (1990). *Processes of Vegetation Change*. Unwin Hyman, London.

Busch, D.E. and Smith, S.D. (1995). Mechanisms associated with the decline of woody species in riparian ecosystems of the southwestern U.S. *Ecological Monographs*, **65**, 347–370.

Busch, D.E., Ingraham, N.L., and Smith, S.D. (1992). Water uptake in woody riparian phreatophytes of the southwestern United States: a stable isotope study. *Ecological Applications*, **2**, 450–459.

Butler, R.A. and Laurance, W.F. (2008). New strategies for conserving tropical forests. *Trends in Ecology and Evolution*, **23**, 469–472.

Butler, W.F. and Acott, T.G. (2007). An inquiry concerning the acceptance of intrinsic value theories of nature. *Environmental Values*, **16**, 149–168.

Caccianiga, M., Luzzaro, A., Pierce, S., Ceriani, R.M., and Cerabolini, B. (2006). The functional basis of a primary succession resolved by CSR classification. *Oikos*, **112**, 10–20.

Cain, A.T., Tuovila, V.R., Hewitt, D.G., and Tewes, M.E. (2003). Effects of a highway and mitigation projects on bobcats in Southern Texas. *Biological Conservation*, **114**, 189–197.

Calderón, F.J. (1993). *The Role of Mycorrhizae in the Nutrient Absorptive Strategy of Important Landslide Colonizers*. M.S. thesis, University of Puerto Rico, Río Piedras.

Cale, P. and Willoughby, N. (2009). An alternative stable state model for landscape-level restoration in South Australia. In R. J. Hobbs and K.N. Suding, eds *New Models for Ecosystem Dynamics and Restoration*, pp. 295–310. Island Press, Washington, DC.

Callaway, R.M. (1995). Positive interactions among plants. *Botanical Review*, **61**, 306–349.

Callaway, R.M. (2007). *Positive Interactions and Interdependence in Plant Communities*. Springer, New York.

Callaway, R.M. and Walker, L.R. (1997). Competition and facilitation: a synthetic approach to interactions in plant communities. *Ecology*, **78**, 1958–1965.

Callaway, R.M., Kidodze, D., Chiboshvili, M., and Khetsuriani, L. (2005). Unpalatable plants protect neighbors from grazing and increase plant community diversity. *Ecology*, **86**, 1856–1862.

Camargo, J.A. and Alonso, A. (2006). Ecological and toxicological effects of inorganic nitrogen pollution in aquatic ecosystems: a global assessment. *Environment International*, **32**, 831–849.

Camilli, R., Reddy, C.M., Yoerger, D.R. *et al.* (2010). Tracking hydrocarbon plume transport and biodegradation at *Deepwater Horizon*. *Science*, **330**: 201–204.

Cardinale, B.J. and Palmer, M.A. (2002). Disturbance moderates biodiversity-ecosystem function relationships: experimental evidence from caddisflies in stream mesocosms. *Ecology*, **83**, 1915–1927.

Cardinale, B.J., Hillebrand, H., and Charles, D.F. (2006). Geographic patterns of diversity in streams are predicted by a multivariate model of disturbance and productivity. *Journal of Ecology*, **94**, 609–618.

Cardinale, B.J., Ives, A.R., and Inchausti, P. (2004). Effects of species diversity on the primary productivity of ecosystems: extending our spatial and temporal scales of inference. *Oikos*, **104**, 437–450.

Cargill, S.M. and Chapin, F.S., III (1987). Application of successional theory to tundra restoration: a review. *Arctic and Alpine Research*, **19**, 366–372.

Carlquist, S. (1966). The biota of long-distance dispersal. I. Principles of dispersal and evolution. *The Quarterly Review of Biology*, **41**, 247–270.

Carlson, K.M., Asner, G.P., Hughes, R.F., Ostertag, R., and Martin, R.E. (2007). Hyperspectral remote sensing of canopy biodiversity in Hawaiian lowland rainforests. *Ecosystems*, **10**, 536–549.

Carlsson, R.B. and Callaghan, T.V. (1991). Positive plant interactions in tundra vegetation and the importance of shelter. *Journal of Ecology*, **79**, 973–984.

Carpenter, R.C. (1986). Partitioning herbivory and its effects on coral reef algal communities. *Ecological Monographs*, **56**, 345–363.

Carpenter, S.R. (2003). *Regime Shifts in Lake Ecosystems: Pattern and Variation*. Ecology Institute, Oldendorf/Luhe, Germany.

Carpenter, S.R. (2005). Eutrophication of aquatic ecosystems: bistability and soil phosphorus. *Proceedings of the National Academy of Sciences USA*, **102**, 10002–10005.

Carpenter, S.R. and Brock, W.A. (2006). Rising variance: a leading indicator of ecological transition. *Ecology Letters*, **9**, 311–318.

Carpenter, S.R. and Scheffer, M. (2009). Critical transitions and regime shifts in ecosystems: consolidating recent advances. In R.J. Hobbs and K.N. Suding, eds *New Models for Ecosystem Dynamics and Restoration*, pp. 22–32. Island Press, Washington, DC.

Carpenter, S.R., Kitchell, J.F., Hodgson, J.R., *et al.* (1987). Regulation of lake primary productivity by food web structure. *Ecology*, **68**, 1863–1876.

Carpenter, S.R, Caraco, N.F., Correll, D.L., Howarth, R.W., Sharpley, A.N., and Smith, V.H. (1998). Nonpoint pollution of surface waters with phosphorus and nitrogen. *Ecological Applications*, **8**, 559–568.

Carreiro, M.M. (2008). Using the urban-rural gradient approach to determine the effects of land use on forest remnants. In M.M. Carreiro, Y.-C. Song, and J. Wu, eds *Ecology, Planning and Management of Urban Forests*, pp. 169–186. Springer, New York.

Carreiro, M.M. and Zipperer, W.C. (in press). Co-adapting societal and ecological interactions following large disturbances in urban park woodlands. *Austral Ecology*, **36**(8), doi: 10.1111/j.1442-9993.2010.02237.x

Carreiro, M.M., Song, Y.-C., and Wu, J., eds (2008). *Ecology, Planning and Management of Urban Forests*. Springer, New York.

Carson, W.P., Banta, J.A., Royo, A.A., and Kirschbaum, C. (2005). Plant communities growing on boulders in the Allegheny National Forest: evidence for boulders as refugia from deer and as a bioassay of overbrowsing. *Natural Areas Journal*, **25**, 10–18.

Case, T.J. (1990). Invasion resistance arises in strongly interacting species-rich model competition communities. *Proceedings of the National Academy of Sciences USA*, **87**, 9610–9614.

Cassel, D.K. and Fryrear, D.W. (1990). Evaluation of productivity changes due to accelerated soil erosion. In W.E. Larson, G.R. Foster, R.R. Allmaras and C.M. Smith, eds *Proceedings of Soil Erosion and Productivity Workshop*, pp. 41–54. University of Minnesota, St. Paul, MN.

Cederholm, C.J., Kunze, M.D., Murota, T., and Sibatani, A. (1999). Pacific salmon carcasses: essential contributions of nutrients and energy for aquatic and terrestrial ecosystems. *Fisheries*, **24**, 6–15.

Chaneton, E.J. and Facelli, J.M. (1991). Disturbance effects on plant community diversity: spatial scales and dominance hierarchies. *Vegetatio*, **93**, 143–155.

Chapin, F.S., III (1993). Physiological controls over plant establishment in primary succession. In J. Miles and D.H. Walton, *Primary Succession on Land*, pp. 161–178. Blackwell, Oxford.

Chapin, F.S., III (1995). New cog in the nitrogen cycle. *Nature*, **377**, 199–200.

Chapin, F.S., III, Vitousek, P.M., and Van Cleve, K. (1986). The nature of nutrient limitation in plant communities. *The American Naturalist*, **127**, 48–58.

Chapin, F.S., III, Walker, L.R., Fastie, C.L., and Sharman, L.C. (1994). Mechanisms of primary succession following deglaciation at Glacier Bay, Alaska. *Ecological Monographs*, **64**, 149–175.

Chapin, F.S., III, Zavaleta, E.S., Eviner, V.T. *et al.* (2000). Consequences of changing biodiversity. *Nature*, **405**, 234–242.

Chapin, F.S., III, Lovecraft, A.L., Zavaleta, E.S., *et al.* (2006). Policy strategies to address sustainability of Alaskan boreal forests in response to directionally changing climate. *Proceedings of the National Academy of Sciences USA*, **7**, 16637–16643.

Chapin, F.S., III, Power, M.E., and Cole, J.J. (2011). Coupled biogeochemical cycles and Earth stewardship. *Frontiers in Ecology and the Environment*, **9**, 3.

Chapman, M.G. (1999). Improving sampling designs for measuring restoration in aquatic habitats. *Journal of Aquatic Ecosystem Stress and Recovery*, **6**, 235–251.

Charley, J.L. and West, N.E. (1975). Plant-induced soil chemical patterns in some shrub-dominated semi-desert ecosystems of Utah. *Journal of Ecology*, **63**, 945–964.

Chaves, M.M., Maroco, J.P., and Pereira, J.S. (2003). Understanding plant responses to drought – from genes to the whole plant. Functional Plant Biology, **30**, 239–264.

Chazdon, R.L. (2003). Tropical forest recovery: legacies of human impact and natural disturbances. *Perspectives in Plant Ecology, Evolution and Systematics*, **6**, 51–71.

Chen, C.R., Condron, L.M., Sinaj, S., Davis, M.R., Sherlock, R.R., and Frossard, E. (2003). Effects of plant species on phosphorus availability in a range of grassland soils. *Plant and Soil*, **256**, 115–130.

Chen, W.Y. and Jim, C.Y. (2008). Assessment and valuation of the ecosystem services provided by urban forests. In M.M. Carreiro, Y.-C. Song, and J. Wu, eds *Ecology, Planning and Management of Urban Forests*, pp. 53–83. Springer, New York.

Cheng, L. and Birch, M.C. (1987). Insect flotsam: an unstudied marine resource. *Ecological Entomology*, **3**, 87–97.

Chengrui, M. and Dregne, H.E. (2001). Review article: silt and the future development of China's Yellow River. *The Geographical Journal*, **167**, 7–22.

Childress, W.M., Crisafulli, C.M., and Rykiel, E.J. (1998). Comparison of Markovian matrix models of a primary successional plant community. *Ecological Modelling*, **107**, 92–102.

Christensen, N.L. (1985). Shrubland fire regimes and their evolutionary consequences. In S.T.A. Pickett and P.S. White, eds *The Ecology of Natural Disturbance and Patch Dynamics*, pp. 86–100. Academic Press, London.

Christensen, N.L. and Peet, R.K. (1984). Convergence during secondary forest succession. *Journal of Ecology*, **72**, 25–36.

Christensen, N.S., Wood, A.W., Voisin, N., Lettenmaier, D.P., and Palmer, R.N. (2004). The effects of climate change on the hydrology and water resources of the Colorado River Basin. *Climate Change*, **62**, 337–363.

Christensen, P., Recher, H., and Hoare, J. (1981). Response of open forests (dry sclerophyll forests) to fire regimes. In A.M. Gill, R.H. Groves, and I.R. Noble, eds *Fire and the Australian Biota*, pp. 367–393. Australian Academy of Science, Canberra.

Christian, D. (1997). *Imperial and Soviet Russia: Power, Privilege, and the Challenge of Modernity*. Palgrave Macmillan, New York.

Clark, C. (1982). *Planet Earth: Flood*. Time-Life Books, Alexandria, Virginia.

Clark, C.J., Poulsen, J.R., Malonga, R., and Elkan, P.W., Jr (2009). Logging concessions can extend the conservation estate for central African tropical forests. *Conservation Biology*, **23**, 1281–1293.

Clark, D.B. (1990). The role of disturbance in the regeneration of Neotropical moist forests. In K.S. Bawa and M. Hadley, eds *Reproductive Ecology of Tropical Forest Plants*, Man and the Biosphere Series, pp. 291–315. United Nations Educational Scientific and Cultural Organization, Paris, France.

Clark, J.S. (1998). Why trees migrate so fast: confronting theory with dispersal biology and the paleorecord. *The American Naturalist*, **152**, 204–224.

Clark, J.S., Silman, M., Kern, R., Macklin, E., and Hille Ris Lambers, J. (1999). Seed dispersal near and far: patterns across temperate and tropical forests. *Ecology*, **80**, 1475–1495.

Clarkson, B.R. and Clarkson, B.D. (1983). Mt. Tarawera: 2. Rates of change in the vegetation and flora of the high domes. *New Zealand Journal of Botany*, **6**, 107–119.

Clarkson, B.R. and Clarkson, B.D. (1995). Recent vegetation changes on Mount Tarawera, Rotorua, New Zealand. *New Zealand Journal of Botany*, **33**, 339–354.

Clements, F.E. (1916). *Plant Succession: an Analysis of the Development of Vegetation*. Carnegie Institute of Washington Publication 242. Carnegie Institute, Washington, DC.

Clements, F.E. (1928). *Plant Succession and Indicators*. H.W. Wilson, New York.

Clements, F.E. (1936). Nature and structure of the climax. *Journal of Ecology*, **24**, 252–284.

Clover, C. (2004). *End of the Line: How Overfishing is Changing the World and What We Eat*. Ebury Press, London.

Coffin, A.W. (2007). From roadkill to road ecology: a review of the ecological effects of roads. *Journal of Transport Geography*, **15**, 396–406.

Cogbill, C.V. (1996). Black growth and fiddlebutts: the nature of old-growth red spruce. In M.B. Davis, ed. *Eastern Old-Growth Forests: Prospects for Rediscovery and Recovery*, pp. 113–125. Island Press, Washington, DC.

Cohen, A.S. (2011). Scientific drilling and biological evolution in ancient lakes: lessons learned and recommendations for the future. *Hydrobiologia*, doi: 10.1007/s10750-010-0546-7.

Cohen, J.E. (1995). *How Many People Can the Earth Support?* Norton, New York.

Cole, J.J., Caraco, N.F., Kling, G.W., and Kratz, T.K. (1994). Carbon dioxide supersaturation in the surface water of lakes. *Science*, **265**, 1568–1570.

Cole, L., Buckland, S.M., and Bardgett, R.D. (2008). Influence of disturbance and nitrogen addition on plant and soil animal diversity in grassland. *Soil Biology and Biochemistry*, **40**, 505–514.

Coley, P.D., Bryant, J.P., and Chapin, F.S., III (1985). Resource availability and plant antiherbivore defense. *Science*, **230**, 895–899.

Colinvaux, P.A. (1973). *Introduction to Ecology*. Wiley, New York.

Collins, S.L., Glenn, S.M., and Gibson, D.J. (1995). Experimental analysis of intermediate disturbance and initial floristic composition: decoupling cause and effect. *Ecology*, 76, 486–492.

Collins, S.L., Suding, K.N., Cleland, E.E., *et al.* (2008). Rank clocks and plant community dynamics. *Ecology*, 89, 3534–3541.

Compton, S.G., Ross, S.J., and Thornton, I.W.B. (1994). Pollinator limitation of fig tree reproduction on the island of Anak Krakatau (Indonesia). *Biotropica*, 26, 180–186.

Connell, J.H. (1978). Diversity in tropical rain forests and coral reefs. *Science*, 199, 1302–1310.

Connell, J.H. and Keough, M.J. (1985). Disturbance and patch dynamics of subtidal marine animals on hard substrata. In S.T.A. Pickett and P.S. White, eds *The Ecology of Natural Disturbance and Patch Dynamics*, pp. 125–152, Academic Press, New York.

Connell, J.H. and Slatyer, R.O. (1977). Mechanisms of succession in natural communities and their roles in community stability and organization. *The American Naturalist*, 111, 1119–1144.

Connell, J.H., Noble, I.R., and Slatyer, R.O. (1987). On the mechanisms producing successional change. *Oikos*, 50, 136–137.

Connell, J.H., Hughes, T.P., and Wallace, C.C. (1997). A 30-year study of coral abundance, recruitment, and disturbance at several scales in space and time. *Ecological Monographs*, 67, 461–488.

Constantine, R., Brunton, D.H., and Dennis, T. (2004). Dolphin-watching tour boats change bottlenose dolphin (*Tursiops truncatus*) behaviour. *Biological Conservation*, 117, 299–307.

Cook, J.E. (1996). Implications of modern succession theory for habitat typing: a review. *Forest Science*, 42, 67–75.

Cooke, B.J., Nealis, V.G., and Régnière, J. (2007). Insect defoliators as periodic disturbances in northern forest ecosystems. In E.A. Johnson and K. Miyanishi, eds *Plant Disturbance Ecology: the Process and the Response*, pp. 487–525. Academic Press, Amsterdam.

Cooke, J.A. (1999). Mining. In L.R. Walker, ed. *Ecosystems of Disturbed Ground*, Ecosystems of the World 16, pp. 365–384. Elsevier, Amsterdam.

Cooper, W.S. (1923). The recent ecological history of Glacier Bay, Alaska: II. The present vegetation cycle. *Ecology*, 4, 223–246.

Correll, D.L. (1998). The role of phosphorus in the eutrophication of receiving waters: a review. *Journal of Environmental Quality*, 27, 261–266.

Costanza, R., d'Arge, R., deGroot, R., *et al.* (1997). The value of the world's ecosystem services and natural capital. *Nature*, 387, 253–260.

Covich, A.P., Crowl, T.A., and Heartsill-Scalley, T. (2006). Effects of drought and hurricane disturbances on headwater distributions of palaemonid river shrimp (*Macrobrachium* spp.) in the Luquillo Mountains, Puerto Rico. *Journal of the North American Benthological Society*, 25, 99–107.

Cowles, H.C. (1899). The ecological relations of the vegetation on the sand dunes of Lake Michigan. *Botanical Gazette*, 27, 95–117, 167–202, 281–308, 361–391.

Cowles, H.C. (1901). The physiographic ecology of Chicago and vicinity: a study of the origin, development, and classification of plant societies. *Botanical Gazette*, 31, 73–108, 145–82.

Cox, P.M., Betts, R.A., Jones, C.D., Spall, S.A., and Totterdell, I.J. (2000). Acceleration of global warming due to carbon cycle feedbacks in a coupled climate model. *Nature*, 408, 184–187.

Craddock, C., Lutz, R.A., and Vrijenhoek, R.C. (1997). Patterns of dispersal and larval development of archaeogastropod limpets at hydrothermal vents in the eastern Pacific. *Journal of Experimental Marine Biology and Ecology*, 210, 37–51.

Cramer, V.A. and Hobbs, R.J., eds (2007). *Old Fields: Dynamics and Restoration of Abandoned Farmland*. Island Press, Washington, DC.

Crawford, R.L., Sugg, P.M., and Edwards, E.S. (1995). Spider arrival and primary establishment on terrain depopulated by volcanic eruption at Mount St. Helens, Washington. *American Midland Naturalist*, 133, 60–75.

Cremene, C., Groza, G., Rakosy, L., *et al.* (2005). Alterations of steppe-like grasslands in Eastern Europe: a threat to regional biodiversity hotspots. *Conservation Biology*, **19**, 1606–1618.

Crews, T., Kitayama, K., Fownes, J., *et al.* (1995). Changes in soil phosphorus fractions and ecosystem dynamics across a long chronosequence in Hawaii. *Ecology*, **76**, 1407–1424.

Crone, T.J. and Tolstoy, M. (2010). Magnitude of the 2010 Gulf of Mexico oil leak. *Science*, **330**, 634.

Crosby, A.W. (1986). *Ecological Imperialism: the Biological Expansion of Europe, 900–1900*. Cambridge University Press, Cambridge [2nd edition 2004].

Crutzen, P.J. (2002). Geology of mankind. *Nature*, **415**, 23.

Cutler, N.A., Belyea, L.R., and Dugmore, A.J. (2008). The spatiotemporal dynamics of a primary succession. *Journal of Ecology*, **96**, 231–246.

Czech, B. (2008). Prospects for reconciling the conflict between economic growth and biodiversity conservation with technological progress. *Conservation Biology*, **22**, 1389–1398.

Dabbert, S., Dubgaard, A., Slangen, L., and Whitby, M., eds (1998). *The Economics of Landscape and Wildlife Conservation*. CAB International, Wallingford, U.K.

Dale, V.H., Joyce, L.A., McNulty, S., *et al.* (2001). Climate change and forest disturbances. *BioScience*, **51**, 723–734.

Dalling, J.W. (1994). Vegetation colonization of landslides in the Blue Mountains, Jamaica. *Biotropica*, **26**, 392–399.

Dando, P.R., Hughes, J.A., Leahy, Y., Taylor, L.J., and Zivanovic, S. (1995). Earthquakes increase hydrothermal venting and nutrient inputs into the Aegean. *Continental Shelf Research*, **15**, 655–662.

Dangerfield, J.M., McCarthy, T.S., and Ellery, W.N. (1998). The mound-building termite *Macrotermes michaelseni* as an ecosystem engineer. *Journal of Tropical Ecology*, **14**, 507–520.

Danin, A. (1991). Plant adaptations in desert dunes. *Journal of Arid Environments*, **21**, 193–212.

D'Antonio, C.M. (1990). *Invasion of Coastal Plant Communities by the Introduced Succulent,* Corpobrotus edulis *(Aizoaceae)*. Ph.D. Dissertation, University of California, Santa Barbara.

D'Antonio, C.M. and Vitousek, P.M. (1992). Biological invasions by exotic grasses, the grass/fire cycle, and global change. *Annual Review of Ecology and Systematics*, **23**, 63–87.

Daskalov, G.M. (2002). Overfishing drives a trophic cascade in the Black Sea. *Marine Ecology Progress Series*, **225**, 53–63.

Daszak, P., Berger, L., Cunningham, A.A., Hyatt, A.D., Green, D.E., and Speare, R. (1999). Emerging infectious diseases and amphibian population declines. *Emerging Infectious Diseases*, **5**, 735–748.

Dauble, D.D., Hanrahan, T.P., Geist, D.R., and Parsley, M.J. (2003). Impacts of the Columbia River hydroelectric system on main-stem habitats of fall Chinook salmon. *North American Journal of Fisheries Management*, **23**, 641–659.

Davidson, D.W. (1993). The effects of herbivory and granivory on terrestrial plant succession. *Oikos*, **68**, 23–35.

Davidson, E.A. and Janssens, I.A. (2006). Temperature sensitivity of soil carbon decomposition and feedbacks to climate change. *Nature*, **440**, 165–173.

Davis, K.B., Barans, C.A., Stender, B.W., Schmidt, D.J., and Pashuk, O. (1991). Shelf water conditions of the South Atlantic Bight six weeks after Hurricane Hugo. *Journal of Coastal Research Special Issue* **8**, 69–82.

Davis, M.A. (2003). Biotic globalization: does competition from introduced species threaten biodiversity? *BioScience*, **53**, 481–489.

Davis, W.M. (1909). The geographical cycle. In D.W. Johnson, ed. *Geographical Essays*, pp. 254–256. Ginn and Company, Oxford.

Davydchuk, V. (1997). Ecosystem remediation in radioactively polluted areas: the Chernobyl experience. *Ecological Engineering*, **8**, 325–336.

Day, R.T., Keddy, P.A., McNeill, J., and Carleton, T. (1988). Fertility and disturbance gradients: a summary model for riverine marsh vegetation. *Ecology*, **69**, 1044–1054.

Dayton, P.K. (1971). Competition, disturbance, and community organization: the provision and subsequent utilization of space in a rocky intertidal community. *Ecological Monographs*, **41**, 351–389.

DeAngelis, D.L. and Waterhouse, J.C. (1987). Equilibrium and non-equilibrium concepts in ecological models. *Ecological Monographs*, **57**, 1–21.

DeGange, A.R., Byrd, G.V., Walker, L.R., and Waythomas, C.F. (2010). Introduction – the impacts of the 2008 eruption of Kasatochi Volcano on terrestrial and marine ecosystems in the Aleutian Islands, Alaska. *Arctic, Antarctic, and Alpine Research*, **42**, 245–249.

Demarais, S., Tazik, D.J., Guertin, P.J., and Jorgensen, E.E. (1999). Disturbance associated with military exercises. In L.R. Walker, ed. *Ecosystems of Disturbed Ground*, Ecosystems of the World 16, pp. 385–396. Elsevier, Amsterdam.

Deng, Y., Ogden, J., Horrocks, M., and Anderson, S. (2006). Application of palynology to describe vegetation succession in estuarine wetlands on Great Barrier Island, northern New Zealand. *Journal of Vegetation Science*, **17**, 765–782.

Dent, C.L. and Grimm, N.B. (1999). Spatial heterogeneity of stream water nutrient concentrations over successional time. *Ecology*, **80**, 2283–2298.

Dethier, M.N. (1984). Disturbance and recovery in intertidal pools: maintenance of mosaic patterns. *Ecological Monographs*, **54**, 99–118.

Diamond, J. (1997). *Guns, Germs, and Steel: the Fates of Human Societies*. Norton, New York.

Diamond, J. (2005). *Collapse: How Societies Choose to Fail or Succeed*. Penguin Group, New York.

Díaz, S., Cabido, M., and Casanoves, F. (1998). Plant functional traits and environmental filters at a regional scale. *Journal of Vegetation Science*, **9**, 113–122.

Didden, W.A.M. (1993). Ecology of terrestrial Enchytraeidae. *Pedobiologia*, **37**, 3–29.

Diemer, M. and Schmid, B. (2001). Effects of biodiversity and disturbance on the survival and performance of two *Ranunculus* species with differing clonal architectures. *Ecography*, **24**, 59–67.

Dobson, A.P., Bradshaw, A.D., and Baker, A.J.M. (1997). Hopes for the future: restoration ecology and conservation biology. *Science*, **277**, 515–522.

Dollar, S.J. and Grigg, R.W. (1981). Impact of a kaolin clay spill on a coral reef in Hawaii. *Marine Biology*, **65**, 269–276.

Done, T.J. (1992). Effects of tropical cyclone waves on ecological and geomorphological structures on the Great Barrier Reef. *Continental Shelf Research*, **12**, 859–872.

Dornelas, M. (2010). Disturbance and change in biodiversity. *Philosophical Transactions of the Royal Society B: Biological Sciences*, **365**, 3719–3727.

Driscoll, C.T., Driscoll, K.M., Mitchell, M.J., and Raynal, D.J. (2003). Effects of acidic deposition on forest and aquatic ecosystems in New York State. *Environmental Pollution*, **123**, 327–336.

Drury, W.H. and Nisbet, I.C.T. (1973). Succession. *Journal of the Arnold Arboretum*, **54**, 331–368.

Duever, M.J., Meeder, J.F., Meeder, L.C., and McCollom, J.M. (1994). The climate of South Florida and its role in shaping the Everglades ecosystem. In S.M. Davis and J.M. Ogden, eds *Everglades: the Ecosystem and its Restoration*, pp. 225–248. St Lucie Press, Delray Beach, Florida, U.S.

Dufrêne, M. and Legendre, P. (1997). Species assemblages and indicator species: the need for a flexible asymmetrical approach. *Ecological Monographs*, **67**, 345–366.

Dyurgerov, E. (2002). *Glacier Mass Balance and Regime: Data of Measurements and Analysis*. Occasional Paper Number 55, Institute of Arctic and Alpine Research, University of Colorado, Boulder, CO, U.S.

Eakin, C.M., Morgan, J.A., Heron, S.F., *et al.* (2010). Caribbean corals in crisis: record thermal stress, bleaching, and mortality in 2005. *PLoS ONE*, **5**, e12969.

Easton, T.A. (2009). *Taking Sides: Clashing Views on Environmental Issues*, 13th edn. McGraw-Hill, Boston.

Edwards, C.A. (2004). *Earthworm Ecology*, 2nd edn. CRC Press, Boca Raton, FL, U.S.

Edwards, J.W. (1988). Life in the allobiosphere. *Trends in Ecology and Evolution*, **3**, 111–114.

Egler, F.E. (1951). A commentary on American plant ecology based on the textbooks of 1947–1949. *Ecology*, **32**, 673–695.

Egler, F.E. (1954). Vegetation science concepts. I. Initial floristic composition, a factor in old-field vegetation development. *Vegetatio*, **4**, 412–417.

Ehrenfeld, J.G. (2000). Defining the limits of restoration: the need for realistic goals. *Restoration Ecology*, **8**, 2–9.

Elderfield, H. and Schultz, A. (1996). Mid-ocean ridge hydrothermal fluxes and the chemical composition of the ocean. *Annual Review of Earth Planetary Science*, **24**, 191–224.

Ellis, E.C., Klein Goldewijk, K., Siebert, S., Lightman, D., and Ramankutty, N. (2010). Anthropogenic transformation of the biomes, 1700 to 2000. *Global Ecology and Biogeography*, **19**, 589–606.

Ellis, J. and Galvin, K.A. (1994). Climate patterns and land-use practices in the dry zones of Africa. *BioScience*, **44**, 340–349.

Elser, J.J., Bracken, M.E.S., Cleland, E.E., *et al.* (2007). Global analysis of nitrogen and phosphorus limitation of primary producers in freshwater, marine and terrestrial ecosystems. *Ecology Letters*, **10**, 1035–1042.

Elton, C.S. (1958). *The Ecology of Invasions by Animals and Plants*. Methuen, London, U.K.

Emanuel, K. (2005). Increasing destructiveness of tropical cyclones over the past 30 years. *Nature*, **436**, 686–688.

Engelmark, O. (1999). Boreal forest disturbances. In L.R. Walker, ed. *Ecosystems of Disturbed Ground*, Ecosystems of the World 16, pp. 161–186. Elsevier, Amsterdam.

Epps, C.W., Palsbøll, P.J., Wehausen, J.D., Roderick, G.K., Ramey, R.R. II, and McCullough, D.R. (2005). Highways block gene flow and cause a rapid decline in genetic diversity of desert bighorn sheep. *Ecology Letters*, **8**, 1029–1038.

Ernst, W.H.O., Van Duin, W.E., and Oolbekking, G.T. (1984). Vesicular-arbuscular mycorrhiza in dune vegetation. *Acta Botanica Neerlandica*, **33**, 151–160.

Ettema, C.H. and Wardle, D.A. (2002). Spatial soil ecology. *Trends in Ecology and Evolution*, **17**, 177–183.

Ettema, C.H., Rathbun, S.L., and Coleman, D.C. (2000). On spatiotemporal patchiness and the coexistence of five species of *Chronogaster* (Nematoda: Chronogasteridae) in a riparian wetland. *Oecologia*, **125**, 444–452.

Evenari, M. (1985). The desert environment. In M. Evenari, I. Noy-Meir, and D.W. Goodall, eds *Hot Deserts and Arid Shrublands*, Ecosystems of the World 12A, pp. 1–22. Elsevier, Amsterdam.

Eviner, V.T. and Chapin, F.S., III (2003). Functional matrix: a conceptual framework for predicting multiple plant effects on ecosystem processes. *Annual Review of Ecology and Systematics*, **34**, 455–485.

Evrendilek, F., Celik, I., and Kilic, S. (2004) Changes in soil organic carbon and other physical soil properties along adjacent Mediterranean forests, grassland, and cropland ecosystems in Turkey. *Journal of Arid Environments*, **59**, 743–752.

Facelli, J.M. and D'Anela, E. (1990). Directionality, convergence and the rate of change during early succession in the Inland Pampa, Argentina. *Journal of Vegetation Science*, **1**, 255–260.

Farber, J.M. (1997). *Ancient Hawaiian Fishponds: Can Restoration Succeed on Moloka'i?* Neptune House Publications, Encinitas, CA, U.S.

Fargione, J.E. and Tilman, D. (2005). Diversity decreases invasion via both sampling and complementarity effects. *Ecology Letters*, **8**, 604–611.

Farrell, T.M. (1991). Models and mechanisms of succession: an example from a rocky intertidal community. *Ecological Monographs*, **61**, 95–113.

Fastie, C.L. (1995). Causes and ecosystem consequences of multiple pathways on primary succession at Glacier Bay, Alaska. *Ecology*, **76**, 1899–1916.

Fenner, M. (1985). *Seed Ecology*. Chapman and Hall, London.

Fernández, D.S. and Fetcher, N. (1991). Changes in light availability following Hurricane Hugo in a subtropical montane forest in Puerto Rico. *Biotropica*, **23**, 393–399.

Ferreira, S.M. and van Aarde, R.J. (1999). Habitat associations and competition in *Mastomys–Saccostomus–Aethomys* assemblages on coastal dune forests. *African Journal of Ecology*, **37**, 121–136.

Fettig, C.J., Klepzig, K.D., Billings, R.F., *et al.* (2007). The effectiveness of vegetation management practices for prevention and control of bark beetle infestations in coniferous forests of the western and southern United States. *Forest Ecology and Management*, **238**, 24–53.

Field, C.B., Behrenfeld, M.J., Randerson, J.T., and Falkowski, P.G. (1998). Primary productivity of the biosphere: integrating terrestrial and oceanic components. *Science*, **281**, 237–240.

Field, C.B., Randerson, J.T., and Malmström, C.M. (1995). Global net primary production: combining ecology and remote sensing. *Remote Sensing of Environment*, **51**, 74–88.

Finkl, C.W., Jr and Pilkey, O.H., eds (1991). Impacts of Hurricane Hugo: September 10–22, 1989. *Journal of Coastal Research Special Issue*, **8**, 1–356.

Finnigan, J.J. (2007). The turbulent wind in plant and forest canopies. In E.A. Johnson and K. Miyanishi, eds *Plant Disturbance Ecology: the Process and the Response*, pp. 15–58. Academic Press, Amsterdam.

Fisher, S.G., Gray, L.J., Grimm, N.B., and Busch, D.E. (1982). Temporal succession in a desert stream ecosystem following flash flooding. *Ecological Monographs*, **52**, 93–110.

Fitter, A.H., and Fitter, R.S.R. (2002). Rapid changes in flowering time in British plants. *Science*, **296**, 1689–1691.

Flanary, W., McGinley, M., and Johnson, G. (2008). Environmental effects of the Cherynobyl accident. In: J. Cutler, ed. *Encyclopedia of Earth*, National Council for Science and the Environment, Washington, DC.

Flannery, T. (2002). *The Future Eaters: an Ecological History of the Australasian Lands and People*. Grove Press, New York.

Flannery, T. (2005). *The Weather Makers*. Grove Press, New York.

Fleming, R.A., Hopkin, A.A., and Candau, J.-N. (2000). Insect and disease disturbance regimes in Ontario's forests. In A.H. Perera, D.L. Euler, and I.D. Thompson, eds *Ecology of a Managed Terrestrial Landscape: Patterns and Processes of Forest Landscapes in Ontario*, pp. 141–162. UBC Press, Vancouver, British Columbia.

Flewelling, L.J., Naar, J.P., Abbott, J.P., *et al.* (2005). Brevetoxicosis: red tides and marine mammal mortalities. *Nature*, **435**, 755–756.

Flinn, K.M. (2007). Microsite-limited recruitment controls fern colonization of post-agricultural forests. *Ecology*, **88**, 3103–3114.

Flöder, S. and Sommer, U. (1999). Diversity in planktonic communities: an experimental test of the intermediate disturbance hypothesis. *Limnology and Oceanography*, **44**, 1114–1119.

Flores-Martinez, A., Ezcurra, E., and Sanchez-Colon, S. (1994). Effect of *Neobuxbaumia tetetzo* on growth and fecundity of its nurse plant *Mimosa luisana*. *Journal of Ecology*, **82**, 325–330.

Forman, R.T.T. (1997). *Land Mosaics: the Ecology of Landscapes and Regions*. Cambridge University Press, Cambridge.

Forman, R.T.T. (2000). Estimate of the area affected ecologically by the road system in the United States. *Conservation Biology*, **14**, 31–35.

Forman, R.T.T. and Alexander, L.E. (1998). Roads and their major ecological effects. *Annual Review of Ecology and Systematics*, **29**, 207–231.

Forman, R.T.T., Sperling, D., Bissonette, J.A., *et al.* (2003). *Road Ecology: Science and Solutions*. Island Press, Washington, DC.

Fort, K.P. and Richards, J.H. (1998). Does seed dispersal limit initiation of primary succession in desert playas? *American Journal of Botany*, **85**, 1722–1731.

Foster, B.L. and Tilman, D. (2000). Dynamic and static views of succession: testing the descriptive power of the chronosequence approach. *Plant Ecology*, **146**, 1–10.

Foster, R., Hagan, A., Perera, N. *et al.* (2006). *Tsunami and earthquake damage to coral reefs of Aceh, Indonesia*. Reef Check Foundation, Pacific Palisades, CA, U.S.

Fox, B.J. (1990). Changes in the structure of mammal communities over successional time scales. *Oikos*, **59**, 321–329.

Franklin, J.F., MacMahon, J.A., Swanson, F.J., and Sedell, J.R. (1985). Ecosystem responses to catastrophic disturbances: lessons from Mount St. Helens. *National Geographic Research*, **1**, 198–216.

Franklin, J.F., Shugart, H.H., and Harmon, M.E. (1987). Tree death as an ecological process: the causes, consequences, and variability of tree mortality. *BioScience*, **37**, 550–556.

Franklin, J.F., Frenzen, P., and Swanson, F.J. (1988). Re-creation of ecosystems at Mount St. Helens: contrasts in artificial and natural approaches. In J. Cairns, Jr, ed. *Rehabilitating Damaged Ecosystems*, Volume II, pp. 1–37. CRC, Boca Raton, FL.

Frelich, L.E. and Reich, P.B. (2010). Will environmental changes reinforce the impact of global warming on the prairie-forest border of central North America? *Frontiers in Ecology and the Environment*, **8**, 371–378.

Friedman, T.L. (2008). *Hot, Flat and Crowded: Why We Need a Green Revolution – and How It Can Renew America*. Farrar, Straus and Giroux, New York.

Fukami, T., Bezemer, T.M., Mortimer, S.R., and van der Putten, W.H. (2005). Species divergence and trait convergence in experimental plant community assembly. *Ecology Letters*, **8**, 1283–1290.

Fukami, K., Sugiura, T., Magome, J., and Kawakami, T. (2009). *Integrated Flood Analysis System (IFAS Version 1.2)*. User's Manual. Public Works Research Institute, Tsukuba, Japan.

Fukami, T., Wardle, D.A., Bellingham, P.J., *et al.* (2006). Above- and below-ground impacts of introduced predators in seabird dominated ecosystems. *Ecology Letters*, 9, 1299–1307.

Fullen, M.A. and Mitchell, D.J. (1994). Desertification and reclamation in North-Central China. *Ambio*, **23**, 131–135.

Fuller, R.N. (1999). *The Role of Refugia in Primary Succession on Mount St. Helens, Washington*. M.S. Thesis, University of Washington, Seattle.

Fuller, R.N. and del Moral, R. (2003). The role of refugia and dispersal in primary succession on Mount St. Helens, Washington. *Journal of Vegetation Science*, **14**, 637–644.

Gadgil, M., and Guha, R. (1992). *This Fissured Land: an Ecological History of India*. Oxford University Press, Oxford.

Gaertner, M., Den Breeyen, A., Hui, C., and Richardson, D.M. (2009). Impacts of alien plant invasions on species richness in Mediterranean-type ecosystems: a meta-analysis. *Progress in Physical Geography*, **33**, 319–338.

Galan, A. (2000). Benthic Amphipoda and Isopoda (Crustacea) from the sublittoral zone off Surtsey and Heimaey south of Iceland. *Surtsey Research*, **11**, 89–96.

Gallagher, E.D., Jumars, P.A., and Trueblood, D.D. (1983). Facilitation of soft-bottom benthic succession by tube builders. *Ecology*, **64**, 1200–1216.

Galloway, J.N., Townsend, A.R., Erisman, J.W., *et al.* (2008). Transformation of the nitrogen cycle: recent trends, questions, and potential solutions. *Science*, **320**, 889–892.

Gardner, T.A., Barlos, J., Sodhi, N.S., and Peres, C.A. (2010). A multi-region assessment of tropical forest biodiversity in a human-modified world. *Biological Conservation*, 143, 2293–2300.

Garner, W. and Steinberger, Y. (1989). A proposed mechanism for the formation of "fertile islands" in the desert ecosystem. *Journal of Arid Environments*, 16, 257–262.

Garrity, S.D. and Levings, S.C. (1985). Interspecific interactions and scarcity of a tropical limpet. *Journal of Molluscan Studies*, 51, 297–308.

Garwood, N., Janos, P., and Brokaw, N. (1979). Earthquake-caused landslides: a major disturbance in tropical forests. *Science*, 205, 997–999.

Gaskin, J.F. and Shafroth, P.B. (2005). Hybridization of *Tamarix ramosissima* and *T. chinensis* (salt-cedars) with *T. aphylla* (athel) (Tamaricaceae) in the southwestern USA determined from DNA sequence data. *Madroño*, 52, 1–10.

Gauthier, S., Bergeron, Y., and Simon, J.-P. (1996). Effects of fire regime on the serotiny level of jack pine. *Journal of Ecology*, 84, 539–548.

Gehrke, P.C., Gilligan, D.M., and Barwick, M. (2002). Changes in fish communities of the Shoal-haven River 20 years after construction of Tallowa Dam, Australia. *River Research and Applications*, 18, 265–286.

Geider, R.J., DeLucia, E.H., Falkowski, P.G., *et al.* (2001). Primary productivity of planet earth: biological determinants and physical constraints in terrestrial and aquatic habitats. *Global Change Biology*, 7, 849–882.

Ghersa, C.M. and León, R.J.C. (1999). Successional changes in agroecosystems of the rolling pampa. In L.R. Walker, ed. *Ecosystems of Disturbed Ground*, Ecosystems of the World 16, pp. 487–502. Elsevier, Amsterdam.

Giampietro, M. (1999). Economic growth, human disturbance to ecological systems, and sustainability. In L.R. Walker, ed. *Ecosystems of Disturbed Ground*, Ecosystems of the World 16, pp. 723–746. Elsevier, Amsterdam.

Gibbons, J.W., Greene, J.L., and Congdon, J.D. (1983). Drought-related responses of aquatic turtle populations. *Journal of Herpetology*, 17, 242–246.

Gibson, D.J., Ely, J.S., and Looney, P.B. (1997). A Markovian approach to modeling succession on a coastal barrier island following beach nourishment. *Journal of Coastal Research*, 13, 831–841.

Gilbert, G.S. and Hubbell, S.P. (1996). Plant diseases and the conservation of tropical forests. *BioScience*, 46, 98–106.

Gilbert, N. (2009). Environment: the disappearing nutrient. *Nature*, 461, 716–718.

Gilbert, O.L. (1989). *The Ecology of Urban Habitats*. Chapman and Hall, London.

Gill, R.A., Kelly, R.H., Parton, W.J., *et al.* (2002). Using simple environmental variables to estimate below-ground productivity in grasslands. *Global Ecology and Biogeography*, 11, 79–86.

Gilman, E.F., Flower, F.B., and Leon, I.A. (1985). Standardized procedures for planting vegetation on completed landfills. *Waste Management and Research*, 3, 65–80.

Gitay, H. and Noble, I.R. (1997). What are functional types and how should we seek them? In T.M. Smith, H.H. Shugart, and F.I. Woodward, eds., *Plant Functional Types*, pp. 3–19. Cambridge University Press, Cambridge.

Gitay, H. and Wilson, J.B. (1995). Post-fire changes in community structure of tall tussock grasslands: a test of alternative models of succession. *Journal of Ecology*, 83, 775–782.

Gleason, H.A. (1917). The structure and development of the plant association. *Bulletin of the Torrey Botanical Club*, 44, 463–481.

Gleason, H.A. (1926). The individualistic concept of the plant association. *Bulletin of the Torrey Botanical Club*, 53, 7–26.

Gleason, H.A. (1939). The individualistic concept of the plant association. *American Midland Naturalist*, 21, 92–110.

Glenn-Lewin, D.C. (1980). The individualistic nature of plant community development. *Vegetatio*, **43**, 141–146.

Glenn-Lewin, D.C. and van der Maarel, E. (1992). Patterns and processes in vegetation dynamics. In D.C. Glenn-Lewin, R.K. Peet, and T.T. Veblen, eds *Plant Succession: Theory and Prediction*, pp. 11–59. Chapman and Hall, London.

Glenn-Lewin, D.C., Peet, R.K., and Veblen, T.T. eds (1992). *Plant Succession: Theory and Prediction*. Chapman and Hall, London.

Glynn, P.W. (1997). Bioerosion and coral reef growth: a dynamic balance. In C. Birkeland, ed. *Life and Death of Coral Reefs*, pp. 68–95. Chapman and Hall, London.

Goldenberg, S.B., Landsea, C.W., Mestas-Nuñez, A.M., and Gray, W.M. (2001). The recent increase in Atlantic hurricane activity: causes and implications. *Science*, **293**, 474–479.

Gomiero, T., Paoletti, M.G., and Pimentel, D. (2010). Biofuels: efficiency, ethics, and limits to human appropriation of ecosystem services. *Journal of Agricultural and Environmental Ethics*, **23**, 403–434.

González, G. and Seastedt, T.R. (2001). Soil fauna and plant litter decomposition in tropical and subalpine forests. *Ecology*, **82**, 955–964.

Gonzalez, P., Neilson, R.P., Lenihan, J.M., and Drapek, R.J. (2010). Global patterns in the vulnerability of ecosystems to vegetation shifts due to climate change. *Global Ecology and Biogeography*, **19**, 755–768.

Goodenough, A.E. (2010). Are the ecological impacts of alien species misrepresented? A review of the "native good, alien bad" philosophy. *Community Ecology*, **11**, 13–21.

Gordon, I.J., Hester, A.J., and Festa-Bianchet, M. (2004). The management of wild large herbivores to meet economic conservation and environmental objectives. *Journal of Applied Ecology*, **41**, 1021–1031.

Goudswaard, K.P.C., Witte, F., and Katunzi, E.F.B. (2008). The invasion of an introduced predator, Nile perch (*Lates niloticus*, L.) in Lake Victoria (East Africa): chronology and causes. *Environmental Biology of Fishes*, **81**, 127–139.

Graham, P.A. (2008). Reparations, self-determination, and the seventh generation. *Harvard Human Rights Journal*, **21**, 47–103.

Grant, A. and Briggs, A.D. (2002). Toxicity of sediments from around a North Sea oil platform: are metals or hydrocarbons responsible for ecological impacts? *Marine Environmental Research*, **53**, 95–116.

Grassle, J.F. and Morse-Porteous, L.S. (1987). Macrofaunal colonization of disturbed deep-sea environments and the structure of deep-sea benthic communities. *Deep-Sea Research*, **34**, 1911–1950.

Gray, J.S. (1997). Marine biodiversity: patterns, threats and conservation needs. *Biodiversity and Conservation*, **6**, 153–175.

Greene, D.F., Zasada, J.C., Sirois, L., *et al*. (1999). A review of the regeneration dynamics of North American boreal forest tree species. *Canadian Journal of Forest Research*, **29**, 824–839.

Greenslade, P. (1999). Long distance migration of insects to a subantarctic island. *Journal of Biogeography*, **26**, 1161–1167.

Grime, J.P. (1973). Competitive exclusion in herbaceous vegetation. *Nature*, **242**, 344–347.

Grime, J.P. (1977). Evidence for the existence of three primary strategies in plants and its relevance to ecological and evolutionary theory. *The American Naturalist*, **111**, 1169–1194.

Grime, J.P. (1979). *Plant Strategies and Vegetation Processes*. Wiley, New York.

Grimm, N.B., Morgan Grove, J., Pickett, S.T.A., and Redman, C.L. (2000). Integrated approaches to long-term studies of urban ecological systems. *BioScience*, **50**, 571–584.

Grimm, N.B., Faeth, S.H., Golubiewski, N.E., *et al*. (2008). Global change and the ecology of cities. *Science*, **319**, 756–760.

Groffman, P.M. and Bohlen, P.J. (1999). Soil and sediment biodiversity. *BioScience*, **49**, 139–148.

de Groot, R.S., Wilson, M., and Boumans, R.M.J. (2002). A typology for the classification, description and valuation of ecosystem functions, goods and services. *Ecological Economics*, **41**, 393–408.

Grootjans, A.P., Ernst, W.H.O., and Stuyfzand, P.J. (1998). European dune slacks: strong interactions of biology, pedogenesis and hydrology. *Trends in Ecology and Evolution*, **13**, 96–100.

Grubb, P.J. (1977). The maintenance of species-richness in plant communities: the importance of the regeneration niche. *Biological Reviews*, **52**, 107–152.

Grubb, P.J. (1986). The ecology of establishment. In A.D. Bradshaw, D.A. Goode, and E. Thorp, eds *Ecology and Design in Landscape*, pp. 83–98. Blackwell, Oxford.

Grubb, P.J. (1987). Some generalizing ideas about colonization and succession in green plants and fungi. In A.J. Gray, M.J. Crawley, and P.J. Edwards, eds. *Colonization, Succession and Stability*, pp. 81–102. Blackwell, Oxford.

Guariguata, M.R. (1990). Landslide disturbance and forest regeneration in the Upper Luquillo Mountains. *Journal of Ecology*, **78**, 814–832.

Guariguata, M.R. and Larsen, M.C. (1990). *Preliminary map showing locations of landslides in El Yunque Quadrangle, Puerto Rico*. US Geological Survey Open-File Report 89–257, scale 1:20,000.

Guliev, I.S. and Feizullayev, A.A. (1996). Geochemistry of hydrocarbon seepages. In D. Schumacher and M.A. Abrams, eds *Hydrocarbon Migration and its Near-Surface Expression*. American Association of Petroleum Geologists Memoir 66, pp. 63–70.

Gunnarsson, K. (2000). Benthic marine algal colonization on the new lava at Heimaey, Vestmannaeyjar archipelago, southern Iceland. *Surtsey Research*, **11**, 69–74.

Gunnarsson, K. and Hauksson, E. (2009). Succession and benthic community development in the sublittoral zone at the recent volcanic island, Surtsey, southern Iceland. *Surtsey Research*, **12**, 161–166.

Gurevitch, J. and Padilla, D.K. (2004). Are invasive species a major cause of extinctions? *Trends in Ecology and Evolution*, **19**, 470–474.

Gutsell, S.L. and Johnson, E.A. (2007). Wildfire and tree population processes. In E.A. Johnson and K. Miyanishi, eds *Plant Disturbance Ecology: the Process and the Response*, pp. 441–485. Academic Press, Amsterdam.

Haeupler, H. (2008). Long-term observations of secondary forests growing on hard-coal mining spoils in the industrial Ruhr region of Germany. In M.M. Carreiro, Song, Y.-C. and Wu, J., eds *Ecology, Planning, and Management of Urban Forests*, pp. 357–368. Springer, New York.

Hagen, J.B. (1992). *An Entangled Bank: the Origins of Ecosystem Ecology*. Rutgers University Press, New Brunswick, NJ, U.S.

Halford, A., Cheal, A.J., Ryan, D., and Williams, D.McB. (2004). Resilience to large-scale disturbance in coral and fish assemblages on the Great Barrier Reef. *Ecology*, **85**, 1892–1905.

Hallock, P. (1988). The role of nutrient availability in bioerosion: consequences to carbonate build-ups. *Palaeogeography, Palaeoclimatology, Palaeoecology*, **63**, 275–291.

Halpern, B.S. and Warner, R.R. (2003). Review paper. Matching marine reserve design to reserve objectives. *Proceedings of the Royal Society B: Biological Sciences*, **270**, 1871–1878.

Hamilton, G. (2011). Aliens to the rescue. *New Scientist*, **209**, 3437.

Hansen, J., Sato, M., Kharecha, P., *et al.* (2008). Target atmospheric CO_2: where should humanity aim? *Open Atmospheric Science Journal*, **2**, 217–231.

Hanski, I. (1995). Effects of landscape pattern on competitive interactions. In L. Hansson, L. Fahrig, and G. Merriam, eds *Mosaic Landscapes and Ecological Processes*, pp. 203–224. Chapman and Hall, New York.

Hanski, I. and Cambefort, Y. (1991). *Dung Beetle Ecology*, Princeton University Press, Princeton, NJ.

Harcombe, P. and Carter, R.M. (2004). Cyclone pumping, sediment partitioning and the development of the Great Barrier Reef shelf system: a review. *Quaternary Science Reviews*, **23**, 107–135.

Harden, B. (1997). *A River Lost: the Life and Death of the Columbia*. Norton, New York.

Harding, J.S., Benfield, E.F., Bolstad, P.V., Helfman, G.S., and Jones, E.B.D., III (1998). Stream biodiversity: the ghost of land use past. *Proceedings of the National Academy of Sciences USA*, **95**, 14843–14847.

Harper, J.L. (1977). *Population Biology of Plants*, Academic Press, New York.

Harris, J.A. and van Diggelen, R. (2006). Ecological restoration as a project for global society. In J. van Andel and J. Aronson, eds *Restoration Ecology*, pp. 3–15. Blackwell, Oxford.

Harris, J.A., Hobbs, R.J., Higgs, E., and Aronson, J. (2006). Ecological restoration and global climate change. *Restoration Ecology*, **14**, 170–176.

Hartshorn, G.S. and Whitmore, J.L. (1999). Anthropogenic disturbance and tropical forestry: implications for sustainable management. In L.R. Walker, ed. *Ecosystems of Disturbed Ground*, Ecosystems of the World 16, pp. 467–486. Elsevier, Amsterdam.

Hauksson, E. (2000). A survey of the benthic coastal fauna of Surtsey, Iceland, in 1997. *Surtsey Research*, **11**, 85–88.

Haven, S.B. (1971). Effects of land-level changes on intertidal invertebrates, with discussion of earthquake ecological succession. In *The Great Alaskan Earthquake of 1964: Biology*, Publication No 1604, pp. 82–126. Committee on the Great Alaskan Earthquake of 1964, Natural Resource Council, National Academy of Sciences, Washington, DC.

Hawken, P. (2009). *Commencement Address to the Class of 2009*. Speech at University of Portland, Portland, Oregon (U.S.). Cited in Brown (2009).

Hayashi, M. and van der Kamp, G. (2007). Water level changes in ponds and lakes: the hydrological process. In E.A. Johnson and K. Miyanishi, eds *Plant Disturbance Ecology: the Process and the Response*, pp. 311–339. Academic Press, Amsterdam.

Hayden, B.P., Santos, M.C.F.V., Shao, G., and Kochel, R.C. (1995). Geomorphological controls on coastal vegetation at the Virginia Coast Reserve. *Geomorphology*, **13**, 283–300.

Heavilin, J., Powell, J., and Logan, J.A. (2007). Dynamics of mountain pine beetle outbreaks. In E.A. Johnson and K. Miyanishi, eds *Plant Disturbance Ecology: the Process and the Response*, pp. 527–553. Academic Press, Amsterdam.

van der Heijden, E.W. and Vosatka, M. (1999). Mycorrhizal associations of *Salix repens* L. communities in succession of dune ecosystems. II. Mycorrhizal dynamics and interactions of ectomycorrhizal and arbuscular mycorrhizal fungi. *Canadian Journal of Botany*, **77**, 7–19.

van der Heijden, M.G.A., Bardgett, R.D., and van Straalen, N.M. (2008). The unseen majority: soil microbes as drivers of plant diversity and productivity in terrestrial ecosystems. *Ecology Letters*, **11**, 296–310.

Heinsohn, G.E. and Spain, A.V. (1974). Effects of a tropical cyclone on littoral and sub-littoral biotic communities and on a population of Dugongs (*Dugong dugon* (Müller)). *Biological Conservation*, **6**, 143–152.

Hesp, P. (2002). Foredunes and blowouts: initiation, geomorphology and dynamics. *Geomorphology*, **48**, 245–268.

Hesp, P.A. and Martínez, M.L. (2007). Disturbance processes and dynamics in coastal dunes. In E.A. Johnson and K. Miyanishi, eds *Plant Disturbance Ecology: the Process and the Response*, pp. 249–282. Academic Press, Amsterdam.

Hewitt, C.D., Broccoli, A.J., Crucifix, M., Gregory, J.M., Mitchell, F.B., and Stouffer, R.J. (2006). The effect of a large freshwater perturbation on the glacial North Atlantic Ocean using a coupled general circulation model. *Journal of Climate*, **19**, 4436–4447.

Heyne, C.M. (2000). *Soil and Vegetation Recovery on Abandoned Paved Roads in a Humid Tropical Rain Forest, Puerto Rico*. M.S. Thesis, University of Nevada Las Vegas.

Hietz, P. (2010). Fern adaptations to xeric environments. In K. Mehltreter, L.R. Walker, and J.M. Sharpe, eds *Fern Ecology*, pp. 140–176. Cambridge University Press, Cambridge.

Higuchi, T. (1983). *The Visual and Spatial Structure of Landscapes*. MIT Press, Cambridge, MA, U.S.

Hill, J. and Woodland, W. (2003). Contrasting water management techniques in Tunisia: towards sustainable agricultural use. *The Geographical Journal*, **169**, 342–357.

Hill, M.O. (1979). *DECORANA – a FORTRAN Program for Detrended Correspondence Analysis and Reciprocal Averaging*. Ecology and Systematics, Cornell University, Ithaca, NY, U.S.

Hill, P.W., Farrar, J., Roberts, P., *et al.* (2011). Vascular plant success in a warming Antarctic may be due to efficient nitrogen acquisition. *Nature Climate Change*, **1**, 50–53.

Hillel, D. (1991). *Out of the Earth: Civilization and the Life of the Soil*. University of California Press, Berkeley, CA.

HilleRisLambers, J., Harpole, W.S., Tilman, D., Knops, J., and Reich, P.B. (2004). Mechanisms responsible for the positive diversity-productivity relationship in Minnesota grasslands. *Ecology Letters*, **7**, 661–668.

Hirata, T. (1992). Succession of sessile organisms on experimental plates immersed in Habeta Bay, Izu Peninsula, Japan. V. An integrated consideration on the definition and prediction of succession. *Ecological Research*, **7**, 31–42.

Hobbie, E.A., Macko, S.A., and Shugart, H.H. (1999). Insights into nitrogen and carbon dynamics of ectomycorrhizal and saprotrophic fungi from isotopic evidence. *Oecologia*, **118**, 353–360.

Hobbs, R.J. and Cramer, V.A. (2007). Old field dynamics: regional and local differences, and lessons for ecology and restoration. In V.A. Cramer and R.J. Hobbs, eds *Old Fields: Dynamics and Restoration of Abandoned Farmland*, pp. 309–318. Island Press, Covalo, WA, U.S.

Hobbs, R.J. and Huenneke, L.F. (1992). Disturbance, diversity, and invasion: implications for conservation. *Conservation Biology*, **6**, 324–337.

Hobbs, R.J. and Norton, D.A. (1996). Towards a conceptual framework for restoration ecology. *Restoration Ecology*, **4**, 93–110.

Hobbs, R.J. and Suding, K.N. (2009a). Synthesis: are new models for ecosystem dynamics scientifically robust and helpful in guiding restoration projects? In Hobbs, R.J. and K.N. Suding, eds *New Models for Ecosystem Dynamics and Restoration*, pp. 325–333. Island Press, Washington, DC.

Hobbs, R.J. and Suding, K.N., eds (2009b). *New Models for Ecosystem Dynamics and Restoration*, Island Press, Washington, DC.

Hobbs, R.J., Arico, S., Aronson, J., *et al.* (2006). Novel ecosystems: theoretical and management aspects of the new ecological world order. *Global Ecology and Biogeography*, **15**, 1–7.

Hobbs, R.J., Walker, L.R., and Walker, J. (2007). Integrating restoration and succession. In L.R. Walker, J. Walker, and R.J. Hobbs, eds *Linking Restoration and Ecological Succession*, pp. 168–179, Springer, New York.

Hobbs, R.J., Cole, D.N., Yung, L., *et al.* (2010). Guiding concepts for park and wilderness stewardship in an era of global environmental change. *Frontiers in Ecology and the Environment*, **8**, 483–490.

Hocking, M.D. and Reimchen, T.E. (2009). Salmon species, density and watershed size predict magnitude of marine enrichment in riparian food webs. *Oikos*, **118**, 1307–1318.

Hocking, M.D. and Reynolds, J.D. (2011). Impacts of salmon on riparian plant diversity. *Science*, **331**, 1609–1612.

Hodkinson, I.D., Webb, N.R., and Coulson, S.J. (2002). Primary community assembly on land – the missing stages: why are the heterotrophic organisms always there first? *Journal of Ecology*, **90**, 569–577.

Hoegh-Guldberg, O., Mumby, P.J., Hooten, A.J., *et al.* (2007). Coral reefs under rapid climate change and ocean acidification. *Science*, **318**, 1737–1742.

Holdaway, R.J. and Sparrow, A.D. (2006). Assembly rules operating along a primary riverbed–grassland successional sequence. *Journal of Ecology*, **94**, 1092–1102.

Hooper, D.U., Chapin, F.S., III, Ewel, J.J., *et al.* (2005). Effects of biodiversity on ecosystem functioning: a consensus of current knowledge. *Ecological Monographs*, **75**, 3–35.

Hopke, P.K., Cohen, D.D., Begum, B.A., *et al.* (2008). Urban air quality in the Asian region. *Science of the Total Environment*, **404**, 103–112.

Hornafius, J.S., Quigley, D., and Luyendyk, B.P. (1999). The world's most spectacular hydrocarbon seeps: Coal Oil Point, Santa Barbara Channel, (California): quantification of emission. *Journal of Geophysical Research*, **104**, 20703–20711.

Horner-Devine, M.C., Carney, K.M., and Bohannan, B.J.M. (2004). An ecological perspective on bacterial biodiversity. *Proceedings of the Royal Society B: Biological Sciences*, **271**, 113–122.

Howarth, R.W., Marino, R., and Lane, J. (1988). Nitrogen fixation in freshwater, estuarine, and marine ecosystems. 1. Rates and importance. *Limnology and Oceanography*, **22**, 669–687.

Howe, H.F. and Miriti, M.N. (2004). When seed dispersal matters. *BioScience*, **54**, 651–660.

Hubbard, D.K., Parsons, K.M., Bythell, J.C., and Walker, N.D. (1991). The effects of Hurricane Hugo on the reefs and associated environments of St. Croix, U.S. Virgin Islands – a preliminary assessment. *Journal of Coastal Research Special Issue*, **8**, 33–48.

Huesemann, M.H., Skillman, A.D., and Crecelius, E.A. (2003). The inhibition of marine nitrification by ocean disposal of carbon dioxide. *Marine Pollution Bulletin*, **44**, 142–148.

Hughes, A. (2010). Disturbance and diversity: an ecological chicken and egg problem. *Nature Education Knowledge*, **1**, 26.

Hughes, A.R., Byrnes, J.E., Kimbro, D.L., and Stachowicz, J.J. (2007). Reciprocal relationships and potential feedbacks between biodiversity and disturbance. *Ecology Letters*, **10**, 849–864.

Hughes, F.M.R. (1997). Floodplain biogeomorphology. *Progress in Physical Geography*, **21**, 501–529.

Hughes, T.P. (1994). Catastrophes, phase shifts, and large-scale degradation of a Caribbean coral reef. *Science*, **265**, 1547–1551.

Hughes, T.P. and Connell, J.H. (1999). Multiple stressors on coral reefs: a long-term perspective. *Limnology and Oceanography*, **44**, 932–940.

Hughes, T.P., Bellwood, D.R., Folke, C., Steneck, R.S., and Wilson, J. (2005). New paradigms for supporting the resilience of marine ecosystems. *Trends in Ecology and Evolution*, **20**, 380–386.

van Hulst, R. (1992). From population dynamics to community dynamics: modeling succession as a species replacement process. In D.C. Glenn-Lewin, R.K. Peet, and T.T. Veblen, *Plant Succession: Theory and Prediction*, pp. 188–214. Chapman and Hall, London.

Hultine, K.R., Nagler, P.L., Morino, K., *et al.* (2010). Sap flux-scaled transpiration by tamarisk (*Tamarix* spp.) before, during and after episodic defoliation by the saltcedar leaf beetle (*Diorhabda carinulata*). *Agricultural and Forest Meteorology*, **150**, 1467–1475.

van Hulzen, J.B., van Soelen, J., Herman, P.M.J., and Bouma, T.J. (2006) The significance of spatial and temporal patterns of algal mat deposition in structuring salt marsh vegetation. *Journal of Vegetation Science*, **17**, 291–298.

Humphries, P. and Baldwin, D.S. (2003). Drought and aquatic ecosystems: an introduction. *Freshwater Biology*, **48**, 1141–1146.

Hunter, J.M. and Arbona, S.I. (1995). Paradise lost: an introduction to the geography of water pollution in Puerto Rico. *Social Science and Medicine*, **40**, 1331–1355.

Hurlbert, A.H., Anderson, T.W., Sturm, K.K., and Hurlbert, S.H. (2007). Fish and fish-eating birds at the Salton Sea: a century of boom and bust. *Lake and Reservoir Management*, **23**, 469–499.

Huston, M. (1979). A general hypothesis of species diversity. *The American Naturalist*, **113**, 81–101.

Huston, M.A. (1994). *Biological Diversity: the Coexistence of Species on Changing Landscapes*. Cambridge University Press, Cambridge.

Huston, M.A. (1997). Hidden treatments in ecological experiments: re-evaluating the ecosystem function of biodiversity. *Oecologia*, **110**, 449–460.

Huston, M.A. and Smith, T. (1987). Plant succession: life history and competition. *The American Naturalist*, **130**, 168–198.

Hutchinson, G.E. (1959). Homage to Santa Rosalia: or, why are there so many kinds of animals? *The American Naturalist*, **93**, 145–159.

Huttenen, J.T., Alm, J., Liikanen, A., *et al.* (2003). Fluxes of methane, carbon dioxide and nitrous oxide in boreal lakes and potential anthropogenic effects on the aquatic greenhouse gas emissions. *Chemosphere*, **52**, 609–621.

Hyatt, L.A., Rosenberg, M.S., Howard, T.G., *et al.* (2003). The distance-dependence prediction of the Janzen–Connell hypothesis: a meta-analysis. *Oikos*, **103**, 590–602.

Ingegnoli, V. (2004). An innovative contribution of landscape ecology to vegetation science. *Israel Journal of Plant Sciences*, **53**, 155–166.

Ismail, K., Kamal, K., Plath, M., and Wronski, T. (2011). Effects of an exceptional drought on daily activity patterns, reproductive behaviour, and reproductive success of reintroduced Arabian oryx (*Oryx leucoryx*). *Journal of Arid Environments*, **75**, 125–131.

IUCN (2004). *2004 IUCN Red List of Threatened Species*. URL: http://www.iucnredlist.org (accessed 18 March 2011).

Ives, A.R. and Carpenter, S.R. (2007). Stability and diversity of ecosystems. *Science*, **317**, 58–62.

Jacoby, G.C., Sheppard, P.R., and Sieh, K.E. (1988). Irregular recurrence of large earthquakes along the San Andreas fault: evidence from trees. *Science*, **241**, 196–200.

Jacquez, G.M. and Patten, D.T. (1996). *Chesneya nubigena* on a Himalayan glacial moraine, a case of facilitation in primary succession? *Mountain Research and Development*, **16**, 265–273.

Janzen, D.H. (1970). Herbivores and the number of tree species in tropical forests. *The American Naturalist*, **104**, 501–528.

Jarrett, R.D. and Malde, H.E. (1987). Paleodischarge of the late Pleistocene Bonneville flood, Snake River, Idaho, computed from new evidence. *Geological Society of America Bulletin*, **99**, 127–134.

Jennings, S. and Polunin, N.V.C. (1996). Impacts of fishing on tropical reef ecosystems. *Ambio*, **25**, 44–49.

Jewett, S.C., Bodkin, J.L., Chenelot, H., Esslinger, G.G., and Hoberg, M.K. (2010). The nearshore benthic community of Kasatochi Island, one year after the 2008 volcanic eruption. *Arctic, Antarctic, and Alpine Research*, **42**, 315–324.

Johnson, E.A. and Miyanishi, K., eds (2007). *Plant Disturbance Ecology: the Process and the Response*. Academic Press, New York.

Johnson, E.A. & Miyanishi, K. (2008). Testing the assumptions of chronosequences in succession. *Ecology Letters*, **11**, 419–431.

Johnson, N.C., Rowland, D.L., Corkidi, L., Egerton-Warburton, L.M., and Allen, E.B. (2003). Nitrogen enrichment alters mycorrhizal allocation at five mesic to semi-natural grasslands. *Ecology*, **84**, 1895–1908.

Jones, D.L. and Kielland, K. (2002). Soil amino acid turnover dominated the nitrogen flux in permafrost-dominated taiga forest soils. *Soil Biology and Biochemistry*, **34**, 209–219.

Jones, E.G., Collins, M.A., Bagley, P.M., Addison, S., and Priede, I.G. (1998). The fate of cetacean carcasses in the deep sea: observations on consumption rates and succession of scavenging species in the abyssal north-east Atlantic Ocean. *Proceedings of the Royal Society B: Biological Sciences*, **265**, 1119–1127.

Jones, J.A., Swanson, F.J., Wemple, B.C., and Snyder, K.U. (2000). Effects of roads on hydrology, geomorphology, and disturbance patches in stream networks. *Conservation Biology*, **14**, 76–85.

Jónsson, S. and Gunnarsson, K. (2000). Seaweed colonization at Surtsey, the volcanic island south of Iceland. *Surtsey Research*, **11**, 59–68.

Jørgensen, S.E. (1997). *Integration of Ecosystem Theories: a Pattern*, 2nd edn. Kluwer, Dordrecht, The Netherlands.

Jumpponen, A., Mattson, K.G., and Trappe, J.M. (1998). Mycorrhizal functioning of *Phialocephala fortinii* with *Pinus contorta* on glacier forefront soil: interactions with soil nitrogen and organic matter. *Mycorrhiza*, **7**, 261–265.

Karl, D.M., McMurtry, G.M., Malahoff, A., and Garcia, M.O. (1988). Loihi Seamount, Hawaii: a mid-plate volcano with a distinctive hydrothermal system. *Nature*, **335**, 532–535.

Kauffman, J.B., Beschta, R.L., and Lytjen, D. (1997). An ecological perspective of riparian and stream restoration in the westerns United States. *Fisheries*, **22**, 12–24.

Kaufmann, R. (2001). Invertebrate succession on an alpine glacier foreland. *Ecology*, **82**, 2261–2278.

Keane, R.M. and Crawley, M.J. (2002). Exotic plant invasions and the enemy release hypothesis. *Trends in Ecology and Evolution*, **17**, 164–170.

Keddy, P.A. (1989). *Competition*. Chapman and Hall, London.

Keever, C. (1950). Causes of succession on old fields of the piedmont, North Carolina. *Ecological Monographs*, **20**, 230–250.

Keever, C. (1979). Mechanisms of succession on old fields of Lancaster County, Pennsylvania. *Bulletin of the Torrey Botanical Club*, **106**, 299–308.

Keller, M., Schimel, D.S., Hargrove, W.W., and Hoffman, F.M. (2008). A continental strategy for the National Ecological Observatory Network. *Frontiers in Ecology and the Environment*, **6**, 282–284.

Kennish, M.J. (1991). *Ecology of Estuaries: Anthropogenic Effects*. CRC Press, Boca Raton, FL, U.S.

Kennish, M.J. (1997). *Practical Handbook of Estuarine and Marine Pollution*. CRC Press, Boca Raton, FL, U.S.

Kijne, J.W. (2003). *Unlocking the Water Potential of Agriculture*. FAO, Rome.

Kilham, K. (1994). *Soil Ecology*. Cambridge University Press, Cambridge.

Kinlan, B.P. and Gaines, S.D. (2003). Propagule dispersal in marine and terrestrial environments: a community perspective. *Ecology*, **84**, 2007–2020.

Kitayama, K., Schuur, E.A.G., Drake, D.R., and Mueller-Dombois, D. (1997). Fate of a wet montane forest during soil aging in Hawaii. *Journal of Ecology*, **85**, 669–679.

Kitzberger, T., Veblen, T.T., and Villalba, R. (1995). Tectonic influences on tree growth in northern Patagonia, Argentina: the roles of substrate stability and climatic variation. *Canadian Journal of Forest Research*, **25**, 1684–1696.

Klironomos, J.N. (2002). Feedback with soil biota contributes to plant rarity and invasiveness in communities. *Nature*, **417**, 67–70.

Kohler, F., Vandenberghe, C., Imstepf, R., and Gillet, F. (2011). Restoration of threatened arable weed communities in abandoned mountainous crop fields. *Restoration Ecology*, **19**, 62–69.

Kotanen, P.M. (1995). Responses of vegetation to a changing regime of disturbance: effects of feral pigs in a Californian coastal prairie. *Ecography*, **18**, 190–199.

Knott, D.M., and Martore, R.M. (1991). The short-term effects of Hurricane Hugo on fishes and decapod crustaceans in the Ashley River and adjacent marsh creeks, South Carolina. *Journal of Coastal Research Special Issue*, **8**, 335–356.

Korte, F.S. and Coulston, F. (2000). The cyanide leaching gold recovery process is a nonsustainable technology with unacceptable impacts on ecosystems and humans: the disaster in Romania. *Ecotoxicology and Environmental Safety*, **46**, 241–245.

Krasny, M.E., Vogt, K.A., and Zasada, J.C. (1984). Root and shoot biomass and mycorrhizal development of white spruce seedlings naturally regenerating in interior Alaskan floodplain communities. *Canadian Journal of Forest Research*, **14**, 554–558.

Krebs, C.J., Boonstra, R., Boutin, S., and Sinclair, A.R.E. (2001). What drives the 10-year cycle of snowshoe hares? *BioScience*, **51**, 25–35.

Kregting, L.T. and Gibbs, M.T. (2006). Salinity controls the upper depth distribution limit of black corals in Doubtful Sound, New Zealand. *New Zealand Journal of Marine and Freshwater Research*, **40**, 43–52.

Kuffner, I.B., Andersson, A.J., Jokiel, P.L., Rodgers, K.S., and Mackenzie, F.T. (2008). Decreased abundance of crustose coralline algae due to ocean acidification. *Nature Geoscience*, **1**, 114–117.

Kurz, W.A., Dymond, C.C., Stinson, G., *et al.* (2008). Mountain pine beetle and forest carbon feedback to climate change. *Nature*, **452**, 987–990.

Kushmaro, A., Rosenberg, E., Fine, M., and Loya, Y. (1997). Bleaching of the coral *Oculina patagonica* by *Vibrio* AK-1. *Marine Ecology Progress Series*, **147**, 159–165.

Kutsch, W., Bahn, M., and Heinemeyer, A., eds (2009). *Soil Carbon Dynamics: an Integrated Methodology*. Cambridge University Press, Cambridge.

Lagerström, A., Nilsson, M.- C., Zackrisson, O., and Wardle, D.A. (2007). Ecosystem input of nitrogen through biological fixation in feather mosses during ecosystem retrogression. *Functional Ecology*, **21**, 1027–1033.

Lake, P.S. (2003). Ecological effects of perturbation by drought in flowing waters. *Freshwater Biology*, **48**, 1161–1172.

Lanta, V. and Lepš, J. (2009). How does surrounding vegetation affect the course of succession: a five-year container experiment. *Journal of Vegetation Science*, **20**, 686–694.

Larsen, M.C. and Parks, J.E. (1997). How wide is a road? The association of roads and mass-wasting in a forested montane environment. *Earth Surface Processes and Landforms*, **22**, 835–848.

Lathrop, E.W. (1983). Recovery of perennial vegetation in military maneuver areas. In R.H. Webb and H.G. Wilshire, eds *Environmental Effects of Off-Road Vehicles: Impacts and Management in Arid Regions*, pp. 266–277. Springer, New York.

Laurance, S.G. and Laurance, W.F. (1999). Tropical wildlife corridors: use of linear rainforest remnants by arboreal mammals. *Biological Conservation*, **91**, 231–239.

Laurance, W.F. (2007). Forest destruction in tropical Asia. *Current Science*, **93**, 1544–1550.

Laurance, W.F. and Luizão, R.C. (2007). Driving a wedge in to the Amazon. *Nature*, **448**, 409–410.

Laurance, W.F. and Williamson, G.B. (2001). Positive feedbacks among forest fragmentation, drought, and climate change in the Amazon. *Conservation Biology*, **15**, 1529–1535.

Lavoie, L. and Sirois, L. (1998). Vegetation changes caused by recent fires in the northern boreal forest of eastern Canada. *Journal of Vegetation Science*, **9**, 483–492.

Lebrija-Trejos, E., Pérez-García, E.A., Meave, J.A., Bongers, F., and Poorter, L. (2010). Functional traits and environmental filtering drive community assembly in a species-rich tropical system. *Ecology*, **91**, 386–398.

Lee, C.S. and Cho, Y. C. (2008). Selection of pollution-tolerant trees for restoration of degraded forests and evaluation of the experimental restoration practices at the Ulsan Industrial Complex, Korea. In M.M. Carreiro, Song, Y.-C., and Wu, J., eds *Ecology, Planning, and Management of Urban Forests*, pp. 369–392. Springer, New York.

Leewis, R. (1991). Environmental impact of shipwrecks in the North Sea. II. Negative aspects: hazardous substances in shipwrecks. *Water Science and Technology*, **4**, 299–300.

Leighton, M. and Wirawan, N. (1986). Catastrophic drought and fire in Borneo tropical rain forest associated with the 1982–1983 El Niño Southern Oscillation event. In G.T. Prance, ed. *Tropical Rain Forests and the World Atmosphere*, pp. 75–102. Westview Press, Boulder, CO, U.S.

Lenz, L., Taylor, J., and Foote, D. (2006). Distribution and abundance of the two spotted leafhopper (*Sophonia rufofascia*) in relation to host plants in Hawaii Volcanoes National Park. *USGS Open File Report 2006–1050*. USGS Pacific Island Ecosystem Research Center, Hilo, HI, U.S.

León, C. and González, M. (1995). Managing the environment in tourism regions: the case of the Canary Islands. *European Environment*, **5**, 171–177.

Lepš, J. and Rejmánek, M. (1991). Convergence or divergence: what should we expect from vegetation succession? *Oikos*, **62**, 261–264.

Lesica, P. and Allendorf, F.W. (2002). Ecological genetics and the restoration of plant communities: mix or match? *Restoration Ecology*, **7**, 42–50.

Lessard, J.L. and Merritt, R.W. (2006). Influence of marine-derived nutrients from spawning salmon on aquatic insect communities in southeast Alaskan streams. *Oikos*, **113**, 334–343.

Lessard, R.R. and DeMarco, G. (2000). The significance of oil spill dispersants. *Spill Science and Technology Bulletin*, **6**, 59–68.

Letnic, M. and Fox, B.J. (1997). The impact of industrial fluoride fallout on faunal succession following sand-mining of dry sclerophyll forest at Tomago, NSW. I. Lizard recolonization. *Biological Conservation*, **80**, 63–81.

Levine, J.M. (2000). Species diversity and biological invasions: relating local process to community pattern. *Science*, **288**, 852–854.

Levine, J.M., Adler, P.B., and Yelenik, S.G. (2004). A meta-analysis of biotic resistance to exotic plant invasions. *Ecology Letters*, **7**, 975–989.

Levins, R. (1969). Some demographic and genetic consequences of environmental heterogeneity for biological control. *Bulletin of the Entomological Society of America*, **15**, 237–240.

Lewis, M.A. (1997). Variability, patchiness, and jump dispersal in the spread of an invading population. In D. Tilman and P. Kareiva, eds *Spatial Ecology: the Role of Space in Population Dynamics and Interspecific Interactions*, pp. 46–69. Princeton University Press, Princeton, NJ, U.S.

Lewis, O.T. (2010). Close relatives are bad news. *Nature*, **466**, 698–699.

Lichter, J. (1998). Primary succession and forest development on coastal Lake Michigan sand dunes. *Ecological Monographs*, **68**, 487–510.

Ligon, F.K., Dietrich, W.E., and Trush, W.J. (1995). Downstream ecological effects of dams. *BioScience*, **45**, 183–192.

Likens, G.E. and Bormann, F.H. (1974). Linkages between terrestrial and aquatic ecosystems. *BioScience*, **24**, 447–456.

Lindenmayer, D.B. (1999). Future directions for biodiversity conservation in managed forests: indicator species, impact studies and monitoring programs. *Forest Ecology and Management*, **115**, 277–287.

Lindenmayer, D.B. and Hobbs, R.J., eds (2007). *Managing and Designing Landscapes for Conservation: Moving from Perspectives to Principles*. Blackwell, Oxford.

Lindenmayer, D.B. and Nix, H.A. (1993). Ecological principles for the design of wildlife corridors. *Conservation Biology*, **7**, 627–631.

Linnaeus, C. (1735). *Systema Naturae*. Theodorum Haak, Leiden.

Littler, M.M., Martz, D.R., and Littler, D.S. (1983). Effects of recurrent sand deposition on rocky intertidal organisms: importance of substrate heterogeneity in a fluctuating environment. *Marine Ecology: Progress Series*, **11**, 129–139.

Lobao, L. and Meyer, K. (2001). The great agricultural transition: crisis, change and social consequences of twentieth century US farming. *Annual Review of Sociology*, **27**, 103–124.

Logan, J.A., Régnière, J., and Powell, J.A. (2003). Assessing the impacts of global climate change on forest pest dynamics. *Frontiers in Ecology and the Environment*, **1**, 130–137.

Loh, R.K. and Daehler, C.C. (2008). Influence of woody invader control methods and seed availability on native and invasive species establishment in a Hawaiian forest. *Biological Invasions*, **10**, 805–819.

Lohrer, A.M., Halliday, N.J., Thrush, S.F., Hewitt, J.E., and Rodil, I.F. (2010). Ecosystem functioning in a disturbance-recovery context: contribution of macrofauna to primary production and nutrient release on intertidal sandflats. *Journal of Experimental Marine Biology and Ecology*, **390**, 6–13.

Long, S.P., Humphries, S., and Falkowski, P.G. (1994). Photoinhibition of photosynthesis in nature. *Annual Review of Plant Physiology and Plant Molecular Biology*, **45**, 633–662.

Lord, J.M. and Norton, D.A. (1990). Scale and the spatial concept of fragmentation. *Conservation Biology*, **4**, 197–202.

Lottermoser, B.G. (2003). *Mine Wastes: Characterization, Treatment, and Environmental Impacts.* Springer, New York.

Louda, S.M. and Stiling, P. (2004). The double-edged sword of biological control in conservation and restoration. *Conservation Biology*, **18**, 50–53.

Lovegrove, B. (1993). *The Living Deserts of Southern Africa.* Fernwood Press, Cape Town, South Africa.

Lovelock, C.E., Jebb, M., and Osmond, C.B. (1994). Photoinhibition and recovery in tropical plant species: response to disturbance. *Oecologia*, **97**, 297–307.

Loya, Y. and Rinkevich, B. (1980). Effects of oil pollution on coral reef communities. *Marine Ecology Progress Series*, **3**, 167–180.

Lugo, A.E. (2010). Let's not forget the biodiversity of the cities. *Biotropica*, **42**, 576–577.

Lugo, A.E. and Gucinski, H. (2000). Function, effects, and management of forest roads. *Forest Ecology and Management*, **133**, 249–262.

Lyons, K.G. and Schwartz, M.W. (2001). Rare species loss alters ecosystem function—invasion resistance. *Ecology Letters*, **4**, 358–365.

van der Maarel, E. and Sykes, M.T. (1993). Small scale plant species turnover in a limestone grassland—the carousel model and some comments on the niche concept. *Journal of Vegetation Science*, **4**, 179–188.

MacArthur, R. (1955). Fluctuations of animal populations and a measure of community stability. *Ecology*, **36**, 533–536.

MacArthur, R. and Wilson, E.O. (1967). *The Theory of Island Biogeography.* Princeton University Press, Princeton, NJ.

McCann, K.S. (2000). The diversity–stability debate. *Nature*, **405**, 228–233.

McClain, M.E., Boyer, E.W., Dent, C.L., *et al.* (2003). Biogeochemical hot spots and hot moments at the interface of terrestrial and aquatic ecosystems. *Ecosystems*, **6**, 301–312.

McClanahan, T.R. (1997). Primary succession of coral-reef algae: different patterns on fished versus unfished reefs. *Journal of Experimental Marine Biology and Ecology*, **218**, 77–102.

McCormick, J. (1968). Succession. *Via*, **1**, 22–35, 131–132.

MacDonald, I.R., Guinasso, N.L., Jr, Ackleson, S.G. *et al.* (1993). Natural oil slicks in the Gulf of Mexico visible from space. *Journal of Geophysical Research*, **98**, 16351–16364.

MacDougall, A.S. and Turkington, R. (2005). Are invasive species the drivers or passengers of change in degraded ecosystems? *Ecology*, **86**, 42–55.

McEvoy, T.J. (2004). *Positive Impact Forestry—a Sustainable Approach to Managing Woodlands.* Island Press, Washington, DC.

McGinley, M.A., Dhillion, S.S., and Neumann, J.C. (1994). Environmental heterogeneity and seedling establishment: ant-plant-microbe interactions. *Functional Ecology*, **8**, 607–615.

McIntosh, R.P. (1985). *The Background of Ecology.* Cambridge University Press, Cambridge.

McIntosh, R.P. (1999). The succession of succession: a lexical chronology. *Bulletin of the Ecological Society of America*, **80**, 256–265.

Mack, M.C. and D'Antonio, C.M. (1998). Impacts of biological invasions on disturbance regimes. *Trends in Ecology and Evolution*, **13**, 195–198.

McKee, K.L. and Baldwin, A.H. (1999). Disturbance regimes in North American wetlands. In L.R. Walker, ed. *Ecosystems of Disturbed Ground*, Ecosystems of the World 16, pp. 331–363. Elsevier, Amsterdam.

Mackey, R.L. and Currie, D.J. (2001). The diversity–disturbance relationship: is it generally strong and peaked? *Ecology*, **82**, 3479–3492.

McKibben, B. (1989). *The End of Nature.* Random House, New York.

MacMahon, J.A. (1981). Empirical and theoretical ecology as a basis for restoration: an ecological success story. In M.L. Pace and P.M. Groffman, eds *Successes, Limitations, and Frontiers in Ecosystem Science*, pp. 220–246. Springer, New York.

MacMahon, J.A. (1999). Disturbance in deserts. In L.R. Walker, ed. *Ecosystems of Disturbed Ground*, Ecosystems of the World 16, pp. 307–330. Elsevier, Amsterdam.

McNaughton, S.J. (1985). Ecology of a grazing ecosystem: the Serengeti. *Ecological Monographs*, **55**, 259–294.

Magnússon, B., Magnússon, S.H., and Fridriksson, S. (2009). Developments in plant colonization and succession on Surtsey during 1999–2008. *Surtsey Research*, **12**, 57–76.

Magurran, A.E. (1988). *Ecological Diversity and its Measurement*. Princeton University Press, Princeton, NJ, U.S.

Magurran, A.E. and McGill, B.J. (2011). *Biological Diversity: Frontiers in Measurement and Assessment*. Oxford University Press, Oxford.

Mahmoud, S.S. and Croteau, R.B. (2002). Strategies for transgenic manipulation of monoterpene biosynthesis in plants. *Trends in Plant Science*, **7**, 366–373.

Mahoney, J.M. and Rood, S.B. (1998). Streamflow requirements for cotton wood seedling recruitment—an integrative model. *Wetlands*, **18**, 634–645.

Majer, J.D., ed. (1989). *Animals in Primary Succession: the Role of Fauna in Reclaimed Lands*. Cambridge University Press, Cambridge.

Majer, J.D. and de Kock, A.E. (1992). Ant recolonization of sand mines near Richards Bay, South Africa: an evaluation of progress with rehabilitation. *South African Journal of Science*, **88**, 31–36.

Malanson, G.P. and Cairns, D.M. (1997). Effects of dispersal, population delays, and forest fragmentation on tree migration rates. *Plant Ecology*, **131**, 67–79.

Maraun, M., Visser, S., and Scheu, S. (1998). Oribatid mites enhance the recovery of the microbial community after a strong disturbance. *Applied Soil Ecology*, **9**, 175–181.

Maraun, M., Salamon, J.-A., Schneider, K., Schaefer, M., and Scheu, S. (2003). Oribatid mite and collembolan diversity, density and community structure in a moder beech forest (*Fagus sylvatica*): effects of mechanical perturbations. *Soil Biology and Biochemistry*, **35**, 1387–1394.

Margalef, R. (1968). *Perspectives in Ecological Theory*. University of Chicago Press, Chicago.

Marleau, J.N., Jin, Y., Bishop, J.G., Fagan, W.F., and Lewis, M.A. (2011). A stoichiometric model of early plant primary succession. *The American Naturalist*, **177**, 233–245.

Marks, P.L. and Bormann, F.H. (1972). Revegetation following forest cutting: mechanisms for return to steady-state nutrient cycling. *Science*, **176**, 914–915.

Maron, J.L., Estes, J.A., Croll, D.A., Danner, E.M., Elmendorf, S.C., and Buckelew, S.L. (2006). An introduced predator alters Aleutian Island plant communities by thwarting nutrient subsidies. *Ecological Monographs*, **76**, 3–24.

Marris, E. (2005). Tsunami damage was enhanced by coral theft. *Nature*, **436**, 1071.

Marrs, R.H., Roberts, R.D., Skeffington, R.A., and Bradshaw, A.D. (1983). Nitrogen and the development of ecosystems. In J.A. Lee, S. McNeill, and I.H. Rorison, eds *Nitrogen as an Ecological Factor*, Symposium of the British Ecological Society, Vol. 22, pp. 113–136. Blackwell, Oxford.

Martínez, M.L., Vázquez, G., and Sánchez Colón, S. (2001). Spatial and temporal variability during primary succession on tropical coastal sand dunes. *Journal of Vegetation Science*, **12**, 361–372.

Martin-López, B., Montes, C., and Benayas, J. (2007). The non-economic motives behind the willingness to pay for biodiversity conservation. *Biological Conservation*, **139**, 67–82.

Mason, B.G., Pyle, D.M., Dade, W.B., and Jupp, T. (2004). Seasonality of volcanic eruptions, *Journal of Geophysical Research*, **109**, B04206, doi: 10.1029/2002JB002293.

Massel, S.R. and Done, T.J. (1993). Effects of cyclone waves on massive coral assemblages on the Great Barrier Reef: meteorology, hydrodynamics and demography. *Coral Reefs*, **12**, 153–166.

Matkin, C.O., Saulitus, E.L., Ellis, G.M., Olesiuk, P., and Rice, S.D. (2008). Ongoing population-level impacts on killer whales *Orcinus orca* following the "Exxon Valdez" oil spill in Prince William Sound, Alaska. *Marine Ecology Progress Series*, **356**, 269–281.

Matthews, J.A. (1992). *The Ecology of Recently-Deglaciated Terrain: a Geoecological Approach to Glacier Forelands and Primary Succession.* Cambridge University Press, Cambridge.

Matthews, J.A. (1999). Disturbance regimes and ecosystem recovery on recently-deglaciated substrates. In L.R. Walker, ed. *Ecosystems of Disturbed Ground,* Ecosystems of the World 16, pp. 17–37. Elsevier, Amsterdam.

Matthews, J.A. and Whittaker, R.J. (1987). Vegetation succession on the Storbreen glacier foreland, Jotunheimen, Norway: a review. *Arctic and Alpine Research,* **19**, 385–395.

Matthews, R.W. and Matthews, J.R. (1978). *Insect Behavior.* Wiley, New York.

Matthews, W.J. and Marsh-Matthews, E. (2003). Effects of drought on fish across axes of space, time and ecological complexity. *Freshwater Biology,* **48**, 1232–1253.

Matthiessen, B., Ptacnik, R., and Hillebrand, H. (2010). Diversity and community biomass depend on dispersal and disturbance in microalgal communities. *Hydrobiologia,* **643**, 65–78.

Matzek, V. and Vitousek, P.M. (2003). Nitrogen fixation in bryophytes, lichens, and decaying wood along a soil-age gradient in Hawaiian montane rain forest. *Biotropica,* **35**, 12–19.

Medellín, R.A., Equihua, M., and Amin, M.A. (2008). Bat diversity and abundance as indicators of disturbance in Neotropical rainforests. *Conservation Biology,* **14**, 1666–1675.

Medina, E., Sternberg, L., and Cuevas, E. (1991). Vertical stratification of $\delta^{13}C$ values in closed natural and plantation forests in the Luquillo mountains, Puerto Rico. *Oecologia* **87**, 369–372.

Meiners, S.J., Cadenasso, M.L., and Pickett, S.T.A. (2007). Succession on the Piedmont of New Jersey and its implications for ecological restoration. In V.A. Cramer and R.J. Hobbs, eds., *Old Fields: Dynamics and Restoration of Abandoned Farmlands,* pp. 145–161. Island Press, Washington, DC.

Menge, B.A. (1978). Predation intensity in a rocky intertidal community: effect of an algal canopy, wave action and desiccation on predator feeding rates. *Oecologia,* **34**, 17–35.

Menge, B.A. and Sutherland, J.A. (1987). Community regulation: variation in disturbance, competition, and predation in relation to environmental stress and recruitment. *The American Naturalist,* **130**, 730–757.

Menge, D.N.L. and Hedin, L.O. (2009). Nitrogen fixation in different biogeochemical niches along a 120 000-year chronosequence in New Zealand. *Ecology,* **90**, 2190–2201.

Menges, E.S. and Waller, D.M. (1983). Plant strategies in relation to elevation and light in floodplain herbs. *The American Naturalist,* **122**, 454–473.

Mercado-Silva, N., Olden, J.D., Maxted, J.T., Hrabik, T.R., and Vander Zanden, M.J. (2006). Forecasting the spread of invasive rainbow smelt in the Laurentian Great Lakes region of North America. *Conservation Biology,* **20**, 1740–1749.

Messer, A.C. (1988). Regional variation in rates of pedogenesis and the influence of climatic factors on moraine chronosequences, southern Norway. *Arctic and Alpine Research,* **20**, 31–39.

Micheli, F. (1999). Eutrophication, fisheries, and consumer-resource dynamics in marine pelagic ecosystems. *Science,* **285**, 1396–1398.

Miles, J. (1979). *Vegetation Dynamics.* Chapman and Hall, London.

Millennium Ecosystem Assessment (2005). *Ecosystems and Human Well-being* (4 volumes). Island Press, Washington, DC.

Miller, K.K. and Wagner, M.R. (1984). Factors influencing pupal distribution of the pandora moth (Lepidoptera: Saturniidae) and their relationship to prescribed burning. *Environmental Entomology,* **13**, 430–431.

Milne, B.T. (1991). Lessons from applying fractal models to landscape patterns. In M.G. Turner and R.H. Gardner, eds *Quantitative Methods in Landscape Ecology,* pp. 199–235. Springer, New York.

Milner, A.M. and Robertson, A.L. (2010). Colonization and succession of stream communities in Glacier Bay, Alaska: what has it contributed to general successional theory? *River Research and Applications,* **26**, 26–35.

Milner, A.M., Fastie, C., Chapin, F.S., III, Engstrom, D.R., and Sharman, L.S. (2007). Interactions and linkages among ecosystems during landscape evolution. *BioScience*, **57**, 237–247.

Milner, A.M., Robertson, A.L., Monaghan, K.A., Veal, A.J., and Flory, E.A. (2008). Colonization and development of an Alaskan stream community over 28 years. *Frontiers in Ecology and the Environment*, **6**, 413–419.

Mining Annual Review (1995). *Mining Annual Review*. Mining Journal Ltd, London.

Mittermeier, R.A., Gil, P.R., Hoffman, M., *et al.* (2005). *Hotspots Revisited: Earth's Biologically Richest and Most Endangered Terrestrial Ecoregions*. Conservation International, Arlington, VA, U.S.

Mix, M.C. (1986). Cancerous diseases in aquatic animals and their association with environmental pollutants: a critical literature review. *Marine Environmental Research*, **20**, 1–141.

Miyanishi, K. and Johnson, E.A. (2007). Coastal dune succession and the reality of dune processes. In E.A. Johnson and K. Miyanishi, eds *Plant Disturbance Ecology: the Process and the Response*, pp. 249–282. Academic Press, Amsterdam.

Moe, S.J., Stelzer, R.S., Forman, R., Harpole, W.S., Daufresne, T., and Yoshida, T. (2005). Recent advances in ecological stoichiometry: insights for population and community ecology. *Oikos*, **109**, 29–39.

Molino, J.-F. and Sabatier, D. (2001). Tree diversity in tropical rain forests: a validation of the intermediate disturbance hypothesis. *Science*, **294**, 1702–1704.

Molles, M.C., Jr (2010). *Ecology: Concepts and Applications*, 5th edn. McGraw-Hill, New York.

Moore, C.J., Moore, S.L., Leecaster, M.K., and Weisberg, S.B. (2001). A comparison of plastic and plankton in the North Pacific Central Gyre. *Marine Pollution Bulletin* **42**, 1297–1300.

Moore, J.C. and DeRuiter, P.C. (1993). Assessment of disturbance on soil ecosystems. *Veterinary Parasitology*, **48**, 75–85.

del Moral, R. (1993). Mechanisms of primary succession on volcanoes: a view from Mount St. Helens. In J. Miles and D.W.H. Walton, eds *Primary Succession on Land*, pp. 79–100. Blackwell, Oxford.

del Moral, R. (2000a). Succession and species turnover on Mount St. Helens, Washington. *Acta Phytogeographica Suecica*, **85**, 53–62.

del Moral, R. (2000b). Local species turnover on Mount St. Helens. In P. White, ed. *Proceeding of the 41st Symposium of the IUVS*, pp. 195–197. Opulus Press, Uppsala, Sweden.

del Moral, R. (2007). Limits to convergence of vegetation during early primary succession. *Journal of Vegetation Science*, **18**, 479–488.

del Moral, R. and Grishin, S.Yu (1999). Volcanic disturbances and ecosystem recovery. In L.R. Walker, ed. *Ecosystems of Disturbed Ground*, Ecosystems of the World 16, pp. 137–160. Elsevier, Amsterdam.

del Moral, R. and Walker, L.R. (2007). *Environmental Disasters, Natural Recovery and Human Responses*. Cambridge University Press, Cambridge.

del Moral, R. and Wood, D. (1993). Early primary succession on a barren volcanic plain at Mount St. Helens, Washington. *American Journal of Botany*, **80**, 981–992.

del Moral, R., Walker, L.R., and Bakker, J.P. (2007). Insights gained from succession for the restoration of landscape structure and function. In L.R. Walker, J. Walker, and R.J. Hobbs, eds *Linking Restoration and Ecological Succession*, pp. 19–44, Springer, New York.

del Moral, R., Saura, J.M., and Emenegger, J.N. (2010). Primary succession trajectories on a barren plain, Mount St. Helens, Washington. *Journal of Vegetation Science*, **21**, 857–867.

Morin, H., Jardon, Y., and Gagnon, R. (2007). Relationship between spruce budworm outbreaks and forest dynamics in eastern North America. In E.A. Johnson and K. Miyanishi, eds *Plant Disturbance Ecology: the Process and the Response*, pp. 555–577. Academic Press, Amsterdam.

Moro, M.J., Pugnaire, F.I., Haase, P., and Puigdefábregas, J. (1997). Mechanisms of interaction between a leguminous shrub and its understory in a semi-arid environment. *Ecography*, **20**, 175–184.

Mote, P.W. and Kaser, G. (2007). The shrinking glaciers of Kilimanjaro: can global warming be blamed? *American Scientist*, **95**, 318–325.

Mueller-Dombois, D. (1973). A non-adapted vegetation interferes with soil water removal in a tropical rain forest area in Hawaii. *Tropical Ecology*, **24**, 1–18.

Mueller-Dombois, D. (1986). Perspectives for an etiology of stand-level dieback. *Annual Review of Ecology and Systematics*, **17**, 221–243.

Mueller-Dombois, D. (2006). Long-term rain forest succession and landscape change in Hawai'i: the "Maui Forest Trouble" revisited. *Journal of Vegetation Science*, **17**, 685–692.

Mueller-Dombois, D. and Ellenberg, H. (1974). *Aims and Methods of Vegetation Ecology*. Wiley, New York.

Mulder, C.P. and Keall, S.N. (2001). Burrowing seabirds and reptiles: impacts on seeds, seedlings and soils in an island forest in New Zealand. *Oecologia*, **127**, 350–360.

Mulder, C.P.H., Koricheva, J., Huss-Danell, K., Hogberg, P., and Joshi, J. (1999). Insects affect relationships between plant species richness and ecosystem processes. *Ecology Letters*, **2**, 237–246.

Mulder, C., Uliassi, D., and Doak, D. (2001). Physical stress and diversity–productivity relationships: the role of positive interactions. *Proceedings of the National Academy of Sciences USA*, **98**, 6704–6708.

Mullineaux, L.S., Fisher, C.R., Peterson, C.H., and Schaeffer, S.W. (2000). Tubeworm succession at hydrothermal vents: use of biogenic cues to reduce habitat selection error? *Oecologia*, **123**, 275–284.

Mullineaux, L.S., Micheli, F., Peterson, C.H., Lenihan, H.S., and Markus, N. (2009). Imprint of past environmental regimes on structure and succession of a deep-sea hydrothermal vent community. *Oecologia*, **161**, 387–400.

Munshower, F.F. (1993). *Practical Handbook of Disturbed Land Revegetation*. Lewis Publishers, Boca Raton, FL, U.S.

Muotka, T., Paavola, R., Haapala, A., Novikmec, M., and Laasonen, P. (2002). Long-term recovery of stream habitat structure and benthic invertebrate communities from in-stream restoration. *Biological Conservation*, **105**, 243–253.

Myers, N. (1993). *Gaia: an Atlas of Planet Management*. Anchor Press/Doubleday, Garden City, NJ, U.S.

Myers, R.A. and Worm, B. (2003). Rapid worldwide depletion of predatory fish communities. *Nature*, **423**, 280–283.

Myster, R.W. and Pickett, S.T.A. (1994). A comparison of the rate of succession over 18 yrs in 10 contrasting old fields. *Ecology*, **75**, 387–392.

Myster, R.W. and Walker, L.R. (1997). Plant successional pathways on Puerto Rican landslides. *Journal of Tropical Ecology*, **13**, 165–173.

Nadkarni, N.M. and Matelson, T.J. (1992). Biomass and nutrient dynamics of epiphytic litterfall in a neotropical montane forest, Costa Rica. *Biotropica*, **24**, 24–30.

Naeem, S., Knops, J.M.H., Tilman, D., Howe, K.M., Kennedy, T., and Gale, S. (2000). Plant diversity increases resistance to invasion in the absence of covarying extrinsic factors. *Oikos*, **91**, 97–108.

Nakamura, F. and Inahara, S. (2007). Fluvial geomorphic disturbances and life history traits of riparian tree species. In E.A. Johnson and K. Miyanishi, eds *Plant Disturbance Ecology: the Process and the Response*, pp.283–310. Academic Press, Amsterdam.

Nara, K. and Hogetsu, T. (2004). Ectomycorrhizal fungi on established shrubs facilitate subsequent seedling establishment of successional plant species. *Ecology*, **85**, 1700–1707.

Naveh, Z. and Lieberman, A.S. (1984). *Landscape Ecology: Theory and Application*. Springer, New York.

Naylor, R.L., Goldburg, R.J., Primavera, J.H. *et al.* (2000). Effect of aquaculture on world fish supplies. *Nature*, **405**, 1017–1024.

Needham, J. (1986). *Science and Civilization in China: Volume 4, Physics and Physical Technology, Part 3, Civil Engineering and Nautics*. Cave Books, Taipei.

Neher, D.A. (2001). Role of nematodes in soil health and their use as indicators. *Journal of Nematology*, **33**, 161–168.

Nelson, D.D. (1991). Factors effecting beach morphology changes caused by Hurricane Hugo, northern South Carolina. *Journal of Coastal Research Special Issue*, **8**, 163–180.

Nepstad, D., Soares-Filho, B.S., Merry, F., *et al.* (2009). The end of deforestation in the Brazilian Amazon. *Science*, **326**, 1350–1351.

New, T.R. and Thornton, I.W. B. (1988). A pre-vegetation population of crickets subsisting on allochthonous aeolian debris on Anak-Krakatau. *Philosophical Transactions of the Royal Society of London Series B, Biological Sciences*, **322**, 481–485.

Nicolson, M. and McIntosh, R.P. (2002). H. A. Gleason and the individualistic hypothesis revisited. *Bulletin of the Ecological Society of America*, **83**, 133–142.

Nicotra, A.B., Chazdon, R.L., and Iriarte, S.V.B. (1999). Spatial heterogeneity of light and woody seedling regeneration in tropical wet forests. *Ecology*, **80**, 1908–1926.

Nielsen, S.P., Iospe, M., and Strand, P. (1997). Collective dose to man from dumping of radioactive waste in the seas. *Science of the Total Environment*, **202**, 136–146.

Niering, W.A. (1987). Vegetation dynamics (succession and climax) in relation to plant community management. *Conservation Biology*, **1**, 287–295.

Niering, W.A., Whittaker, R.H., and Lowe, C.H. (1963). The saguaro: a population in relation to environment. *Science*, **142**, 15–23.

Nilsson, C. and Berggren, K. (2000). Alterations of riparian ecosystems caused by river regulation. *BioScience*, **50**, 783–792.

Nilsson, C. and Wilson, S.D. (1991). Convergence in plant community structure along a disparate gradient: are lakeshores inverted mountainsides? *The American Naturalist*, **137**, 774–790.

Noble, I.R. and Slatyer, R.O. (1980). The use of vital attributes to predict successional changes in plant-communities subject to recurrent disturbances. *Vegetatio*, **43**, 5–21.

Nordin, A., Schmidt, I.K., and Shaver, G.R. (2004). Nitrogen uptake by arctic soil microbes and plants in relation to soil nitrogen supply. *Ecology*, **85**, 955–962.

Nordstrom, D.K. and Alpers, C.N. (1999). Negative pH, efflorescent mineralogy, and consequences for environmental restoration at the Iron Mountain Superfund site, California. *Proceedings of the National Academy of Sciences USA*, **96**, 3455–3462.

Nosi, C. and Zanni, L. (2004). Moving from "typical products" to "food-related services": the Slow Food case as a new business paradigm. *British Food Journal*, **106**, 779–792.

Noss, R.F., La Roe E.T., III, and Scott, J.M. (1995). *Endangered Ecosystems of the United States: a Preliminary Assessment of Loss and Degradation*. Biological Report 28, U.S. Department of the Interior, National Biological Service, Washington, DC.

Nussbaum, R., Anderson, J., and Spencer, T. (1995). Effects of selective logging on soil characteristics and growth of planted dipterocarp seedlings in Sabah. In R. Primack and T. Lovejoy, eds *Ecology, Conservation, and Management of Southeast Asian Rainforests*, pp. 105–115. Yale University Press, New Haven, CT, U.S.

Oberndorfer, E., Lundholm, J., Bass, B., *et al.* (2007). Green roofs as urban ecosystems: ecological structures, functions, and services. *BioScience*, **57**, 823–833.

Oboyski, P. (1995). *Macroarthropod Communities on Vine Maple, Red Alder and Sitka Alder Along Riparian Zones in the Central Western Cascade Range, Oregon*. M.S. Thesis, Oregon State University, Corvallis, OR, U.S.

O'Connell, K.E.B., Gower, S.T., and Norman, J.M. (2003). Net ecosystem production of two contrasting boreal black spruce forest communities. *Ecosystems*, **6**, 248–260.

O'Dowd, D.J., Green, P.T., and Lake, P.S. (2003). Invasional "meltdown" on an oceanic island. *Ecology Letters*, **6**, 812–817.

Odum, E.P. (1950). Organic production and turnover in old field succession. *Ecology*, **41**, 34–49.

Odum, E.P. (1959). *Fundamentals of Ecology*, 2nd edn. Saunders, Philadelphia, PA, U.S.

Odum, E.P. (1969). The strategy of ecosystem development. *Science*, **164**, 262–270.

Odum, H.T. and Pigeon, R.F., eds (1970). *A Tropical Rain Forest*. U.S. Atomic Energy Commission, Division of Technical Information, Oak Ridge, TN.

Oesterheld, M. and Semmartin, M. (in press). Impact of grazing on species composition: adding complexity to a generalized model. *Austral Ecology*, **36**(8), doi: 10.1111/j.1442-9993.2010.02235.x

Oesterheld, M., Loreti, J., Semmartin, M., and Paruelo, J.M. (1999). Grazing, fire, and climate effects on primary productivity of grasslands and savannas. In L.R. Walker, ed. *Ecosystems of Disturbed Ground*, Ecosystems of the World 16, pp. 287–306. Elsevier, Amsterdam.

Ólafsson, E. and Ingimarsdóttir, M. (2009). The land-invertebrate fauna on Surtsey during 2002–2006. *Surtsey Research*, **12**, 113–128.

Olff, H. and Bakker, J.P. (1991). Long-term dynamics of standing crop and species composition after cessation of fertilizer application to mown grassland. *Journal of Applied Ecology*, **28**, 1040–1052.

Olff, H. and Ritchie, M.E. (1998). Effects of herbivores on grassland plant diversity. *Trends in Ecology and Evolution*, **13**, 261–265.

Olson, J.S. (1958). Rates of succession and soil changes on southern Lake Michigan sand dunes. *Botanical Gazette*, **119**, 125–170.

O'Neill, R.V., DeAngelis, D.L., Waide, J.B., and Allen, T.F.H. (1986). *A Hierarchical Concept of Ecosystems*. Princeton University Press, Princeton, NJ, U.S.

Onipchenko, V.G. and Zobel, M. (2000). Mycorrhiza, vegetative mobility and responses to disturbance of alpine plants in the northwestern Caucasus. *Folia Geobotanica*, **35**, 1–11.

Oosting, H.J. (1948). *The Study of Plant Communities*. Freeman, San Francisco.

O'Reilly, C.M., Alin, S.R., Plisnier, P.-D., Cohen, A.S., and McKee, B.A. (2003). Climate change decreases aquatic ecosystem productivity of Lake Tanganyika, Africa. *Nature*, **424**, 766–768.

Orlove, B., Wiegandt, E., and Luckman, B.H. (2008). The place of glaciers in natural and cultural landscapes. In B. Orlove, E. Wiegandt, and B.H. Luckman, eds *Darkening Peaks: Glacier Retreat, Science, and Society*, pp. 3–22. University of California Press, Berkeley, CA, U.S.

Orr, D.W. (2004). *Earth in Mind: on Education, Environment, and the Human Prospect*. Island Press, Washington, DC.

Orr, J.C., Fabry, V.J., Aumont, O., *et al.* (2005). Anthropogenic ocean acidification over the twenty-first century and its impact on calcifying organisms. *Nature*, **437**, 681–686.

Ortlieb, L., Barrientos, S., and Guzman, N. (1996). Coseismic coastal uplift and coralline algae record in Northern Chile: the 1995 Antofagasta earthquake case. *Quaternary Science Reviews*, **15**, 949–960.

Pacala, S.W., Canham, C.D., and Silander, J.A., Jr (1993). Forest models defined by field measurements. 1. The design of a northeastern forest simulator. *Canadian Journal of Forest Science*, **23**, 1980–1988.

Pace, M.L. and Groffman, P.M., eds (1998). *Successes, Limitations, and Frontiers in Ecosystem Science*. Springer, New York.

Packer, A. and Clay, K. (2004). Development of negative feedback during successive growth cycles of black cherry. *Proceedings of the Royal Society B: Biological Sciences*, **271**, 317–324.

Paine, R.T. (1966). Food web complexity and species diversity. *The American Naturalist*, **100**, 65–75.

Paine, R.T. (1980). Food webs: linkage, interaction strength and community structure. *Journal of Animal Ecology*, **49**, 667–685.

Paine, R.T., and Levin, S.A. (1981). Intertidal landscapes: disturbance and the dynamics of pattern. *Ecological Monographs*, **51**, 145–178.

Paine, R.T., Ruesink, J.L., Sun, A. *et al.* (1996). Trouble on oiled waters: lessons from the *Exxon Valdez* oil spill. *Annual Review of Ecology and Systematics*, **27**, 197–235.

Palik, B.J., Mitchell, R.J., and Hiers, J.K. (2002). Modeling silviculture after natural disturbance to sustain biodiversity in the longleaf pine (*Pinus palustris*) ecosystem: balancing complexity and implementation. *Forest Ecology and Management*, **155**, 347–356.

Palmer, M.A., Bernhardt, E.S., Allan, J.D., *et al.* (2005). Standards for ecologically successful river restoration. *Journal of Applied Ecology*, **42**, 208–217.

Palmer, M.W. (1994). Variation in species richness: toward a unification of hypotheses. *Folia Geobotanica et Phytotaxonomica*, **29**, 511–530.

Parker, J.D., Burkepile, D.E., and Hay, M.E. (2006). Opposing effects of native and exotic herbivores on plant invasions. *Science*, **311**, 1459–1461.

Parker, M.A. (2001). Mutualism as a constraint on invasion success for legumes and rhizobia. *Diversity and Distributions*, **7**, 125–136.

Parmenter, R.P. and MacMahon, J.A. (1987). Early successional patterns of arthropod recolonization on reclaimed strip mines in southwestern Wyoming: the ground dwelling beetle fauna (Coleoptera). *Environmental Entomology*, **16**, 168–177.

Parmenter, R.P. and MacMahon, J.A. (2009). Carrion decomposition and nutrient cycling in a semi-arid shrub-steppe ecosystem. *Ecological Monographs*, **79**, 637–661.

Parmenter, R.P., MacMahon, J.A., Waaland, M.E., Stuebe, M.M., Landres, P., and Crisafulli, C. (1985). Reclamation of surface coal mines in western Wyoming for wildlife habitat: a preliminary analysis. *Reclamation and Revegetation Research*, **4**, 98–115.

Parnell, J.J., Crowl, T.A., Weimer, B.C., and Pfrender, M.E. (2009). Biodiversity in microbial communities: system scale patterns and mechanisms. *Molecular Ecology*, **18**, 1455–1462.

Parry, M.A.J. and Hawkesford, M.J. (2010). Food security: increasing yield and improving resource use efficiency. *Proceedings of the Nutrition Society*, **69**, 592–600.

Parsons, W.F.J., Ehrenfeld, J.G., and Handle, S.N. (1998). Vertical growth and mycorrhizal infection of woody plant roots as potential limits to the restoration of woodlands on landfills. *Restoration Ecology*, **6**, 280–289.

Pastor N., López-Lázaro M., Tella J.L., *et al.* (2001). DNA damage in birds after the mining waste spill in southwestern Spain: a Comet assay evaluation. *Journal of Environmental Pathology, Toxicology and Oncology*, **20**, 317–324.

Pastorok, R.A. and Bilyard, G.R. (1985). Effects of sewage pollution on coral reef communities. *Marine Ecology Progress Series*, **21**, 175–189.

Patten, B.C. and Jorgenson, S.E. (1995). *Complex Ecology: the Part–Whole Relation in Ecosystem*, Prentice Hall, Upper Saddle River, NJ, U.S.

Pausas, J.G. and Bradstock, R.A. (2007). Fire persistence traits of plants along a productivity and disturbance gradient in mediterranean shrublands of south-east Australia. *Global Ecology and Biogeography*, **16**, 330–340.

Pausas, J.G. and Lavorel, S. (2003). A hierarchical deductive approach for functional types in disturbed ecosystems. *Journal of Vegetation Science*, **14**, 409–416.

Pearlstine, L., McKellar, H., and Kitchens, W. (1985). Modelling the impacts of a river diversion on bottomland forest communities in the Santee River floodplain, South Carolina. *Ecological Modelling*, **29**, 283–302.

Peet, R.K. (1992). Community structure and ecosystem function. In D.C. Glenn-Lewin, R.K. Peet, and T.T. Veblen, eds *Plant Succession: Theory and Prediction*, pp. 103–151. Chapman and Hall, London.

Peh, K.S.-H., Lewis, S.L., and Lloyd, J. (2011). Mechanisms of monodominance in diverse tropical tree-dominated systems. *Journal of Ecology*, **99**, 891–898.

Peltzer, D.A., Bellingham, P.J., Kurokawa, H., Walker, L.R., Wardle, D.A., and Yeates, G.W. (2009). Punching above their weight: low-biomass non-native plant species alter soil properties during primary succession. *Oikos*, **118**, 1001–1014.

Peltzer, D.A., Wardle, D.A., Allison, V.J., *et al.* (2010). Understanding ecosystem retrogression. *Ecological Monographs*, **80**, 509–529.

Peng, C. and Apps, M.J. (1999). Modelling the response of net primary productivity (NPP) of boreal forest ecosystems to changes in climate and fire disturbance regimes. *Ecological Modelling*, **122**, 175–193.

Perfecto, I., Vandermeer, J., Hanson, P., and Cartín, V. (1997). Arthropod biodiversity loss and the transformation of a tropical agro-ecosystem. *Biodiversity and Conservation*, **6**, 935–945.

Peterson, C.H. (1991). Intertidal zonation of marine invertebrates in sand and mud. *American Scientist*, **79**, 236–249.

Peterson, C.H., Rice, S.D., Short, J.W. *et al.* (2003). Long-term ecosystem response to the *Exxon Valdez* oil spill. *Science*, **302**, 2082–2086.

Petrides, G.A. (1975). The importation of wild ungulates into Latin American, with remarks on their environmental effects. *Environmental Conservation*, **2**, 47–51.

Pianka, E.R. (1970). On r and K selection. *The American Naturalist*, **104**, 592–597.

Pickett, S.T.A. (1976). Succession: an evolutionary perspective. *The American Naturalist*, **110**, 107–119.

Pickett, S.T.A. (1989). Space-for-time substitutions as an alternative to long-term studies. In G.E. Likens, ed. *Long-term Studies in Ecology*, pp. 110–135. Springer, New York.

Pickett, S.T.A. and Cadenasso, M.L. (1995). Landscape ecology: spatial heterogeneity in ecological systems. *Science*, **269**, 331–334.

Pickett, S.T.A. and White, P.S., eds (1985). *The Ecology of Natural Disturbance and Patch Dynamics*. Academic Press, New York.

Pickett, S.T.A., Collins, S.L., and Armesto, J.J. (1987a). Models, mechanisms and pathways of succession. *Botanical Review*, **53**, 335–371.

Pickett, S.T.A., Collins, S.L., and Armesto, J.J. (1987b). A hierarchical consideration of causes and mechanisms of succession. *Vegetatio*, **69**, 109–114.

Pickett, S.T.A., Wu, J., and Cadenasso, M.L. (1999). Patch dynamics and the ecology of disturbed ground: a framework for synthesis. In L.R. Walker, ed. *Ecosystems of Disturbed Ground*, Ecosystems of the World 16, pp. 707–722. Elsevier, Amsterdam.

Pickett, S.T.A., Cadenasso, M.L., Grove, J.M., *et al.* (2001). Urban ecological systems: linking terrestrial ecological, physical, and socioeconomic components of metropolitan areas. *Annual Review of Ecology and Systematics*, **32**, 127–157.

Pickett, S.T.A., Cadenasso, M.L., and Meiners, S.J. (2009). Ever since Clements: from succession to vegetation dynamics and understanding to intervention. *Applied Vegetation Science*, **12**, 9–21.

Pickford, N. (2006). *Lost Treasure Ships of the Northern Seas: a Guide and Gazetteer to 2000 Years of Shipwreck*, Chatham, London.

Pimentel, D., ed. (1993). *Soil Erosion and Conservation*. Cambridge University Press, Cambridge.

Pimentel, D. (2006). Soil erosion: a food and environmental threat. *Environment, Development, and Sustainability*, **8**, 119–137.

Pimentel, D. and Harvey, C. (1999). Ecological effects of erosion. In L.R. Walker, ed. *Ecosystems of Disturbed Ground*, Ecosystems of the World 16, pp. 123–135. Elsevier, Amsterdam.

Pimentel, D., Lach, L., Zuniga, R., and Morrison, D. (2000). Environmental and economic costs of nonindigenous species in the United States. *BioScience*, **50**, 53–65.

Piper, D.J.W., Shor, A.N., and Clarke, J.H. (1988). The 1929 Grand Banks earthquake, slump, and turbidity current. *Geological Society of America Special Papers*, **229**, 77–92.

Pitcher, G.C., Figueiras, F.G., Hickey, B.M., and Moita, M.T. (2010). The physical oceanography of upwelling systems and the development of harmful algal blooms. *Progress in Oceanography*, **85**, 5–32.

Platt, W.J. and Connell, J.H. (2003). Natural disturbances and directional replacement of species. *Ecological Monographs*, **73**, 507–522.

Platt, W.J. and Weis, I.M. (1977). Resource partitioning and competition within a guild of fugitive prairie plants. *The American Naturalist*, **111**, 479–513.

Poff, N.L. and Allan, J.D. (1995). Functional organization of stream fish assemblages in relation to hydrological variability. *Ecology*, **76**, 606–627.

Polis, G.A. and Hurd, S.D. (1996). Linking marine and terrestrial food webs: allochthonous input from the ocean supports high secondary productivity on small islands and coastal land communities. *The American Naturalist*, **147**, 396–423.

Porazinska, D.L., Bardgett, R.D., Blaauw, M.B., *et al.* (2003). Relationships at the aboveground–belowground interface: plants, soil biota and soil processes. *Ecological Monographs*, **73**, 377–395.

Pörtner, H.O., Langenbuch, M., and Reipschläger, A. (2004). Biological impact of elevated ocean CO_2 concentrations: lessons from animal physiology and earth history. *Journal of Oceanography*, **60**, 705–718.

Pounds, J.A. and Crump, M.L. (2002). Amphibian declines and climate disturbance: the case of the golden toad and the harlequin frog. *Conservation Biology*, **8**, 72–85.

Power, M.E. and Dietrich, W.E. (2002). Food webs in river networks. *Ecological Research*, **17**, 451–471.

Power, M.E., Dietrich, W.E., and Finlay, J.C. (1996). Dams and downstream aquatic biodiversity: potential food web consequences of hydrologic and geomorphic change. *Environmental Management*, **20**, 887–895.

Prach, K. (2003). Spontaneous vegetation succession in central European man-made habitats: what information can be used in restoration practice? *Applied Vegetation Science*, **6**, 125–129.

Prach, K. and Walker, L.R. (2011). Four opportunities for studies of ecological succession. *Trends in Ecology and Evolution*, **26**, 119–123.

Prach, K., Marrs, R., Pyšek, P., and van Diggelen, R. (2007). Manipulation of succession. In L.R. Walker, J. Walker, and R.J. Hobbs, eds *Linking Restoration and Ecological Succession*, pp. 121–149. Springer, New York.

Prach, K., Řehounková, K., Řehounek, J., and Konvalinková, P. (2011). Ecological restoration of central European mining sites: a summary of a multi-site analysis. *Landscape Research*, **36**, 263–268.

Price, L.W. (1971). Vegetation, microtopography, and depth of active layer on different exposures in subarctic alpine tundra. *Ecology*, **52**, 638–647.

Primack, R.B. (2010). *Essentials of Conservation Biology*, 5th edn. Sinauer, Sunderland, MA, U.S.

Prober, S.M., Lunt, I.D., and Morgan, J.W. (2009). Rapid internal plant-soil feedbacks lead to alternative stable states in temperate Australian grassy woodlands. In R.J. Hobbs and K.N. Suding, eds *New Models for Ecosystem Dynamics and Restoration*, pp. 156–168. Island Press, Washington, DC.

Pugh, P.J.A. and Davenport, J. (1997). Colonisation vs. disturbance: the effects of sustained ice-scouring on intertidal communities. *Journal of Experimental Marine Biology and Ecology*, **210**, 1–21.

Pugnaire, F., ed. (2010). *Positive Plant Interactions and Community Dynamics*. CRC Press, Boca Raton, FL, U.S.

Purcell, J.E., Uye, S., and Lo, W.-T. (2007). Anthropogenic causes of jelly fish blooms and their direct consequences for humans: a review. *Marine Ecology Progress Series*, **350**, 153–174.

Quine, C.P. and Gardiner, B.A. (2007). Understanding how the interaction of wind and trees results in wind-throw, stem breakage, and canopy gap formation. In E.A. Johnson and K. Miyanishi, eds., *Plant Disturbance Ecology: the Process and the Response*, pp. 103–155. Academic Press, New York.

Quist, M.C., Fay, P.A., Guy, C.S., Knapp, A.K., and Rubenstein, B.N. (2003). Military training effects on terrestrial and aquatic communities on a grassland military installation. *Ecological Applications*, **13**, 432–442.

Rabalais, N.N. (2002). Nitrogen in aquatic ecosystems. *Ambio*, **31**, 102–112.

Raeburn, P. (1996). *The Last Harvest: the Genetic Gamble that Threatens to Destroy American Agriculture*. Bison Books, Winnipeg, Manitoba.

Raffa, K.F., and Berryman, A.A. (1983). The role of host plant resistance in the colonization behavior and ecology of bark beetles (Coleoptera: Scolytidae). Ecological Monographs, **53**, 27–49.

Ragnarsdóttir, K.V. (2008). Rare metals getting rarer. *Nature Geoscience*, **1**, 720–721.

Rahmstorf, S. (2010). A new view on sea level rise. *Nature Reports Climate Change*, **4**, 44–45.

Rapport, D.J., Costanza, R., and McMichael, A.J. (1998). Assessing ecosystem health. *Trends in Ecology and Evolution*, **13**, 397–402.

Raup, H.M. (1981). Physical disturbance in the life of plants. In M.H. Nitecki, ed. *Biotic Crises in Ecological and Evolutionary Time*, pp. 39–52. Academic Press, New York.

Read, P. and Fernandes, T. (2003). Management of environmental impacts of marine aquaculture in Europe. *Aquaculture*, **226**, 139–163.

Rebele, F. and Lehmann, C. (2002). Restoration of a landfill site in Berlin, Germany by spontaneous and directed succession. *Restoration Ecology*, **10**, 340–347.

Reddy, C.M., Eglinton, T.I., Hounshell, A., *et al.* (2002). The West Falmouth soil spill after thirty years: the persistence of petroleum hydrocarbons in marsh sediments. *Environmental Science and Technology*, **36**, 4754–4760.

Reece, E. (2006). *Lost Mountain: Radical Strip Mining and the Devastation of Appalachia*, Riverhead Books, New York.

Reed, D.C., Raimondi, P.T., Carr, M.H., and Goldwasser, L. (2000). The role of dispersal and disturbance in determining spatial heterogeneity in sedentary organisms. *Ecology*, **81**, 2011–2026.

Reed, R.A., Johnson-Barnard, J., and Baker, W.L. (1996). Contribution of roads to forest fragmentation in the Rocky Mountains. *Conservation Biology*, **10**, 1098–1106.

Řehounková, K. and Prach, K. (2010). Life-history traits and habitat preferences of colonizing plant species in long-term spontaneous succession in abandoned gravel-sand pits. *Basic and Applied Ecology*, **11**, 45–53.

Reice, S.R. (2003). *The Silver Lining: the Benefits of Natural Disasters*. Princeton University Press, Princeton, NJ, U.S.

Reinhart, K.O., Packer, A., van der Putten, W.H., and Clay, K. (2003). Escape from natural soil pathogens enables a North American tree to invade Europe. *Ecology Letters*, **6**, 1046–1050.

Reissek, S. (1856). Vortrag ueber die Bildungsgeschichte der Donauinseln immittleren Laufe dieses Stromes. *Flora*, **39**, 609–624 (not seen but cited in Clements 1928).

Restrepo, C., Walker, L.R., Shiels, A.B., *et al.* (2009). Landsliding and its multiscale influence on mountainscapes. *BioScience*, **59**, 685–698.

Ricciardi, A. (2001). Facilitative interactions among aquatic invaders: is an "invasional meltdown" occurring in the Great Lakes? *Canadian Journal of Fisheries and Aquatic Science*, **58**, 2513–2525.

Rice, E.L. and Pancholy, S.K. (1972). Inhibition of nitrification by climax ecosystems. *American Journal of Botany*, **59**, 1033–1040.

Richards, E.N. and Goff, M.L. (1997). Arthropod succession on exposed carrion in three contrasting tropical habitats on Hawaii Island, Hawaii. *Journal of Medical Entomology*, **34**, 328–339.

Rindos, D. (1987). *The Origins of Agriculture: an Evolutionary Perspective*. Academic Press, New York.

Robigou, V., Delaney, J.R., and Stakes, D.S. (1993). Large massive sulfide deposits in a newly discovered active hydrothermal system, the High Rise Field, Endeavour Segment, Juan de Fuca Ridge. *Geophysical Research Letters*, **20**, 1887–1890.

Robinson, G.R. and Handel, S.N. (1993). Forest restoration on a closed landfill: rapid addition of new species by bird dispersal. *Conservation Biology*, **7**, 271–278.

Robinson, K. (1997). *Where Dwarfs Reign: a Tropical Rain Forest in Puerto Rico*. University of Puerto Rico Press, San Juan, Puerto Rico.

Robinson, R.C., Sheffield, E., and Sharpe, J.M. (2010). Problem ferns: their impact and management. In K. Mehltreter, L.R. Walker, and J.M. Sharpe, eds *Fern Ecology*, pp. 255–322. Cambridge University Press, Cambridge.

Robles, C.D. (1982). Disturbance and predation in an assemblage of herbivorous Diptera and algae on rocky shores. *Oecologia*, **54**, 23–31.

Rogers, C.S. (1993). Hurricanes and coral reefs: the intermediate disturbance hypothesis revisited. *Coral Reefs*, **12**, 127–137.

Rogers, P. (2002). *Losing Control: Global Security in the 21st century*. Pluto Press, London.

Rohde, K. (2005). *Nonequilibrium Ecology*. Cambridge University Press, Cambridge.

Rongo, T., Bush, M., and Van Woesik, R. (2009). Did ciguatera prompt the late Holocene Polynesian voyages of discovery? *Journal of Biogeography*, **36**, 1423–1432.

Roland, J. (1993). Large-scale forest fragmentation increases the duration of tent caterpillar outbreak. *Oecologia*, **93**, 25–30.

Rosenberg, E. and Ben-Haim, Y. (2002). Microbial diseases of coral and global warming. *Environmental Microbiology*, **4**, 318–326.

Rossiter, N.A., Setterfield, S.A., Douglas, M.M., and Hutley, L.B. (2003). Testing the grass-fire cycle: alien grass invasion in the tropical savannas of northern Australia. *Diversity and Distributions*, **9**, 169–176.

Roxburgh, S.H., Shea, K., and Wilson, J.B. (2004). The intermediate disturbance hypothesis: patch dynamics and mechanisms of species coexistence. *Ecology*, **85**, 359–371.

Rudgers, J.A., Holah, J., Orr, S.P., and Clay, K. (2007). Forest succession suppressed by an introduced plant–fungal symbiosis. *Ecology*, **88**, 18–25.

Ruffo, S., and Kareiva, P.M. (2009). Using science to assign value to nature. *Frontiers in Ecology and the Environment*, **7**, 3.

van Ruijven, J., De Deyn, G.B., and Berendse, F. (2003). Diversity reduces invasibility in experimental plant communities: the role of plant species. *Ecology Letters*, **6**, 910–918.

de Ruiter, P.C., Neutel, A.N., and Moore, J.C. (1995). Energetics, patterns of interaction strengths, and stability in real ecosystems. *Science*, **269**, 1257–1260.

Ruiz, G.M. and Carlton, J.T., eds (2003). *Invasive Species: Vectors and Management Strategies*. Island Press, Washington, DC.

Rundel, P.W. (1981). Structural and chemical components of flammability. In H.A. Mooney, T.A. Bonnicksen, N.L. Christensen, J.E. Lotan, and W.A. Reiners, eds *Fire Regimes and Ecosystem Properties*, pp. 183–207. U.S. Department of Agriculture Forest Service General Technical Report WO-26.

Rundel, P.W. (1999). Disturbance in Mediterranean-climate shrublands and woodlands. In L.R. Walker, ed. *Ecosystems of Disturbed Ground*, Ecosystems of the World 16, pp. 271–285. Elsevier, Amsterdam.

Rusch, G. (1992). Spatial patterns of seedling recruitment at two different scales in a limestone grassland. *Oikos*, **65**, 433–442.

Rydin, H. and Borgegård, S.-O. (1991). Plant characteristics over a century of primary succession on islands, Lake Hjälmaren. *Ecology*, **72**, 1089–1101.

Sachs, J.D., Baillie, J.E.M., Sutherland, W.J., *et al.* (2009). Biodiversity conservation and the Millenium Development Goals. *Science*, **325**, 1502–1503.

Sagarin, R. and Pauchard, A. (2010). Observational approaches in ecology open new ground in a changing world. *Frontiers in Ecology and the Environment*, **8**, 379–386.

Sala, A., Devitt, D.E., and Smith, S.D. (1996). Water use by *Tamarix ramosissima* and associated phreatophytes in a Mojave Desert floodplain. *Ecological Applications*, **6**, 888–898.

Samuels, C.L. and Drake, J.A. (1997). Divergent perspectives on community convergence. *Trends in Ecology and Evolution*, **12**, 427–432.

Sandmeier, F.C., Tracey, R., duPré, S., and Hunter, K. (2009). Upper respiratory tract disease (URTD) as a threat to desert tortoise populations: a reevaluation. *Biological Conservation*, **142**, 1255–1268.

Santas, R., Lianou, C., and Danielidis, D. (1997). UVB radiation and depth interaction during primary succession of marine diatom assemblages of Greece. *Limnology and Oceanography*, **42**, 986–991.

Sarrazin, F. and Barbault, R. (1996). Reintroduction: challenges and lessons for basic ecology. *Trends in Ecology and Evolution*, **11**, 474–478.

Sarrazin, J., Robigou, V., Juniper, S.K., and Delaney, J. (1997). Biological and geological dynamics over four years on a high-temperature sulfide structure at the Juan de Fuca Ridge hydrothermal observatory. *Marine Ecology Progress Series*, **153**, 5–24.

Sauer, C.O (1952). *Agricultural Origins and Dispersals*. MIT Press, Cambridge, MA, U.S.

Savidge, W.B. and Taghon, G.L. (1988). Passive and active components of colonization following two types of disturbance on intertidal sandflat. *Journal of Experimental Marine Biology and Ecology*, **115**, 137–155.

Scarth, A. (1999). *Vulcan's Fury: Man Against the Volcano*. Yale University Press, New Haven, CT, U.S.

Scatena, F.N. and Larsen, M.C. (1991). Physical aspects of Hurricane Hugo in Puerto Rico. *Biotropica*, **23**, 317–323.

Scatena, F.N. and Lugo, A.E. (1995). Geomorphology, disturbance, and the soil and vegetation of two subtropical wet steepland watersheds of Puerto Rico. *Geomorphology*, **13**, 199–213.

Scheffer, M., Straile, D., van Nes, E.H., and Hosper, H. (2001). Climatic warming causes regime shifts in lake food webs. *Limnology and Oceanography*, **46**, 1780–1783.

Scheffer, N. (1998). *Ecology of Shallow Lakes*. Chapman and Hall, London.

Scheidat, M., Castro, C., Gonzalez, J., and Williams, R. (2004). Behavioural responses of humpback whales (*Megaptera novaeangliae*) to whalewatching boats near Isla de la Plata, Machalilla National Park, Ecuador. *Journal of Cetacean Research and Management*, **6**, 63–68.

Schenk, H.J. (1999). Clonal splitting in desert shrubs. *Plant Ecology*, **141**, 41–52.

Schimel, J.P., Cates, R.G., and Ruess, R. (1998). The role of balsam poplar secondary chemicals in controlling soil nutrient dynamics through succession in the Alaskan taiga. *Biogeochemistry*, **42**, 221–234.

Schipper, L.A., Degens, B.P., Sparling, G.P., and Duncan, L. (2001). Changes in microbial heterotrophic diversity along five plant successional sequences. *Soil Biology and Biochemistry*, **33**, 2093–2103.

Schlesinger, W.H. (1997). *Biogeochemistry: an Analysis of Global Change*, 2nd edn. Academic Press, San Diego.

Schlesinger, W.H. (2009). On the fate of anthropogenic nitrogen. *Proceedings of the National Academy of Sciences USA*, **106**, 203–208.

Schlesinger, W.H., Raikes, J.A., Hartley, A.E., and Cross, A.E. (1996). On the spatial pattern of soil nutrients in desert ecosystems. *Ecology*, **77**, 364–374.

Schneider, K., Migge, S., Norton, R.A., *et al.* (2004). Trophic niche differentiation in soil microarthropods (Oribatida, Acari): evidence from stable isotope ratios ($^{15}N/^{14}N$). *Soil Biology and Biochemistry*, **36**, 1769–1774.

Schoeman, D.S., McLachlan, A., and Dugan, J.E. (2000). Lessons from a disturbance experiment in the intertidal zone of an exposed sandy beach. *Estuarine, Coastal and Shelf Science*, **50**, 869–884.

Schoenly, K. and Reid, W. (1987). Dynamics of heterotrophic succession in carrion arthropod assemblages: discrete seres or a continuum of change? *Oecologia*, **73**, 192–202.

Schowalter, T.D. (1985). Adaptations of insects to disturbance. In S.T.A. Pickett and P.S. White, eds *The Ecology of Natural Disturbance and Patch Dynamics*, pp. 235–252. Academic Press, New York.

Schowalter, T.D. and Lowman, M.D. (1999). Forest herbivory: insects. In L.R. Walker, ed. *Ecosystems of Disturbed Ground*, Ecosystems of the World 16, pp. 253–269. Elsevier, Amsterdam.

Schowalter, T.D. and Turchin, P. (1993). Southern pine beetle infestation development: interaction between pine and hardwood basal areas. *Forest Science*, **39**, 201–210.

Schrautzer, J., Rinker, A., Jensen, K., Müller, F., Schwartze, P. and Dierßen, K. (2007). Succession and restoration of drained fens: perspectives from northwestern Europe. In L.R. Walker, J. Walker, and R.J. Hobbs, eds *Linking Restoration and Ecological Succession*, pp. 90–120.

Schwartz, D.M. and Bazzaz, F.A. (1973). *In situ* measurements of carbon dioxide gradients in a soil–plant–atmosphere system. *Oecologia*, **12**, 161–167.

Schwilk, D.W., Keeley, J.E., and Bond, W.J. (1997). The intermediate disturbance hypothesis does not explain fire and diversity pattern in fynbos. *Plant Ecology*, **132**, 77–84.

Schwimmer, H. and Haim, A. (2009). Physiological adaptations of small mammals to desert ecosystems. *Integrative Zoology*, **4**, 357–366.

Seager, R., Ting, M., Held, I., *et al.* (2007). Model projections of an imminent transition to a more arid climate in southwestern North America. *Science*, **316**, 1181–1184.

Searle, M. (2006). Co-seismic uplift of coral reefs along the western Andaman Islands during the December 26th 2004 earthquake. *Coral Reefs*, **25**, 2.

Seddon, P.J., Armstrong, D.P., and Maloney, R.F. (2007). Developing the science of reintroduction biology. *Conservation Biology*, **21**, 303–312.

Sekercioglu, C.H. (2006). Increasing awareness of avian ecological function. *Trends in Ecology and Evolution*, **21**, 465–471.

Sexton, W.J. and Hayes, M.O. (1991). The geologic impact of Hurricane Hugo and post-storm shoreline recovery along the undeveloped coastline of South Carolina, Dewees Island to the Santee Delta. *Journal of Coastal Research Special Issue*, **8**, 275–290.

Shafer, C.L. (1990). *Nature Reserves: Island Theory and Conservation Practice*. Smithsonian Institution Press, Washington, DC.

Shanahan, T.M., Overpeck, J.T., Anchukaitis, K.J., *et al.* (2009). Atlantic forcing of persistent drought in West Africa. *Science*, **324**, 377–380.

Shank, T.M., Fornari, D.J., Von Damm, K.L., Lilley, M.D., Haymon, R.M., and Lutz, R.A. (1998). Temporal and spatial patterns of biological community development at nascent deep-sea hydrothermal vents (9°N, East Pacific Rise). *Deep-Sea Research*, **45**, 465–516.

Shapiro, A.M. (2002). The Californian urban butterfly fauna is dependent on alien plants. *Diversity and Distributions*, **8**, 31–40.

Shea, K., Roxburgh, S.H., and Rauschert, E.S.J. (2004). Moving from pattern to process: coexistence mechanisms under intermediate disturbance regimes. *Ecology Letters*, **7**, 491–508.

Sheil, D. and Burslem, D.F.R.P. (2003). Disturbing hypotheses in tropical forests. *Trends in Ecology and Evolution*, **18**, 18–26.

Sheldon, P.J., Knox, J.M., and Lowry, K. (2005). Sustainability in a mature mass tourism destination: the case of Hawaii. *Tourism Review International*, **9**, 47–59.

Shelford, V.E. (1911). Ecological succession. II. Pond fishes. *Biological Bulletin*, **21**, 127–151.

Shepherd, K.D. and Walsh, M.G. (2002). Development of reflectance spectral libraries for characterization of soil properties. *Soil Science Society of America Journal*, **66**, 988–998.

Shiels, A.B. and Walker, L.R. (2003). Bird perches increase forest seeds on Puerto Rican landslides. *Restoration Ecology*, **11**, 457–465.

Shiels, A.B., Zimmerman, J.K., García-Montiel, D.C., *et al.* (2010). Plant responses to simulated hurricane impacts in a subtropical wet forest, Puerto Rico. *Journal of Ecology*, **98**, 659–673.

Short, F.T. and Neckles, H.A. (1998). The effects of global climate change on seagrasses. *Aquatic Botany*, **63**, 169–196.

Short, F.T. and Wyllie-Echeverria, S. (1996). Natural and human-induced disturbance of seagrasses. *Environmental Conservation*, **23**, 17–27.

Shumway, S.W. (2000). Facilitative effects of a sand dune shrub on species growing beneath the shrub canopy. *Oecologia*, **124**, 138–148.

Sidle, R.C. and Ochiai, H. (2006). *Landslides: Processes, Prediction and Land Use*. American Geophysical Union, San Francisco, CA.

Siepel, H. (1996a). The importance of unpredictable and short-term environmental extremes for biodiversity in oribatid mites. *Biodiversity Letters*, **3**, 26–34.

Siepel, H. (1996b). Biodiversity of soil microarthropods: the filtering of species. *Biodiversity and Conservation*, **5**, 251–260.

Sikes, D.S. and Slowik, J. (2010). Terrestrial arthropods of pre- and post-eruption Kasatochi Island, Alaska, 2008–2009: a shift from a plant-based to a necromass-based food web. *Arctic, Antarctic, and Alpine Research*, **42**, 297–305.

Silver, W.L., Brown, S., and Lugo, A.E. (1996a). Effects of changes in biodiversity on ecosystem function in tropical forests. *Conservation Biology*, **10**, 17–24.

Silver, W.L., Brown, S., and Lugo, A.E. (1996b). Biodiversity and biogeochemistry in tropical forests. In G. Orians, R. Dirzo, and H. Cushman, eds *Biodiversity and Ecosystem Processes in Tropical Forests*, pp. 49–67. Springer, New York.

Simberloff, D. (2006). Invasional meltdown 6 years later: important phenomenon, unfortunate metaphor, or both? *Ecology Letters*, **9**, 912–919.

Simberloff, D. and Von Holle, B. (1999). Positive interactions of nonindigenous species: invasional meltdown? *Biological Invasions*, **1**, 21–32.

Simonett, D.S. (1967). Landslide distribution and earthquakes in Bewani and Torricelli Mountains, New Guinea—a statistical analysis. In J.N. Jennings and J.A. Mabbutt, eds *Landform Studies from Australia and New Guinea*, pp. 64–85. Australian National University Press, Canberra, Australia.

Skaggs, J.M. (1995). *The Great Guano Rush: Entrepreneurs and American Overseas Expansion*. St Martin's Press, New York.

Slocum, M.G., Aide, T.M., Zimmerman, J.K., and Navarro, L. (2004). Natural regeneration of subtropical montane forest after clearing fern thickets in the Dominican Republic. *Journal of Tropical Ecology*, **20**, 483–486.

Smetacek, V. and Naqvi, S.W.A. (2008). The next generation of iron fertilization experiments in the Southern Ocean. *Philosophical Transactions of the Royal Society A: Mathematical, Physical and Engineering Sciences*, **366**, 3947–3967.

Smith, C.K., Kukert, H., Wheatcroft, R.A., Jumars, P.A., and Deming, J.W. (1989). Vent fauna on whale remains. *Nature*, **341**, 27–28.

Smith, C.R., Matbaum, H.L., Baco, A.R. *et al.* (1998). Sediment community structure around a whale skeleton in the deep northeast Pacific: macrofaunal, microbial and bioturbation effects. *Deep-Sea Research*, **45**, 335–364.

Smith, D.W., Peterson, R.O., and Houston, D.B. (2003). Yellowstone after wolves. *BioScience*, **53**, 330–340.

Smith, R.A., Alexander, R.B., and Schwarz, G.E. (2003). Natural background concentrations of nutrients in streams and rivers of the conterminous United States. *Environmental Science and Technology*, **37**, 3039–3047.

Smith, S.D., Huxman, T.E., Zitzer, S.F., *et al.* (2000). Elevated CO_2 increases productivity and invasive species success in an arid ecosystem. *Nature*, **408**, 79–82.

Smith, S.V. (1984). Phosphorus versus nitrogen limitation in the marine environment. *Limnology and Oceanography*, **29**, 1149–1160.

Smith, S.V. and Atkinson, M.J. (1984). Phosphorus limitation of net production in a confined aquatic ecosystem. *Nature*, **307**, 626–627.

Smith, T.M. and Smith, R.L. (2009). *Elements of Ecology*, 7th edn. Benjamin Cummings, San Francisco, CA.

Smol, J.P. and Cumming, B.F. (2000). Tracking long-term changes in climate using algal indicators in lake sediments. *Journal of Phycology*, **36**, 986–1011.

Sobey, D.G. and Kenworthy, J.B. (1979). The relationship between herring gulls and the vegetation of their breeding colonies. *Journal of Ecology*, **67**, 469–496.

Sojka, R.E. (1999). Physical aspects of soils of disturbed ground. In L.R. Walker, ed. *Ecosystems of Disturbed Ground*, Ecosystems of the World 16, pp. 503–519. Elsevier, Amsterdam.

Solomon, B.D., Corey-Luse, C.M., and Halvorsen, K.E. (2004). The Florida manatee and eco-tourism: toward a safe minimum standard. *Ecological Economics*, **50**, 101–115.

Solomon, S., Plattner, G.-K., Knutti, R., and Friedlingstein, P. (2009). Irreversible climate change due to carbon dioxide emissions. *Proceedings of the National Academy of Sciences USA*, **106**, 1704–1709.

Sousa, W.P. (1979a). Experimental investigations of disturbance and ecological succession in a rocky intertidal algal community. *Ecological Monographs*, **49**, 227–254.

Sousa, W.P. (1979b). Disturbance in marine intertidal boulder fields: the nonequilibrium maintenance of species diversity. *Ecology*, **60**, 1225–1239.

Sousa, W.P. (1985). Disturbance and patch dynamics on rocky intertidal shores. In S.T.A. Pickett and P.S. White, eds *The Ecology of Natural Disturbance and Patch Dynamics*, pp. 101–124. Academic Press, New York.

Specht, R.L. (1997). Ecosystem dynamics in coastal dunes of eastern Australia. In E. van der Maarel, ed. *Dry Coastal Ecosystems: General Aspects, Ecosystems of the World 2C*, pp. 483–495. Elsevier, Amsterdam.

Spedding, F. and Daane, A.H. (1961). *The Rare Earths*. Wiley, New York.

Spellerberg, I. (1998). Ecological effects of roads and traffic: a literature review. *Global Ecology and Biogeography*, **7**, 317–333.

Spieles, D.J. (2010). *Protected Land: Disturbance, Stress, and American Ecosystem Management*. Springer, New York.

Sprent, J.I. (1987). *The Ecology of the Nitrogen Cycle*. Cambridge University Press, Cambridge.

Sprent, J.I. and Sprent, P. (1990). *Nitrogen Fixing Organisms: Pure and Applied Aspects*. Chapman and Hall, London.

Stachowicz, J.J., Whitlatch, R.B., and Osman, R.W. (1999). Species diversity and invasion resistance in a marine ecosystem. *Science*, **286**, 1577–1579.

Stahl, P.D., Williams, S.E., and Christensen, M. (1988). Efficacy of native vesicular-arbuscular mycorrhizal fungi after severe soil disturbance. *New Phytologist*, **110**, 347–354.

Stampe, E.D. and Daehler, C.C. (2003). Mycorrhizal species identity affects plant community structure and invasion: a microcosm study. *Oikos*, **100**, 362–372.

Stein, B.A., Kutner, L.S., and Adams, J.S., eds (2000). *Precious Heritage: the Status of Biodiversity in the United States*. Oxford University Press, Oxford.

Stevens, M.H.H. and Carson, W.P. (2002). Resource quantity, not resource heterogeneity, maintains plant diversity. *Ecology Letters*, **5**, 420–426.

Stohlgren, T.J., Barnett, D.T., Jarnevich, C.S., Flather, C., and Kartesz, J. (2008). The myth of plant species saturation. *Ecology Letters*, **11**, 313–322.

Stohlgren, T.J., Binkley, D., Chong, G.W., *et al.* (1999). Exotic plant species invade hot spots of native plant diversity. *Ecological Monographs*, **69**, 25–46.

Strom, A.L. (2008). Catastrophic slides and avalanches. In K. Sassa and P. Canuti, eds *Landslides: Disaster Risk Reduction*, pp. 377–399. Springer, New York.

Suarez, A.V. and Tsutsui, N.D. (2007). The evolutionary consequences of biological invasions. *Molecular Ecology*, **17**, 351–360.

Suding, K.N. and Hobbs, R.J. (2009). Models of ecosystem dynamics as frameworks for restoration ecology. In Hobbs, R.J. and K.N. Suding, eds *New Models for Ecosystem Dynamics and Restoration*, pp. 3–21. Island Press, Washington, DC.

Suding, K.N., Gross, K.L., and Houseman, G.R. (2004). Alternative states and positive feedbacks in restoration ecology. *Trends in Ecology and Evolution*, **19**, 46–53.

Sugg, P.M. and Edwards, J.S. (1998). Pioneer aeolian community development on pyroclastic flows after the eruption of Mount St. Helens, Washington, U.S.A. *Arctic and Alpine Research*, **30**, 400–407.

Sugihara, G. and May, R.M. (1990). Applications of fractals in ecology. *Trends in Ecology and Evolution*, **5**, 79–86.

Sukopp, H., Hejny, S., and Kowarik, I., eds (1990). *Urban Ecology*. SPB Academic, The Hague.

Svenning, J.-C. and Skov, F. (2007). Ice age legacies in the geographical distribution of tree species richness in Europe. *Global Ecology and Biogeography*, **16**, 234–245.

Swart, J.A.A., van der Windt, H.J., and Keulartz, J. (2001). Valuation of nature in conservation and restoration. *Restoration Ecology*, **9**, 230–238.

Swetnam, T.W., Allen, C.D., and Betancourt, J.L. (1999). Applied historical ecology: using the past to manage for the future. *Ecological Applications*, **9**, 1189–1206.

Swift, M.J., Heal, O.W., and Anderson, J.M. (1979). *Decomposition in Terrestrial Ecosystems*. University of California Press, Berkeley.

Szentiks, C.A., Kondgen, S., Silinski, S., Speck, S., and Leendertz, F.H. (2009). Lethal pneumonia in a captive juvenile chimpanzee (*Pan troglodytes*) due to human-transmitted human respiratory syncytial virus (HRSV) and infection with *Streptococus pneumoniae*. *Journal of Medical Primatology*, **38**, 236–240.

Tabarelli, M. (2010). Tropical biodiversity in human-modified landscapes: what is our trump card? *Biotropica*, **42**, 553–554.

Talbot, S.S., Talbot, S.L., and Walker, L.R. (2010). Post-eruption legacy effects and their implications for long-term recovery of the vegetation on Kasatochi Island, Alaska. *Arctic, Antarctic, and Alpine Research*, **42**, 285–296.

Tanner, R.A. and Gange, A.C. (2005). Effects of golf courses on local biodiversity. *Landscape and Urban Planning*, **71**, 137–146.

Tansley, A.G. (1935). The use and abuse of vegetational concepts and terms. *Ecology*, **16**, 284–307.

Tarasov, V.G., Gebruk, A.V., Mironov, A.N., and Moskalev, L.I. (2005). Deep-sea and shallow-water hydrothermal vent communities: two different phenomena? *Chemical Geology*, **224**, 5–39.

Teal, J.M. and Howarth, R.W. (1984). Oil spill studies: a review of ecological effects. *Environmental Management*, **8**, 27–44.

Terborgh, J. (1999). *Requiem for Nature*. Island Press, Washington, DC.

Thoreau, H.D. (1860). *Succession of Forest Trees*. Massachuetts Board of Agriculture Eighth Annual Report. William White, Printer, Boston. (Not seen but cited in McIntosh 1999.)

Thornton, I.W.B., New, T.R., McLaren, D.A., Sudarman, H.K. and Vaughan, P.J. (1998). Air-borne arthropod fallout on Anak Krakatau and a possible pre-vegetation pioneer community. *Philosophical Transactions of the Royal Society B: Biological Sciences*, **322**, 471–479.

Thrush, S.F., Hewitt, J.E., Funnell, G.A. *et al.* (2001). Fishing disturbance and marine biodiversity: the role of habitat structure in simple soft-sediment systems. *Marine Ecology Progress Series*, **223**, 277–286.

Tilman, D. (1985). The resource-ratio hypothesis of plant succession. *The American Naturalist*, **125**, 827–852.

Tilman, D. (1988). *Plant Strategies and the Dynamics and Structure of Plant Communities*. Princeton University Press, Princeton, NJ, U.S.

Tilman, D. (1994). Competition and biodiversity in spatially structured habitats. *Ecology*, **75**, 2–16.

Tilman, D. (1997). Community invasibility, recruitment limitation, and grassland biodiversity. *Ecology*, **78**, 81–92.

Tilman, D. (1999). The ecological consequences of changes in biodiversity: a search for general principles. *Ecology*, **80**, 1455–1474.

Tilman, D. and Kareiva, P., eds (1997). *Spatial Ecology: the Role of Space in Population Dynamics and Interspecific Interactions*. Princeton University Press, Princeton, NJ, U.S.

Tilman, D., Cassman, K.G., Matson, P.A., Naylor, R., and Polasky, S. (2002). Agricultural sustainability and intensive production practices. *Nature*, **418**, 671–677.

Titus, J.H. and del Moral, R. (1998). The role of mycorrhizae in primary succession on Mount St. Helens. *American Journal of Botany*, **85**, 370–375.

Tolliver, K.S., Colley, D.M., and Young, D.R. (1995). Inhibitory effects of *Myrica cerifera* on *Pinus taeda*. *American Midland Naturalist*, **133**, 256–263.

Townsend, C.R., Scarsbrook, M.R., and Dolédec, S. (1997). The intermediate disturbance hypothesis, refugia, and biodiversity in streams. *Limnology and Oceanography*, **42**, 938–949.

Townsend, C.R., Thompson, R.M., McIntosh, A.R., Kilroy, C., Edwards, E., and Scarsbrook, M.R. (1998). Disturbance, resource supply, and food-web architecture in streams. *Ecology Letters*, **1**, 200–209.

Townsend, C., Dolédec, S., and Scarsbrook, M. (2003). Species traits in relation to temporal and spatial heterogeneity in streams: a test of habitat templet theory. *Freshwater Biology*, **37**, 367–387.

Trimble, S.W. and Crosson, P. (2000). U.S. soil erosion rates—myth and reality. *Science*, **280**, 248–250.

Tsutsui, W. (2003). Landscapes in the dark: toward an environmental history of wartime Japan. *Environmental History*, **8**, 294–311.

Tucker, C.J., Dregne, H.E., and Newcomb, W.W. (1991). Expansion and contraction of the Sahara Desert from 1980 to 1990. *Science*, **253**, 299–300.

Turley, C.M., Roberts, J.M., and Guinotte, J.M. (2007). Corals in deep-water: will the unseen hand of ocean acidification destroy cold-water ecosystems? *Coral Reefs*, **26**, 445–448.

Turner, B.L. (2008). Resource partitioning for soil phosphorus: a hypothesis. *Journal of Ecology*, **96**, 698–702.

Turner, B.L., Matson, P.A., McCarthy, J.J., *et al.* (2003). Illustrating the coupled human-environment system for vulnerability analysis: three case studies. *Proceedings of the National Academy of Sciences USA*, **100**, 8080–8085.

Turner, B.L., Condron, L.M., Richardson, S.J., Peltzer, D.A., and Allison, V.J. (2007). Soil organic phosphorus transformation during pedogenesis. *Ecosystems*, **10**, 1166–1181.

Turner, M.G. (1989). Landscape ecology: the effect of pattern on process. *Annual Review of Ecology and Systematics*, **20**, 171–197.

Turner, M.G. (2010). Disturbance and landscape dynamics in a changing world. *Ecology*, **91**, 2833–2849.

Turner, M.G., Gardner, R.H., Dale, V.H., and O'Neill, R.V. (1989). Predicting the spread of disturbance across heterogeneous landscapes. *Oikos*, **55**, 121–129.

Turner, R.E. and Rabalais, N.N. (1994). Coastal eutrophication near the Mississippi River Delta. *Nature*, **364**, 619–621.

Turner, T. (1983). Facilitation as a successional mechanism in a rocky intertidal community. *The American Naturalist*, **121**, 729–738.

Uehlinger, U. and Naegeli, M.W. (1998). Ecosystem metabolism, disturbance, and stability in a pre-alpine gravel bed river. *Journal of the North American Benthological Society*, **17**, 165–178.

Underhill, J.E. and Angold, P.G. (2000). Effects of roads on wildlife in an intensively modified landscape. *Environmental Reviews*, **8**, 21–39.

Underwood, A.J. (1999). Physical disturbances and their direct effect on an indirect effect: responses of an intertidal assemblage to a severe storm. *Journal of Experimental Marine Biology and Ecology*, **232**, 125–140.

Underwood, A.J. (2000). Experimental ecology of rocky intertidal habitats: what are we learning. *Journal of Experimental Marine Biology and Ecology*, **250**, 51–76.

Urban, D.L. and Shugart, H.H. (1992). Individual-based models of forest succession. In D.C. Glenn-Lewin, R.K. Peet, and T.T. Veblen, eds *Plant Succession: Theory and Prediction*, pp. 249–292. Chapman and Hall, London.

USGS Earthquake Hazard Program (2010). URL: http://earthquake.usgs.gov/earthquakes/eqarchives/year/eqstats.php (accessed 24 September 2010).

Usher, M.B. (1981). Modeling ecological succession, with particular reference to Markovian models. *Vegetatio*, **46**, 11–18.

Usher, M.B. (1992). Statistical models of succession. In D.C. Glenn-Lewin, R.K. Peet, and T.T. Veblen, eds *Plant Succession: Theory and Prediction*, pp. 215–248. Chapman and Hall, London.

Valett, H.M., Fisher, S.G., Grimm, N.B., and Camill, P. (1994). Vertical hydrologic exchange and ecological stability of a desert stream ecosystem. *Ecology*, **75**, 548–560.

Valiela, I. and Rietsma, C.S. (1995). Disturbance of salt marsh vegetation by wrack mats in Great Sippewissett Marsh. *Oecologia*, **102**, 106–112.

Valiela, I., McClelland, J., Hauxwell, J., Behr, P.J., Hersh, D., and Foreman, K. (1997). Macroalgal blooms in shallow estuaries: controls and ecophysiological and ecosystem consequences. *Limnology and Oceanography*, **42**, 1105–1118.

Valiente-Banuet, A. and Ezcurra, E. (1991). Shade as a cause of the association between the cactus *Neobuxbaumia tetezo* and the nurse plant *Mimosa luisana* in the Tehuacán Valley, Mexico. *Journal of Ecology*, **79**, 961–971.

Vandermeer, J., Boucher, D., Perfecto, I., and Granzow de la Cerda, I. (1996). A theory of disturbance and species diversity: evident from Nicaragua after Hurricane Joan. *Biotropica*, **28**, 600–613.

Van der Putten, W.H. (2005). Plant-soil feedback and plant diversity affect the composition of plant communities. In R.D. Bardgett, M.B. Usher, and D.W. Hopkins, eds *Biological Diversity and Function in Soils*, pp. 250–272. Cambridge University Press, Cambridge.

Van Dover, C.L. (2000). *The Ecology of Deep-Sea Hydrothermal Vents*. Princeton University Press, Princeton, NJ, U.S.

Vasek, F.C. (1980). Creosote bush: long-lived clones in the Mojave Desert. *American Journal of Botany*, **67**, 246–255.

Vasek, F.C. and Lund, L.J. (1980). Soil characteristics associated with a primary plant succession on a Mojave Desert dry lake. *Ecology*, **61**, 1013–1018.

Vaughan, N.E. and Lenton, T.M. (in press). A review of climate geo-engineering proposals. *Climatic Change*, doi: 10.1007/s10584-011-0027-7.

Vavrus, S., Ruddiman, W.F., and Kutzbach, J.E. (2008). Climate model tests of the anthropogenic influence on greenhouse-induced climate change: the role of early human agriculture, industrialization, and vegetation feedbacks. *Quaternary Science Reviews*, **27**, 1410–1425.

Veblen, T.T. (1992). Regeneration dynamics. In D.C. Glenn-Lewin, R.K. Peet, and T.T. Veblen, eds *Plant Succession: Theory and Prediction*, pp. 152–187. Chapman and Hall, London.

Veblen, T.T. and Ashton, D.H. (1978). Catastrophic influence on the vegetation of the Valdivian Andes, Chile. *Vegetatio*, **36**, 149–167.

Veblen, T.T., Holz, A., Paritsis, J., Raffaele, E., Kitzberger, T., and Blackhall, M. (in press). Adapting to global environmental change in Patagonia: what role for disturbance ecology? *Austral Ecology*, **36**(8), doi: 10.1111/j.1442-9993.2010.02236.x.

Velázquez, E. and Gómez-Sal, A. (2007). Environmental control of early succession on a large land-slide in a tropical dry ecosystem (Casita Volcano, Nicaragua). *Biotropica*, **35**, 601–609.

Vellinga, M. and Wood, R.A. (2002). Global climatic impacts of a collapse of the Atlantic thermohaline circulation. *Climatic Change*, **54**, 251–267.

Vermonden, K., Leuven, R.S.E.W., van der Velde, G., van Katwijk, M., Roelofs, J.G.M., and Hendriks, A.J. (2009). Urban drainage systems: an undervalued habitat for aquatic macroinvertebrates. *Biological Conservation*, **142**, 1104–1115.

Versfeld, D.B. and van Wilgen, B.W. (1986). Impact of woody aliens on ecosystem properties. In I.A.W. Macdonald, F.J. Kruger, and A.A. Ferrar, eds *The Ecology and Management of Biological Invasions in Southern Africa*, pp. 239–246. Oxford University Press, Oxford.

Vine, F.J. and Matthews, D.H. (1963). Magnetic anomalies over oceanic ridges. *Nature*, **199**, 947–949.

Visser, J.H. (1986). Host odor perception in phytophagous insects. *Annual Review of Entomology*, **31**, 121–144.

Vitousek, P.M. (1994). Potential nitrogen fixation during primary succession in Hawai'i Volcanoes National Park. *Biotropica*, **26**, 234–240.

Vitousek, P. (2004). *Nutrient Cycling and Limitation: Hawai'i as a Model System*. Princeton University Press, Princeton, NJ, U.S.

Vitousek, P.M. and Matson, P.A. (1990). Gradient analysis of ecosystems. In J.J. Cole, G.M. Lovett, and S.E.G. Findlay, eds *Comparative Analysis of Ecosystems: Patterns, Mechanisms and Theories*, pp. 287–298. Springer, New York.

Vitousek, P.M. and Reiners, W.A. (1975). Ecosystem succession and nutrient retention: a hypothesis. *BioScience*, **25**, 376–381.

Vitousek, P.M. and Walker, L.R. (1987). Colonization, succession and resource availability: Ecosystem-level interactions. In A.J. Gray, M.J. Crawley, and P.J. Edwards, eds *Colonization, Succession and Stability*, Symposium of the British Ecological Society, Vol. 26, pp. 207–224. Blackwell, Oxford.

Vitousek, P.M. and Walker, L.R. (1989). Biological invasion by *Myrica faya* in Hawaii: Plant demography, nitrogen fixation, and ecosystem effects. *Ecological Monographs*, **59**, 247–265.

Vitousek, P.M., Ehrlich, P.R., Ehrlich, A.H., and Matson, P.A. (1986). Human appropriation of the products of photosynthesis. *BioScience*, **36**, 368–373.

Vitousek, P.M., Walker, L.R., Whiteaker, L.D., Mueller-Dombois, D., and Matson, P.A. (1987). Biological invasion by *Myrica faya* alters ecosystem development in Hawaii. *Science*, **238**, 802–804.

Vitousek, P.M., D'Antonio, C.M., Loope, L.L., and Westbrooks, R. (1996). Biological invasions as global environmental change. *American Scientist*, **84**, 468–478.

Vitousek, P.M., Aber, J.D., Howarth, R.W. *et al.* (1997a). Human alteration of the global nitrogen cycle: sources and consequences. *Ecological Applications*, **7**, 737–750.

Vitousek, P.M., Mooney, H.A., Lubchenco, J., and Melillo, J.M. (1997b). Human domination of Earth's ecosystems. *Science*, **277**, 494–499.

Vitousek, P.M., Porder, S., Houlton, B.Z., and Chadwick, O.A. (2010). Terrestrial phosphorus limitation: mechanisms, implications, and nitrogen-phosphorus interactions. *Ecological Applications*, **20**, 5–15.

Vizzini, S., Tomasello, A., Di Maida, G., Pirrotta, M., Mazzola, A., and Calvo, S. (2010). Effect of explosive shallow hydrothermal vents on δ¹³C and growth performance in the seagrass *Posidonia oceanica*. *Journal of Ecology*, **98**, 1284–1291.

Volney, W.J.A. and McCullough, D. (1994). Jack pine budworm population behaviour in northwestern Wisconsin. *Canadian Journal of Forest Research*, **24**, 502–510.

Vörösmarty, C.J. and Sahagian, D. (2000). Anthropogenic disturbance of the terrestrial water cycle. *BioScience*, **50**, 753–765.

Vougioukalakis, G., Sbrana, A., and Mitropoulos, D. (1995). The 1649–50 Kolumbo submarine volcano activity, Santorini, Greece. In F. Barberi, R. Casale, and M. Fratta, eds *The European Laboratory Volcanoes: Workshop Proceedings*, pp. 189–192. EC European Science Commission, Luxembourg.

Wackernagel, M. and Rees, W. (1996). *Our Ecological Footprint*. New Society Publishers, Philadelphia, PA.

Waide, R.B. and A.E. Lugo (1992). A research perspective on disturbance and recovery of a tropical montane forest. In J.G. Goldhammer, ed. *Tropical Forests in Transition: Ecology of Natural and Anthropogenic Disturbance Processes*, pp. 173–189. Birkahuser Verlag, Basel, Switzerland.

Waide, R.B., Willig, M.R., Steiner, C.F., *et al.* (1999). The relationship between productivity and species richness. *Annual Review of Ecology and Systematics*, **30**, 257–300.

Wali, M.K. (1999). Ecology today: beyond the bounds of science. *Nature and Resources*, **35**, 38–50.

Wali, M.K., Evrendilek, F., West, T.O., *et al.* (1999). Assessing terrestrial ecosystem sustainability. *Nature and Resources*, **35**, 21–33.

Walker, G.P.L. (1994). Geology and volcanology of the Hawaiian Islands. In E.A. Kay, ed. *A Natural History of the Hawaiian Islands*, pp. 53–85. University of Hawaii Press, Honolulu.

Walker, J. and Reddell, P. (2007). Retrogressive succession and restoration on old landscapes. In L.R. Walker, J. Walker, and R.J. Hobbs, eds *Linking Restoration and Ecological Succession*, pp. 69–89. Springer, New York.

Walker, J., Thompson, C.H., Fergus, I.F., and Tunstall, B.R. (1981). Plant succession and soil development in coastal sand dunes of subtropical eastern Australia. In D.C. West, H.H. Shugart, and D.B. Botkin, eds *Forest Succession: Concepts and Applications*, pp. 107–131. Springer, New York.

Walker, J., Thompson, C.H., and Lacey, C.J. (1987). Morphological differences in lignotubers of *Eucalyptus intermedia* R.T. Bak and *E. signata* F. Muell associated with different stages of podzol development on coastal dunes, Cooloola, Queensland. *Australian Journal of Botany*, **35**, 301–310.

Walker, J., Thompson, C.H., Reddell, P., and Olley, J. (2000). Retrogressive succession on an old landscape. In P. White, ed. *Proceedings of the 41st Symposium of the IUVS*, pp. 21–23. Opulus Press, Uppsala, Sweden.

Walker, L. (2010). *Choosing a Sustainable Future: Ideas and Inspiration from Ithaca, NY*. New Society Publishers, Gabriola Island, British Columbia, Canada.

Walker, L.R. (1993). Nitrogen fixers and species replacements in primary succession. In J. Miles and D.W.H. Walton, eds *Primary Succession on Land*, pp. 249–272. Blackwell, Oxford.

Walker, L.R., ed. (1999a). *Ecosystems of Disturbed Ground*, Ecosystems of the World 16. Elsevier, Amsterdam.

Walker, L.R. (1999b). Patterns and processes in primary succession. In L.R. Walker, ed. *Ecosystems of Disturbed Ground*, Ecosystems of the World 16, pp. 585–610. Elsevier, Amsterdam.

Walker, L.R. (2000). Seedling and sapling dynamics of treefall pits in Puerto Rico. *Biotropica*, **32**, 262–275.

Walker, L.R. (in press). Integration of the study of natural and anthropogenic disturbances using severity gradients. *Austral Ecology*, **36**(8), doi: 10.1111/j.1442-9993.2011.02238.x.

Walker, L.R. and Bellingham, P.J. (2011). *Island Environments in a Changing World*. Cambridge University Press, Cambridge.

Walker, L.R. and Chapin, F.S., III (1987). Interactions among processes controlling successional change. *Oikos*, **50**, 131–135.

Walker, L.R. and del Moral, R. (2003). *Primary Succession and Ecosystem Rehabilitation*. Cambridge University Press, Cambridge.

Walker, L.R. and del Moral, R. (2009a). Lessons from primary succession for restoration of severely damaged habitats. *Applied Vegetation Science*, **12**, 55–67.

Walker, L.R. and del Moral, R. (2009b). Transition dynamics in succession: implications for rates, trajectories, and restoration. In R.J. Hobbs and K.N. Suding, eds *New Models for Ecosystem Dynamics and Restoration*, pp. 33–49. Island Press, Washington, DC.

Walker, L.R. and Powell, E.A. (1999). Regeneration of the Mauna Kea silversword *Argyroxiphium sandwicense* (Asteraceae) Hawaii. *Biological Conservation*, **89**, 61–70.

Walker, L.R. and Shiels, A.B. (2008). Post-disturbance erosion impacts carbon fluxes and plant succession on recent tropical landslides. *Plant and Soil*, **313**, 205–216.

Walker, L.R. and Smith, S.D. (1997). Impacts of invasive plants on community and ecosystem properties. In J.O. Luken and J.W. Thieret, eds *Assessment and Management of Plant Invasions*, pp. 69–86. Springer, New York.

Walker, L.R. and Vitousek, P.M. (1991). An invader alters germination and growth of a native dominant tree in Hawai'i. *Ecology*, **72**, 1449–1455.

Walker, L.R. and Willig, M. R. (1999). An introduction to terrestrial disturbances. In L.R. Walker, ed. *Ecosystems of Disturbed Ground*, Ecosystems of the World 16, pp. 1–16. Elsevier, Amsterdam.

Walker, L.R., Brokaw, N.V.L., Lodge, D.J., and Waide, R.B., eds (1991). Ecosystem, plant and animal responses to hurricanes in the Caribbean. *Biotropica*, **23**, 313–521.

Walker, L.R., Voltzow, J., Ackerman, J.D., Fernández, D.S., and Fetcher, N. (1992). Immediate impact of Hurricane Hugo on a Puerto Rican rain forest. *Ecology*, **73**, 691–694.

Walker, L.R., Silver, W.L., Willig, M.R., and Zimmerman, J.K., eds (1996a). Long term responses of Caribbean ecosystems to disturbance. *Biotropica*, **28**, 414–614.

Walker, L.R., Zarin, D.J., Fetcher, N., Myster, R.W., and Johnson, A.H. (1996b). Ecosystem development and plant succession on landslides in the Caribbean. *Biotropica*, **28**, 566–576.

Walker, L.R., Clarkson, B.D., Silvester, W., and Clarkson, B.R. (2003). Colonization dynamics and facilitative impacts of a nitrogen-fixing shrub in post-volcanic primary succession. *Journal of Vegetation Science*, **14**, 277–290.

Walker, L.R., Barnes, P.L., and Powell, E.A. (2006a). *Tamarix aphylla*: a newly invasive tree in southern Nevada. *Western North American Naturalist*, **66**, 191–201.

Walker, L.R., Bellingham, P.B., and Peltzer, D.A. (2006b). Plant characteristics are poor predictors of microsite colonization during the first two years of primary succession. *Journal of Vegetation Science*, **17**, 397–406.

Walker, L.R., Walker, J., and Hobbs, R.J., eds (2007). *Linking Restoration and Ecological Succession*. Springer, New York.

Walker, L.R., Landau, F.H., Velázquez, E., Shiels, A.B., and Sparrow, A.D. (2010a). Early successional woody plants facilitate and ferns inhibit forest development on Puerto Rican landslides. *Journal of Ecology*, **98**, 625–635.

Walker, L.R., Wardle, D.A., Bardgett, R.D., and Clarkson, B.D. (2010b). The use of chronosequences in studies of ecological succession and soil development. *Journal of Ecology*, **98**, 725–736.

Walker, S.J., Schlacher, T.A., and Schlacher-Hoenlinger, M.A. (2007). Spatial heterogeneity of epibenthos on artificial reefs: fouling communities in the early stages of colonization on an East Australian shipwreck. *Marine Ecology*, **28**, 435–445.

Walker, T.W. and Syers, J.K. (1976). The fate of phosphorus during pedogenesis. *Geoderma*, **15**, 1–19.

Walter, K.M., Zimov, S.A., Chanton, J.P., Verbyla, D., and Chapin, F.S., III (2006). Methane bubbling from Siberian thaw lakes as a positive feedback to climate warming. *Nature*, **443**, 71–75.

Wang, B., Michaelson, G., Ping, C.-L., Plumlee, G., and Hageman, P. (2010). Characterization of pyroclastic deposits and pre-eruptive soils following the 2008 eruption of Kasatochi Island volcano, Alaska. *Arctic, Antarctic, and Alpine Research*, **42**, 276–284.

Wang, B.C. and Smith, T.B. (2002). Closing the seed dispersal loop. *Trends in Ecology and Evolution*, **17**, 379–386.

Ward, J.V. (1998). Riverine landscapes: biodiversity patterns, disturbance regimes, and aquatic conservation. *Biological Conservation*, **83**, 269–278.

Ward, J.V., Tockner, K., and Scheimer, F. (1999). Biodiversity of floodplain river ecosystems: ecotones and connectivity. *Regulated Rivers: Research and Management*, **15**, 125–139.

Wardle, D.A. (1995). Impacts of disturbance on detritus food webs in agro-ecosystems of contrasting tillage and weed management practices. *Advances in Ecological Research*, **26**, 105–185.

Wardle, D.A. (2001). Experimental demonstration that plant diversity reduces invasibility—evidence of a biological mechanism or a consequence of sampling effect? *Oikos*, **95**, 161–170.

Wardle, D.A. (2004). *Communities and Ecosystems: Linking the Aboveground and Belowground Components*. Princeton University Press, Princeton, NJ, U.S.

Wardle, D.A. and Bardgett, R.D. (2004). Human-induced changes in large herbivorous mammal density: the consequences for decomposers. *Frontiers in Ecology and the Environment*, **2**, 145–153.

Wardle, D.A. and Ghani, A. (1995). A critique of the microbial metabolic quotient (qCO_2) as a bioindicator of disturbance and ecosystem development. *Soil Biology and Biochemistry*, **27**, 1601–1610.

Wardle, D.A. and Peltzer, D.A. (2007). Aboveground-belowground linkages, ecosystem development, and ecosystem restoration. In L.R. Walker, J. Walker, and R.J. Hobbs, eds *Linking Restoration and Ecological Succession*, pp. 45–68.

Wardle, D.A., Yeates, G.W., Watson, R.N., and Nicholson, K.S. (1995a). The detritus food-web and the diversity of soil fauna as indicators of disturbance regimes in agro-ecosystems. *Plant and Soil*, **170**, 35–43.

Wardle, D.A., Yeates, G.W., Watson, R.N., and Nicholson, K.S. (1995b). Development of the decomposer food-web, trophic relationships, and ecosystem properties during a 3-year primary succession in sawdust. *Oikos*, **73**, 155–166.

Wardle, D.A., Walker, L.R., and Bardgett, R.D. (2004). Ecosystem properties and forest decline in contrasting long-term chronosequences. *Science*, **305**, 509–513.

Wardle, D.A., Bellingham, P.J., Fukami, T., and Mulder, C.P.H. (2007). Promotion of ecosystem carbon sequestration by invasive predators. *Biology Letters*, **3**, 479–482.

Wardle, D.A., Bardgett, R.D., Walker, L.R., Peltzer, D.A., and Lagerstrom, A. (2008). The response of plant diversity to ecosystem retrogression: evidence from contrasting long-term chronosequences. *Oikos*, **117**, 93–103.

Wardle, D.A., Walker, L.R., Bardgett, R.D., and Bonner, K.I. (2009). Among- and within-species variation in plant litter decomposition in contrasting long-term chronosequences. *Functional Ecology*, **23**, 442–453.

Warming, E. (1895). *Plantesamfund: Grundträk af den Ökologiska Plantegeografi*. Philipsen, Copenhagen.

Warming, E. (1909). *Oecology of Plants: an Introduction to the Study of Plant Communities*. Clarendon Press, Oxford.

Warnken, J. and Byrnes, T. (2004). Impact of tourboats in marine environments. In R. Buckley, ed. *Environmental Impacts of Ecotourism*. CABI, Wallingford, U.K.

Wassenaar, T.D., van Aarde, R.J., Pimm, S.L., and Ferreira, S.M. (2005). Community convergence in disturbed subtropical dune forests. *Ecology*, **86**, 655–666.

Watt, A.S. (1947). Pattern and process in the plant community. *Journal of Ecology*, **35**, 1–22.

Waythomas, C.F., Scott, W.E., and Nye, C.J. (2010). The geomorphology of an Aleutian volcano following a major eruption: the 7–8 August 2008 eruption of Kasatochi Volcano, Alaska, and its aftermath. *Arctic, Antarctic, and Alpine Research*, **42**, 260–275.

Webb, S.L. (1999). Disturbance by wind in temperate-zone forests. In L.R. Walker, ed. *Ecosystems of Disturbed Ground*, Ecosystems of the World 16, pp. 187–222. Elsevier, Amsterdam.

Weber, M.G. and Flannigan, M.D. (1997). Canadian boreal forest ecosystem structure and function in a changing climate: impact on fire regimes. *Environmental Reviews*, **5**, 145–166.

Weiher, E. and Keddy, P., eds (1999). *Ecological Assembly Rules: Perspectives, Advances, Retreats.* Cambridge University Press, Cambridge.

Weisberg, P.J. and Baker, W.L. (1995). Spatial variation in tree seedling and krummholz growth in the forest-tundra ecotones of Rocky Mountain National Park, Colorado, U.S.A. *Arctic and Alpine Research*, **27**, 116–129.

Weisman, A. (2007). *The World Without Us.* St Martin's Press, New York.

Wells, A., Duncan, R.P., and Stewart, G.H. (2001). Forest dynamics in Westland, New Zealand: the importance of large infrequent earthquake-induced disturbance. *Journal of Ecology*, **89**, 1006–1018.

Weltzin, J.F., Belote, R.T., and Sanders, N.J. (2003). Biological invaders in a greenhouse world: will elevated CO_2 fuel plant invasions? *Frontiers in Ecology and the Environment*, **1**, 146–153.

Wen, D. (1993). Soil erosion and conservation in China. In D. Pimentel, ed. *Soil Erosion and Conservation*, pp. 63–86. Cambridge University Press, Cambridge.

Westing, A.H. (1971). Ecological effects of military defoliation on the forests of South Vietnam. *BioScience*, **21**, 893–898.

Westing, A.H. (1988). The military sector *vis-à-vis* the environment. *Journal of Peace Research*, **25**, 257–264.

Westman, W.E. (1986). Resilience: concepts and measures. In B. Dell, A.J.M. Hopkins, and B.B. Lamont. *Resilience in Mediterranean-type Ecosystems*, pp. 5–19. Junk, Dordrecht.

Whigham, D.F., Olmsted, I., Cabrera Cano, E., and Harmon, M.E. (1991). The impact of Hurricane Gilbert on trees, litterfall, and woody debris in a dry tropical forest in the northeastern Yucatan Peninsula. *Biotropica*, **23**, 434–441.

Whigham, D.F., Dickinson, M.B., and Brokaw, N.V.L. (1999). Background canopy gap and catastrophic wind disturbance in tropical forests. In L.R. Walker, ed. *Ecosystems of Disturbed Ground*, Ecosystems of the World 16, pp. 223–252. Elsevier, Amsterdam.

White, L., Jr (1962). *Medieval Technology and Social Change.* Oxford University Press, Oxford.

White, P.S. and Jentsch, A. (2001). The search for generality in studies of disturbance and ecosystem dynamics. *Progress in Botany*, **62**, 399–449.

White, P.S. and Pickett, S.T.A. (1985). Natural disturbance and patch dynamics: an introduction. In S.T.A. Pickett and P.S. White, eds *The Ecology of Natural Disturbance and Patch Dynamics*, pp. 3–16. Academic Press, New York.

White, R.P., Murray, S., and Rohweder, M. (2003). *The State of Food Insecurity in the World 2003.* FAO, Rome.

Whitehead, D., Boelman, N.T., Turnbull, M.H., *et al.* (2005). Photosynthesis and reflectance indices for rainforest species in ecosystems undergoing progression and retrogression along a soil fertility chronosequence in New Zealand. *Oecologia*, **144**, 233–244.

Whittaker, R.H. (1967). Gradient analysis of vegetation. *Biological Review*, **49**, 207–264.

Whittaker, R.H., ed. (1973). *Ordination and Classification of Communities.* Junk, The Hague.

Whittaker, R.H. (1977). Evolution of species diversity in land communities. *Evolutionary Biology*, **10**, 1–67.

Whittaker, R.J. (1992). Stochasticism and determinism in island ecology. *Journal of Biogeography*, **19**, 587–591.

Whittaker, R.J. (2010). Meta-analyses and mega-mistakes: calling time on meta-analysis of the species richness–productivity relationship. *Ecology*, **91**, 2522–2533.

Whittaker, R.J., Partomihardjo, T., and Jones, S.H. (1999). Interesting times on Krakatau: stand dynamics in the 1990s. *Philosophical Transactions of the Royal Society B: Biological Sciences*, **354**, 1857–1867.

Whittaker, R.J., Triantis, K.A. and Ladle, R.J. (2008). A general dynamic theory of oceanic island biogeography. *Journal of Biogeography*, **35**, 977–994.

WHO (1992). World Health Organization, Commission on Health and Environment. *Our Planet, Our Health*. World Health Organization, Geneva.

Wiggins, G.B., Mackay, R.B., and Smith, I.M. (1980). Evolutionary and ecological strategies of animals in annual pools. *Archiv für Hydrobiologie Supplement*, **58**, 97–206.

Wigley, T.M.L. (2006). A combined mitigation/geoengineering approach to climate stabilization. *Science*, **314**, 452–454.

Wikberg, S. and Svensson, B.M. (2003). Ramet demography in a ring-forming clonal sedge. *Journal of Ecology*, **91**, 847–854.

Wilcove, D.A., and Master, L.L. (2005). How many endangered species are there in the United States? *Frontiers in Ecology and the Environment,* **3**, 414–420.

Wilcox, B.A. and Murphy, D.D. (1985). Conservation strategy: the effects of fragmentation on extinction. *The American Naturalist*, **125**, 879–887.

Wilkinson, C., Souter, D., and Goldberg, J. (2006). *Status of Coral Reefs in Tsunami Affected Countries*. Global Coral Reef Monitoring Network. Australian Institute of Marine Science, Townsville, Queensland, Australia.

Wilkinson, D.M. (1997). Plant colonization: are wind dispersed seeds really dispersed by birds at larger spatial and temporal scales? *Journal of Biogeography,* **24**, 61–65.

Williams, J.C., Drummond, B.A., and Buxton, R.T. (2010). Initial effects of the August 2008 volcanic eruption on breeding birds and marine mammals at Kasatochi Island, Alaska. *Arctic, Antarctic, and Alpine Research*, **42**, 302–314.

Williams, K.S. and Simon, C. (1995). The ecology, behavior, and evolution of periodical cicadas. *Annual Review of Entomology*, **40**, 269–295.

Williams, R.S., Jr and Moore, J.G. (2008). *Man Against Volcano: the Eruption on Heimaey, Vestmannaeyjar, Iceland*. Diane Publishing, Darby, PA, U.S.

Williamson, N.A., Johnson, M.S., and Bradshaw, A.D. (1982). *Mine Wastes Reclamation. The Establishment of Vegetation on Metal Mine Wastes*. Mining Journal Books, London.

Willig, M.R. and Camilo, G.R. (1991). The effect of Hurricane Hugo on six invertebrate species in the Luquillo Experimental Forest of Puerto Rico. *Biotropica*, **23**, 455–461.

Willig, M.R. and McGinley, M.A. (1999). The response of animals to disturbance and their roles in patch generation. In L.R. Walker, ed. *Ecosystems of Disturbed Ground*, Ecosystems of the World 16, pp. 633–657. Elsevier, Amsterdam.

Willig, M.R. and Walker, L.R. (1999). Disturbance in terrestrial ecosystems: salient themes, synthesis and future directions. In L.R. Walker, ed. *Ecosystems of Disturbed Ground*, Ecosystems of the World 16, pp. 747–767. Elsevier, Amsterdam.

Wilson, J.B. and Agnew, A.D. Q. (1992). Positive-feedback switches in plant communities. *Advances in Ecological Research*, **23**, 263–336.

Wilson, R.D., Monaghan, P.H., Osanik, A., Price, L.C., and Roger, M.A. (1974). Natural marine oil seepage. *Science*, **184**, 857–865.

Wilson, R.R. and Smith, K.L., Jr (1984). Effects of near-bottom currents on detection of bait by the abyssal grenadier fishes *Corphenoides* spp., recorded *in situ* with a video camera on a free vehicle. *Marine Biology*, **84**, 83–91.

Wilson, S.D. (1988). The effects of military tank traffic on prairie: a management model. *Environmental Management*, **12**, 397–403.

Wilson, S.D. and Tilman, D. (1993). Plant competition and resource availability in response to disturbance and fertilization. *Ecology*, **74**, 599–611.

Witte, F., Goldschmidt, T., Goudswaard, P.C., Ligtvoet, W., Van Oijen, M.J.P., and Wanink, J.H. (1991). Species extinction and concomitant ecological changes in Lake Victoria. *Netherlands Journal of Zoology*, **42**, 214–232.

Wood, D.M. and del Moral, R. (1987). Mechanisms of early primary succession in subalpine habitats on Mount St. Helens. *Ecology*, **68**, 780–790.

Wooten, J.T. (1998). Effects of disturbance on species diversity: a multitrophic perspective. *The American Naturalist*, **152**, 803–825.

Worley, I.A. (1973). The "black crust" phenomenon in upper Glacier Bay, Alaska. *Northwest Science*, **47**, 20–29.

Worm, B. and Duffy, J.E. (2003). Biodiversity, productivity and stability in real food webs. *Trends in Ecology and Evolution*, **18**, 628–632.

Worster, D.E. (1979). *Dustbowl: the Southern Plains in the 1930s*. Oxford University Press, Oxford.

Wright, J.P., Jones, C.G., and Flecker, A.S. (2002). An ecosystem engineer, the beaver, increases species richness at the landscape scale. *Oecologia*, **132**, 96–101.

Wright, R.A. and Mueller-Dombois, D. (1988). Relationships among shrub population structure, species associations, seedling root form and early volcanic succession, Hawaii. In M.J.A. Werger, P.J.M. van der Aart, H.J. During, and J.T.A. Verhoeven, eds *Plant Form and Vegetation Structure*, pp. 87–104. SPB Academic, The Hague.

Wu, J. (2008). Toward a landscape ecology of cities: beyond buildings, tree, and urban forests. In Carreiro, M.M., Song, Y.-C. and Wu, J., eds *Ecology, Planning and Management of Urban Forests*, pp. 10–28. Springer, New York.

Wu, J. and Loucks, O.L. (1995). From balance of nature to hierarchical patch dynamics: a paradigm shift in ecology. *Quarterly Review of Biology*, **70**, 439–466.

Wurgler, F.E. and Kramers, P.G.N. (1992). Environmental effects of genotoxins (ecogenotoxicology). *Mutagenesis*, **7**, 321–327.

Wynn-Williams, D.D. (1993). Microbial processes and initial stabilization of fellfield soil. In J. Miles and D.W.H. Walton, eds *Primary Succession on Land*, pp. 17–32. Blackwell, Oxford.

Yang, L.H. (2004). Periodical cicadas as resource pulses in North American forests. *Science*, **306**, 1565–1567.

Yarranton, G. and Morrison, R. (1974). Spatial dynamics of a primary succession: nucleation. *Journal of Ecology*, **62**, 417–428.

Yoffee, N., ed. (2009). *Questioning Collapse*. Cambridge University Press, Cambridge.

Young, I.R., Zieger, S., and Babanin, A.V. (2011). Global trends in wind speed and wave height. *Science*, **332**, 451–455.

Zarin, D.J. and Johnson, A.H. (1995a). Base saturation, nutrient cation, and organic matter increases during early pedogenesis on landslide scars in the Luquillo Experimental Forest, Puerto Rico. *Geoderma*, **65**, 317–330.

Zarin, D.J. and Johnson, A.H. (1995b). Nutrient accumulation during primary succession in a montane tropical forest, Puerto Rico. *Soil Science Society of America Journal*, **59**, 1444–1452.

Zedler, J.B. (2003). Wetlands at your service: reducing impacts of agriculture at the watershed scale. *Frontiers in Ecology and the Environment*, **1**, 65–72.

Zedler, P.H. (2007). Fire effects on grasslands. In E.A. Johnson and K. Miyanishi, eds *Plant Disturbance Ecology: the Process and the Response*, pp. 397–439. Academic Press, Amsterdam.

Zimmerman, J.K., Pulliam, W.M., Lodge, D.J., *et al.* (1995). Nitrogen immobilization by decomposing woody debris and the recovery of tropical wet forest from hurricane damage. *Oikos*, **72**, 314–322.

Zimmerman, J.K., Willig, M.R., Walker, L.R., and Silver, W.L. (1996). Introduction: disturbance and Caribbean ecosystems. *Biotropica*, **28**, 414–423.

Zobel, D.B. and Antos, J.A. (1997). A decade of recovery of understory vegetation buried by volcanic tephra from Mount St. Helens. *Ecological Monographs*, **67**, 317–344.

Zwally, H.J., Abdalati, W., Herring, T., Larson, K., Saba, J., and Steffen, K. (2002). Surface melt-induced acceleration of Greenland ice-sheet flow. *Science*, **297**, 218–222.

Index

Printed and bound by CPI Group (UK) Ltd, Croydon, CR0 4YY